Edited by
R. Morris Bullock

Catalysis Without Precious Metals

Further Reading

Plietker, B. (Ed.)

Iron Catalysis in Organic Chemistry

Reactions and Applications

2008
Hardcover
ISBN: 978-3-527-31927-5

Jackson, S. D., Hargreaves, J. S. J. (Eds.)

Metal Oxide Catalysis

2008
Hardcover
ISBN: 978-3-527-31815-5

Drauz, K., Gröger, H., May, O. (Eds.)

Enzyme Catalysis in Organic Synthesis

2011
Hardcover
ISBN: 978-3-527-32547-4

Pihko, P. M. (Ed.)

Hydrogen Bonding in Organic Synthesis

2009
Hardcover
ISBN: 978-3-527-31895-7

Bäckvall, J.-E. (Ed.)

Modern Oxidation Methods

2011
Hardcover
ISBN: 978-3-527-32320-3

Mohr, F. (Ed.)

Gold Chemistry

Applications and Future Directions in the Life Sciences

2009
Hardcover
ISBN: 978-3-527-32086-8

Dupont, J, Pfeffer, M. (Eds.)

Palladacycles

Synthesis, Characterization and Applications

2008
Hardcover
ISBN: 978-3-527-31781-3

Sheldon, R. A., Arends, I., Hanefeld, U.

Green Chemistry and Catalysis

2007
Hardcover
ISBN: 978-3-527-30715-9

Dalko, P. I. (Ed.)

Enantioselective Organocatalysis

Reactions and Experimental Procedures

2007
Hardcover
ISBN: 978-3-527-31522-2

Cybulski, A., Moulijn, J. A., Stankiewicz, A. (Eds.)

Novel Concepts in Catalysis and Chemical Reactors

Improving the Efficiency for the Future

2010
Hardcover
ISBN: 978-3-527-32469-9

Yudin, A. K. (Eds.)

Catalyzed Carbon-Heteroatom Bond Formation

From Biomimetic Concepts to Applications in Asymmetric Synthesis

2011
Hardcover
ISBN: 978-3-527-32428-6

Edited by R. Morris Bullock

Catalysis Without Precious Metals

WILEY-VCH Verlag GmbH & Co. KGaA

The Editor

Dr. R. Morris Bullock
Pacific Northwest National
Laboratory
P.O. Box 999, K2-57
Richland, WA 99352
USA

All books published by **Wiley-VCH** are carefully produced. Nevertheless, authors, editors, and publisher do not warrant the information contained in these books, including this book, to be free of errors. Readers are advised to keep in mind that statements, data, illustrations, procedural details or other items may inadvertently be inaccurate.

Library of Congress Card No.: applied for

British Library Cataloguing-in-Publication Data
A catalogue record for this book is available from the British Library.

Bibliographic information published by the Deutsche Nationalbibliothek
The Deutsche Nationalbibliothek lists this publication in the Deutsche Nationalbibliografie; detailed bibliographic data are available on the Internet at <http://dnb.d-nb.de>.

© 2010 Wiley-VCH Verlag & Co. KGaA, Boschstr. 12, 69469 Weinheim, Germany

All rights reserved (including those of translation into other languages). No part of this book may be reproduced in any form – by photoprinting, microfilm, or any other means – nor transmitted or translated into a machine language without written permission from the publishers. Registered names, trademarks, etc. used in this book, even when not specifically marked as such, are not to be considered unprotected by law.

Cover Design   Graphik-Design Schulz, Fußgönheim
Typesetting   Toppan Best-set Premedia Limited, Hong Kong
Printing and Binding   Fabulous Printers Pte Ltd, Singapore

Printed in Singapore
Printed on acid-free paper

ISBN: 978-3-527-32354-8

To Cindy, Claude and Lindsay

Contents

Preface *XIII*
List of Contributors *XVII*

1 Catalysis Involving the H· Transfer Reactions of First-Row Transition Metals *1*
John Hartung and Jack R. Norton
1.1 H· Transfer Between M–H Bonds and Organic Radicals *2*
1.2 H· Transfer Between Ligands and Organic Radicals *4*
1.3 H· Transfer Between M–H and C–C Bonds *7*
1.4 Chain Transfer Catalysis *11*
1.5 Catalysis of Radical Cyclizations *15*
1.6 Competing Methods for the Cyclization of Dienes *19*
1.7 Summary and Conclusions *20*
References *21*

2 Catalytic Reduction of Dinitrogen to Ammonia by Molybdenum *25*
Richard R. Schrock
2.1 Introduction *25*
2.2 Some Characteristics of Triamidoamine Complexes *26*
2.3 Possible [HIPTN$_3$N]Mo Intermediates in a Catalytic Reduction of Molecular Nitrogen *30*
2.3.1 MoN$_2$ and MoN$_2^-$ *30*
2.3.2 Mo–N=NH *33*
2.3.3 Conversion of Mo(N$_2$) into Mo–N=NH *33*
2.3.4 [Mo=N–NH$_2$]$^+$ *35*
2.3.5 Mo≡N and [Mo=NH]$^+$ *36*
2.3.6 Mo(NH$_3$) and [Mo(NH$_3$)]$^+$ *37*
2.4 Interconversion of Mo(NH$_3$) and Mo(N$_2$) *38*
2.5 Catalytic Reduction of Dinitrogen *39*
2.6 MoH and Mo(H$_2$) *41*
2.7 Ligand and Metal Variations *44*
2.8 Comments *47*
Acknowledgements *48*
References *48*

3	**Molybdenum and Tungsten Catalysts for Hydrogenation, Hydrosilylation and Hydrolysis** *51*
	R. Morris Bullock
3.1	Introduction *51*
3.2	Proton Transfer Reactions of Metal Hydrides *52*
3.3	Hydride Transfer Reactions of Metal Hydrides *54*
3.4	Stoichiometric Hydride Transfer Reactivity of Anionic Metal Hydride Complexes *56*
3.5	Catalytic Hydrogenation of Ketones with Anionic Metal Hydrides *58*
3.6	Ionic Hydrogenation of Ketones Using Metal Hydrides and Added Acid *59*
3.7	Ionic Hydrogenations from Dihydrides: Delivery of the Proton and Hydride from One Metal *64*
3.8	Catalytic Ionic Hydrogenations With Mo and W Catalysts *65*
3.9	Mo Phosphine Catalysts With Improved Lifetimes *69*
3.10	Tungsten Hydrogenation Catalysts with N-Heterocyclic Carbene Ligands *70*
3.11	Catalysts for Hydrosilylation of Ketones *71*
3.12	Cp_2Mo Catalysts for Hydrolysis, Hydrogenations and Hydrations *73*
3.13	Conclusion *78*
	Acknowledgements *78*
	References *79*

4	**Modern Alchemy: Replacing Precious Metals with Iron in Catalytic Alkene and Carbonyl Hydrogenation Reactions** *83*
	Paul J. Chirik
4.1	Introduction *83*
4.2	Alkene Hydrogenation *86*
4.2.1	Iron Carbonyl Complexes *86*
4.2.2	Iron Phosphine Compounds *89*
4.2.3	Bis(imino)pyridine Iron Complexes *93*
4.2.4	α-Diimine Iron Complexes *99*
4.3	Carbonyl Hydrogenation *101*
4.3.1	Hydrosilylation *101*
4.3.2	Bifunctional Complexes *103*
4.4	Outlook *105*
	References *106*

5	**Olefin Oligomerizations and Polymerizations Catalyzed by Iron and Cobalt Complexes Bearing Bis(imino)pyridine Ligands** *111*
	Vernon C. Gibson and Gregory A. Solan
5.1	Introduction *111*
5.2	Precatalyst Synthesis *112*

5.2.1	Ligand Preparation *112*
5.2.2	Complexation with MX$_2$ (M = Fe, Co) *113*
5.3	Precatalyst Activation and Catalysis *115*
5.3.1	Olefin Polymerization *115*
5.3.1.1	Catalytic Evaluation *116*
5.3.1.2	Steric Versus Electronic Effects *116*
5.3.1.3	Effect of MAO Concentration *119*
5.3.1.4	Effects of Pressure and Temperature *120*
5.3.1.5	α-Olefin Monomers *121*
5.3.2	Olefin Oligomerization *122*
5.3.2.1	Catalytic Evaluation *122*
5.3.2.2	Substituent Effects *122*
5.3.2.3	Schulz–Flory Distributions *124*
5.3.2.4	Poisson Distributions *124*
5.3.2.5	α-Olefin Monomers *125*
5.4	The Active Catalyst and Mechanism *125*
5.4.1	Active Species *125*
5.4.1.1	Iron Catalyst *126*
5.4.1.2	Cobalt Catalyst *127*
5.4.2	Propagation and Chain Transfer Pathways/Theoretical Studies *127*
5.4.3	Well-Defined Iron and Cobalt Alkyls *129*
5.5	Other Applications *133*
5.5.1	Immobilization *133*
5.5.2	Reactor Blending and Tandem Catalysis *134*
5.6	Conclusions and Outlook *134*
	References *136*

6	**Cobalt and Nickel Catalyzed Reactions Involving C–H and C–N Activation Reactions** *143*
	Renee Becker and William D. Jones
6.1	Introduction *143*
6.2	Catalysis with Cobalt *143*
6.3	Catalysis with Nickel *154*
	References *163*

7	**A Modular Approach to the Development of Molecular Electrocatalysts for H$_2$ Oxidation and Production Based on Inexpensive Metals** *165*
	M. Rakowski DuBois and Daniel L. DuBois
7.1	Introduction *165*
7.2	Concepts in Catalyst Design Based on Structural Studies of Hydrogenase Enzymes *166*
7.3	A Layered or Modular Approach to Catalyst Design *170*
7.4	Using the First Coordination Sphere to Control the Energies of Catalytic Intermediates *171*

7.5	Using the Second Coordination Sphere to Control the Movement of Protons between the Metal and the Exterior of the Molecular Catalyst *173*
7.6	Integration of the First and Second Coordination Spheres *174*
7.7	Summary *178*
	Acknowledgements *179*
	References *179*

8 Nickel-Catalyzed Reductive Couplings and Cyclizations *181*
Hasnain A. Malik, Ryan D. Baxter, and John Montgomery

8.1	Introduction *181*
8.2	Couplings of Alkynes with α,β-Unsaturated Carbonyls *182*
8.2.1	Three-Component Couplings via Alkyl Group Transfer – Methods Development *182*
8.2.2	Reductive Couplings via Hydrogen Atom Transfer – Methods Development *184*
8.2.3	Mechanistic Insights *186*
8.2.3.1	Metallacycle-Based Mechanistic Pathway *186*
8.2.4	Use in Natural Product Synthesis *189*
8.3	Couplings of Alkynes with Aldehydes *191*
8.3.1	Three-Component Couplings via Alkyl Group Transfer – Method Development *192*
8.3.2	Reductive Couplings via Hydrogen Atom Transfer – Method Development *193*
8.3.2.1	Simple Aldehyde and Alkyne Reductive Couplings *194*
8.3.2.2	Directed Processes *196*
8.3.2.3	Diastereoselective Variants: Transfer of Chirality *197*
8.3.2.4	Asymmetric Variants *199*
8.3.3	Mechanistic Insights *200*
8.3.4	Cyclocondensations via Hydrogen Gas Extrusion *204*
8.3.5	Use in Natural Product Synthesis *205*
8.4	Conclusions and Outlook *210*
	Acknowledgements *210*
	References *210*

9 Copper-Catalyzed Ligand Promoted Ullmann-type Coupling Reactions *213*
Yongwen Jiang and Dawei Ma

9.1	Introduction *213*
9.2	C–N Bond Formation *213*
9.2.1	Arylation of Amines *213*
9.2.1.1	Arylation of Aliphatic Primary and Secondary Amines *213*
9.2.1.2	Arylation of Aryl Amines *215*
9.2.1.3	Arylation of Ammonia *215*
9.2.2	Arylation and Vinylation of N-Heterocycles *217*

9.2.2.1	Coupling of Aryl Halides and N-Heterocycles	217
9.2.2.2	Coupling of Vinyl Bromides and N-Heterocycles	218
9.2.3	Aromatic Amidation	218
9.2.3.1	Cross-Coupling of aryl Halides with Amides and Carbamates	219
9.2.3.2	Cross-Coupling of Vinyl Halides with Amides or Carbamates	220
9.2.3.3	Cross-Coupling of Alkynl Halides with Amides or Carbamates	220
9.2.4	Azidation	221
9.3	C–O Bond Formation	222
9.3.1	Synthesis of Diaryl Ethers	222
9.3.2	Aryloxylation of Vinyl Halides	223
9.3.3	Cross-Coupling of Aryl Halides with Aliphatic Alcohols	223
9.4	C–C Bond Formation	224
9.4.1	Cross-Coupling with Terminal Acetylene	224
9.4.2	The Arylation of Activated Methylene Compounds	225
9.4.3	Cyanation	227
9.5	C–S Bond Formation	228
9.5.1	The Formation of Bisaryl- and Arylalkyl-Thioethers	228
9.5.2	The Synthesis of Alkenylsulfides	229
9.5.3	Assembly of aryl Sulfones	229
9.6	C–P Bond Formation	230
9.7	Conclusion	230
	References	231
10	**Copper-Catalyzed Azide-Alkyne Cycloaddition (CuAAC)**	**235**
	M.G. Finn and Valery V. Fokin	
10.1	Introduction	235
10.2	Azide–Alkyne Cycloaddition: Basics	237
10.3	Copper-Catalyzed Cycloadditions	238
10.3.1	Catalysts and Ligands	238
10.3.2	CuAAC with In Situ Generated Azides	244
10.3.3	Mechanistic Aspects of the CuAAC	244
10.3.4	Reactions of Sulfonyl Azides	250
10.3.5	Copper-Catalyzed Reactions with Other Dipolar Species	251
10.3.6	Examples of Application of the CuAAC Reaction	252
10.3.6.1	Synthesis of Compound Libraries for Biological Screening	252
10.3.6.2	Copper-Binding Adhesives	253
10.3.7	Representative Experimental Procedures	255
	Acknowledgements	256
	References	257
11	**"Frustrated Lewis Pairs": A Metal-Free Strategy for Hydrogenation Catalysis**	**261**
	Douglas W. Stephan	
11.1	Phosphine-Borane Activation of H_2	263
11.2	"Frustrated Lewis Pairs"	264

11.3	Metal-Free Catalytic Hydrogenation	*267*
11.4	Future Considerations	*273*
	Acknowledgements	*273*
	References	*273*

Index *277*

Preface

Many of the greatest success stories of organometallic and inorganic chemistry are in the application of metal complexes to catalytic reactions. In many cases, precious metals perform the heavy lifting, breaking H–H bonds, forming C–H or C–C bonds, etc. Precious metals have become so familiar in these roles that in some cases the precious metal and their catalytic reactivity seem almost inextricably linked. Wilkinson's catalyst, a rhodium complex, played a pivotal role in our understanding of hydrogenations. More recently, Noyori and co-workers developed remarkably reactive ruthenium complexes for asymmetric catalytic hydrogenations of C=O bonds. Over 150 years have passed since the discovery of a fuel cell that oxidizes hydrogen, yet modern low-temperature fuel cells still require platinum. Many carbon-carbon coupling reactions used extensively in organic synthesis function efficiently with extremely low loadings of palladium catalysts.

Kicking old habits is never easy, despite the allure of significant rewards for making the desired change. Yet we now know that the use of precious metals in catalysis is not always required. The research presented in this book shows how new catalysts that do not require precious metals may ultimately supplant the use of precious metals in some types of reactions. This book also highlights the challenges remaining in the development of catalysts that do not require precious metals. The pathway to devising new types of catalysts using abundant metals often involves scrutiny of reaction mechanisms that could potentially accomplish the desired goal, and finding ways to coerce inexpensive, abundant metals into accomplishing that task. In many cases those mechanisms are altogether different from those used in traditional precious metal catalysts. As can be seen in different chapters in this book, some of the catalytic reactions that use cheap metals are already competitive with well-known reactions that use precious metals. Even in new catalysts that do not yet exhibit rates or lifetimes that compare favorably to long-established and well-optimized precious metal catalysts, fundamentally new reactivity patterns have been discovered, and new classes of catalysts have been developed. This book provides detailed information on many types of reactions that can be catalyzed without the need for precious metals. I hope that these chapters may inspire others to join in the pursuit of "cheap metals for noble tasks."

Research on alternatives to precious metal catalysts has been growing rapidly in recent years, and expected to experience increased growth in the future. The most obvious reason for replacing precious metals is that they are very expensive, often costing more than 100 or 1000 times the cost of base metals. The high cost is obviously connected to the low abundance of these metals. High cost alone is not the only reason, however; in some cases specialized organic ligands (used in asymmetric catalysis, for example) cost more than the metal. Substantial costs are involved in industrial processes when recovery and recycle of the metal is required. Another attribute of avoiding precious metals is that some metals like iron have a minimal environmental and toxicological impact. Importantly, some large-scale uses in energy storage and conversion currently being considered would require large amounts of precious metals. In automotive transportation, for example, conversion to a "hydrogen economy" based entirely on fuel cells that require platinum would not be feasible, not only due to the high cost, but because there is not enough platinum available to accommodate such a huge scale of usage.

The order of the chapters in this book follows their order in periodic table, starting at the first row of group 6 (Norton's chapter on chromium catalysts) and continuing to Group 6 metals molybdenum and tungsten. While most of the inexpensive, abundant metals are from the first row of the periodic table, molybdenum and tungsten (from the second and third row of the periodic table) are exceptions, as they are much less expensive than precious metals. Subsequent chapters focus on catalysis by iron (Group 8), cobalt (Group 9), nickel (Group 10) and copper (Group 11). The last chapter highlights new catalysts that have no transition metals at all, using the main group elements phosphorus and boron. The cover highlights the inexpensive, abundant metals that are discussed in this book, with those metals being highlighted in green, and the precious metals of low abundance and high cost being shown in red. Manganese is abundant and inexpensive, and offers appealing opportunities for development into catalytic reactions. But since no chapters in this book focus on Mn, so it was not shown in green on the cover.

This book focuses on homogeneous (molecular) catalysts. There is a need to replace precious metals used in heterogeneous catalysis as well, but that topic is beyond the scope of material that can be covered in one book.

I sincerely thank all of the authors of the chapters in this book. They contributed their expertise and time in the writing of their chapters, and gracefully put up with annoying e-mails and editorial suggestions from me. I appreciate the enthusiasm they share for developing the chemistry of abundant, inexpensive metals as attractive alternatives to precious metals. Paul Chirik and Jack Norton gave me very helpful advice in the planning of this book.

I am deeply indebted to many scientific colleagues who have influenced my thinking, and who helped teach me chemistry over the years. In particular, Carol Creutz (Brookhaven National Laboratory) and Dan DuBois (Pacific Northwest National Laboratory) have both been extremely generous with their time and patient with my questions. I thank my scientific mentors, Chuck Casey and Jack Norton, for invaluable advice on many topics for more than twenty-five years.

It was a pleasure to work with Dr. Heike Nöthe at Wiley-VCH, and with Dr. Manfred Köhl in the early stages of preparations and planning for this book.

I dedicate this book to my wife, Cindy, to my son, Claude, and to my daughter, Lindsay. I thank them for being immensely supportive, including times when I was in the lab or my office rather than at home.

June 2010   *R. Morris Bullock*

List of Contributors

Ryan D. Baxter
University of Michigan
Department of Chemistry
930 North University Avenue
Ann Arbor, MI 48109-1055
USA

Renee Becker
University of Rochester
Department of Chemistry
Rochester, NY 14627
USA

R. Morris Bullock
Pacific Northwest National Laboratory
Chemical and Materials Sciences
Division
P.O. Box 999
K2-57
Richland, WA 99352
USA

Paul J. Chirik
Cornell University
Department of Chemistry and
Chemical Biology
Baker Laboratory
Ithaca, NY 14853
USA

Daniel L. DuBois
Pacific Northwest National Laboratory
Chemical and Materials Sciences
Division
Richland, WA 99352
USA

M. Rakowski DuBois
Pacific Northwest National Laboratory
Chemical and Materials Sciences
Division
Richland, WA 99352
USA

M. G. Finn
The Scripps Research Institute
Department of Chemistry
10550 North Torrey Pines Road
La Jolla, CA 92037
USA

Valery V. Fokin
The Scripps Research Institute
Department of Chemistry
10550 North Torrey Pines Road
La Jolla, CA 92037
USA

Catalysis Without Precious Metals. Edited by R. Morris Bullock

ISBN: 978-3-527-32354-8

Vernon C. Gibson
Imperial College
Department of Chemistry
South Kensington Campus
London SW7 2AZ
UK

John Hartung
Columbia University
Department of Chemistry
3000 Broadway
New York, NY 10027
USA

Yongwen Jiang
Chinese Academy of Sciences
Shanghai Institute of Organic Chemistry
State Key Laboratory of Bioorganic and Natural Products Chemistry
354 Fenglin Lu
Shanghai 200032
China

William D. Jones
University of Rochester
Department of Chemistry
Rochester, NY 14627
USA

Dawei Ma
Chinese Academy of Sciences
Shanghai Institute of Organic Chemistry
State Key Laboratory of Bioorganic and Natural Products Chemistry
354 Fenglin Lu
Shanghai 200032
China

Hasnain A. Malik
University of Michigan
Department of Chemistry
930 North University Avenue
Ann Arbor, MI 48109-1055
USA

John Montgomery
University of Michigan
Department of Chemistry
930 North University Avenue
Ann Arbor, MI 48109-1055
USA

Jack R. Norton
Columbia University
Department of Chemistry
3000 Broadway
New York, NY 10027
USA

Richard R. Schrock
Massachusetts Institute of Technology
Department of Chemistry
Cambridge, MA 02139
USA

Gregory A. Solan
University of Leicester
Department of Chemistry
University Road
Leicester LE1 7RH
UK

Douglas W. Stephan
University of Toronto
Department of Chemistry
80 St. George St.
Toronto
Ontario
Canada M5S 3H6

1
Catalysis Involving the H• Transfer Reactions of First-Row Transition Metals

John Hartung and Jack R. Norton

The M–H bonds of transition-metal hydride complexes may be cleaved heterolytically (H^+, H^- transfer) or homolytically (H• transfer). ΔG for the H^+ transfer in Equation 1.1 is readily quantified by pK_a measurements (see Chapter 3). Analogous measurements for H^- transfer, or "hydricities", are difficult because the loss of H^- generates a vacant coordination site. However, ΔG for Equation 1.2 can be determined indirectly, from electrochemical and pK_a measurements in the appropriate solvent [1, 2], and we can thus compare the "hydricities" of various hydride complexes (see Chapter 3). The lowest values of ΔG_{H^-} (corresponding to the complexes most eager to transfer H^-) are found for second- and third-row transition metals[1] [3], which is why those (relatively expensive) metals are good H^- donors and effective catalysts for reactions like ionic hydrogenation [5–10].

$$M-H \rightleftharpoons M^- + \text{solvent}H^+ \qquad (1.1)$$

$$[M-H]^- \rightleftharpoons M + H^- \qquad (1.2)$$

The thermodynamics of the homolytic cleavage of an M–H bond (1.3) are also available from electrochemical and pK_a measurements (the thermodynamic cycle in Equations 1.4–1.6). The oxidation potential is that of the one-electron process in Equation 1.5. If the pK_a is measured in CH_3CN, and the potential is measured relative to ferrocene/ferrocenium in that solvent, ΔG for Equation 1.3 in CH_3CN is given by Equation 1.7 [11, 12], while the bond dissociation energy (BDE) for Equation 1.3 (the traditional gas phase "bond strength") is given by Equation 1.8[2].

$$M-H \rightleftharpoons M\bullet + H\bullet \qquad (1.3)$$

$$M-H \rightleftharpoons M^- + H^+ \qquad (1.4)$$

$$M^- \rightleftharpoons M\bullet + e^- \qquad (1.5)$$

$$H^+ + e^- \rightleftharpoons H\bullet \qquad (1.6)$$

---

1) For $[HM(P–P)_2]^+$, experiments show that ΔG_{H^-} decreases in the order Ni > Pt > Pd; see reference [4].

2) The relative and absolute uncertainties in bond strengths determined in this way are discussed in reference [13].

Catalysis Without Precious Metals. Edited by R. Morris Bullock
© 2010 WILEY-VCH Verlag GmbH & Co. KGaA, Weinheim
ISBN: 978-3-527-32354-8

Table 1.1 Bond dissociation energies of some chromium and vanadium hydrides.

Hydride	M–H BDE, kcal mol^{-1}
CpCr(CO)$_3$H	62.2
dppm(CO)$_4$VH	57.9
dppe(CO)$_4$VH	57.5
dppp(CO)$_4$VH	56.0
dppb(CO)$_4$VH	54.9

$$\Delta G(\text{M–H})(\text{kcal mol}^{-1}) = 1.37 \text{p}K_a(\text{M–H}) + 23.06(\text{M}^-/\text{M}^\bullet) + 53.6 \quad (1.7)$$

$$\text{BDE}(\text{M–H})(\text{kcal mol}^{-1}) = 1.37 \text{p}K_a = 23.06 E^\circ + 59.5 \quad (1.8)$$

Such measurements show the weakest M–H bonds to be those of first-row transition metals [4], and suggest that these (relatively abundant and cheap) metals are best for catalyzing reactions that involve H$^\bullet$ transfer. Because most of these metals are nontoxic, their H$^\bullet$ transfer reactions offer an attractive alternative to the tin-mediated radical chemistry that has become ubiquitous in organic synthesis.

Bond dissociation energies of most M–H bonds lie between 60 and 65 kcal mol^{-1} [4]. A few are much stronger: the Os–H bond of Cp(CO)$_2$OsH has a BDE ≥82 kcal mol^{-1} [14]. On the other hand V–H bonds are particularly weak. Calculations at the B3LYP level of theory on the hypothetical VH$_5$ give it the weakest M–H bond (43 kcal mol^{-1}) among neutral "valency-saturated" MH$_n$ (i.e., among complexes where M forms the maximum number of M–H bonds) [15]. Experimentally Table 1.1 [13] shows very weak V–H bonds for (P–P)(CO)$_4$VH (P–P = Ph$_2$P(CH$_2$)$_n$PPh$_2$, with n = 1 (dppm), n = 2 (dppe), n = 3 (dppp), and n = 4 (dppb) [16, 17].

An H$^\bullet$ transfer, or hydrogen atom transfer (HAT), reaction has been defined by Mayer as "the concerted movement of a proton and an electron ... in a single kinetic step where both ... originate from the same reactant and travel to the same product." [18] Mayer considers HAT to be "one type of the broad class of proton-coupled electron transfer (PCET) reactions, which also includes reactions where the proton and electron are separated." The distinction is a matter of ongoing discussion [19, 20], and other acronyms have been proposed [19, 21], but all the reactions to be considered in this chapter can be satisfactorily described as "H$^\bullet$ transfer".

1.1
H$^\bullet$ Transfer Between M–H Bonds and Organic Radicals

HAT reactions from transition-metal hydrides to organic radicals R$^\bullet$ (1.9) are characterized by second-order kinetics. Second-order kinetics have been

Table 1.2 Bond dissociation energies and rates of H˙ transfer to tris(p-tert-butylphenyl)methyl radical (extrapolated to room temperature) of several hydrides.

Hydride	BDE, kcal mol^{-1}	k_H, M^{-1}s^{-1}
Cp(CO)$_2$Fe–H	57[a]	1.2 × 10^4
Cp(CO)$_3$Cr–H	62[b]	335
Cp(CO)$_2$Ru–H	65[a]	1.03 × 10^3
(CO)$_4$Co–H	59,[c] 67[a]	1.6 × 10^3
(CO)$_5$Mn–H	68[a]	741
Cp(CO)$_3$Mo–H	69[a]	514
Cp(CO)$_3$W–H	72[a]	91
(CO)$_4$OsH$_2$	78[d]	15.7

a) Ref [120].
b) Ref [13].
c) Ref [22].
d) Ref [23].

established (the rate constants are shown in Table 1.2) for the transfer of H˙ from a variety of hydrides to the tris(p-tert-butylphenyl)methyl radical [24–26]. The bulky t-Bu substituents keep it entirely monomeric in solution [27, 28]. Additional evidence that H˙ transfers to R˙ obey second-order kinetics is provided by studies of the last step in radical hydrogenation reactions (see Equations 1.14 and 1.15 below). If the hydride L$_n$M–H in a transfer such as Equation 1.9 is coordinatively saturated (18 electrons), the metalloradical L$_n$M˙ will have a 17 electron configuration.

$$L_nM-H + R \rightleftharpoons L_nM˙ + R-H \tag{1.9}$$

Second-order kinetics have also been established for H˙ transfer in the reverse direction, that is, from R–H to M˙. The rate constant for Equation 1.10, from 1,4-cyclohexadiene to Cp(CO)$_2$Os˙, has been measured by time-resolved IR spectroscopy as 2.1 × 10^6 M^{-1}s^{-1} (23°C, hexane) [14]. Intriguingly, H˙ can be abstracted by a photogenerated osmium metalloradical from even stronger C–H bonds, such as those of toluene and THF; photolysis of [Tp(CO)$_2$Os]$_2$ (Tp = tris(pyrazolyl) borate) in either of these solvents gives the osmium hydride. (The C–H bonds of toluene (90 kcal mol^{-1} for the methyl group) and THF (92 kcal mol^{-1}) are considerably stronger than that of cyclohexadiene (77 kcal mol^{-1}) [14, 29].)

$$Cp(CO)_2Os˙ + \underset{}{\text{(cyclohexadiene)}} \longrightarrow Cp(CO)_2Os-H + \underset{}{\text{(cyclohexadienyl˙)}} \tag{1.10}$$

Steric factors can be important in H˙ transfer reactions. The rate constant (k_H) of H˙ transfer to tris(p-tert-butylphenyl)methyl radical decreases by a factor of 37 from Cp(CO)$_3$MoH to Cp*(CO)$_3$MoH (which has an Mo–H bond of comparable strength) [24, 25]. There is a smaller rate difference between Cp/Cp* with the more reactive 5-hexenyl radical; radical clock methods have shown that the rate constant for H· transfer decreases by a factor of two from $Cp(CO)_3MoH$ to $Cp*(CO)_3MoH$ at 298 K [30]. The considerable primary:secondary:tertiary selectivity of $Cp*(CO)_3MoH$ in these radical clock studies (26:7:1) arises from steric interactions with the radical substituents [30].

However, the strength of an M–H bond is particularly important in determining the rate of its H· transfer reactions. Moving from $Cp(CO)_3MoH$ to the (structurally similar) $Cp(CO)_3WH$ in Table 1.2 produces a 5.6-fold decrease in k_H to tris(*p*-*tert*-butylphenyl)methyl as the bond strength increases by $3\,\text{kcal}\,\text{mol}^{-1}$. Consideration of M–H bond strengths has enabled O'Connor and coworkers to employ an H· donor that works effectively with their Ru cycloaromatization catalyst. In Scheme 1.1 $Cp(CO)_3WH$ donates H· only to the diradical intermediate, whereas the weaker M–H bond in $Cp(CO)_3CrH$ transfers H· to the enediyne before cycloaromatization [31].

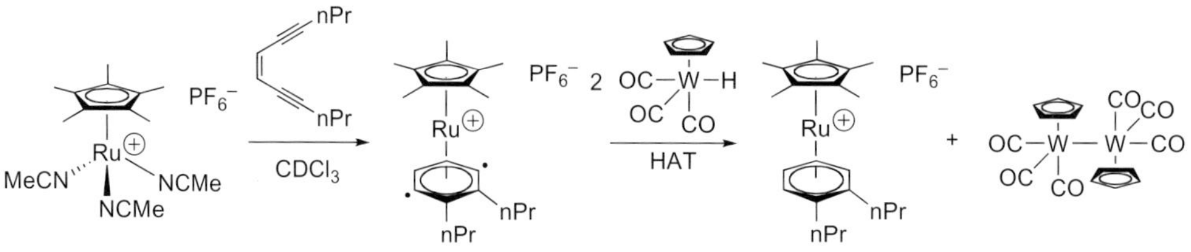

Scheme 1.1 Ruthenium mediated Bergman cycloaromatization.

Gansäuer and coworkers have developed a tandem catalytic approach to reductive epoxide opening that employs H· transfer from a dihydride complex (Scheme 1.2) [32, 33]. The carbon-centered radical that arises from rate-determining epoxide opening with Cp_2TiCl abstracts H· from the H_2 adduct of Wilkinson's catalyst. Manganese or zinc metal regenerates the active Ti^{III} species and makes the reaction catalytic.

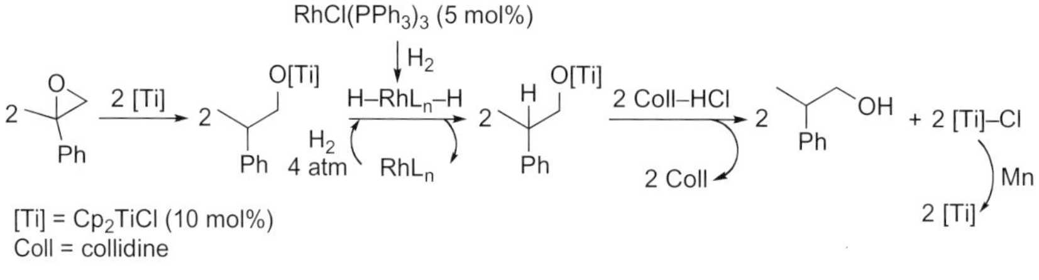

[Ti] = Cp_2TiCl (10 mol%)
Coll = collidine

Scheme 1.2 Tandem catalytic epoxide opening.

1.2
H· Transfer Between Ligands and Organic Radicals

The transfer of H· to organic radicals is not limited to hydride complexes. Although free N–H and O–H bonds have high bond dissociation enthalpies, they are weak-

1.2 H· Transfer Between Ligands and Organic Radicals

ened substantially by coordination of the N or O to a metal center. Such complexes, often of inexpensive transition metals, have been able to accomplish synthetically useful transformations.

Newcomb has measured the rate of H· transfer to carbon-centered radicals from the water and methanol complexes of Cp_2TiCl [34]. The rate constant (k_H) for HAT to the radical below (determined by the radical clock method, Scheme 1.3) in THF at room temperature is $1.0 \times 10^5 \, M^{-1} s^{-1}$.

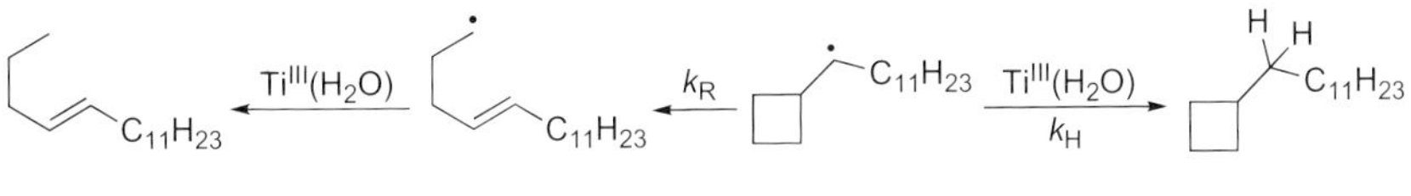

Scheme 1.3 Determination of k_H from $Cp_2TiCl(H_2O)$ to a secondary alkyl radical.

In an investigation related to the total synthesis of (+)-3α-hydroxyreynosin, Cuerva and Oltra discovered that the products of a Ti(III)-mediated epoxide-opening/cyclization cascade varied with the presence or absence of water. In dry THF they obtained the product originally desired, with an exocyclic Δ [3, 14] double bond, but the addition of water resulted in the formation of the reductively cyclized product in Scheme 1.4. Apparently H· is transferred from the titanium–water complex to the cyclized radical [35–37].

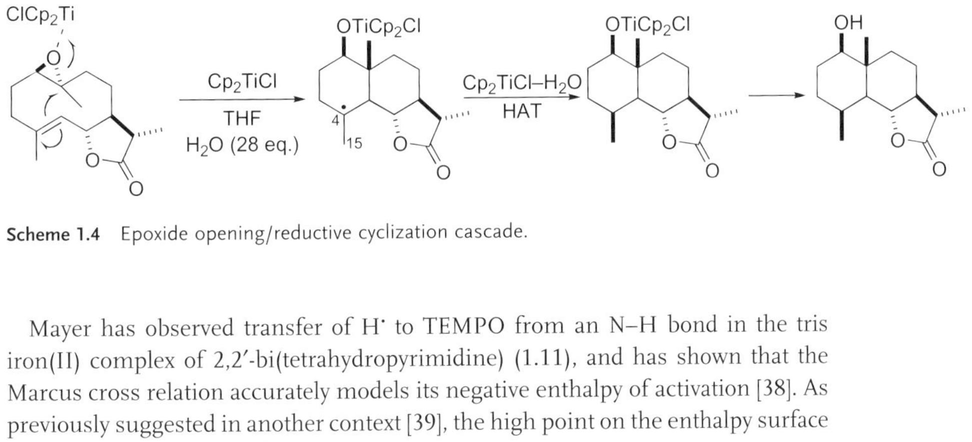

Scheme 1.4 Epoxide opening/reductive cyclization cascade.

Mayer has observed transfer of H· to TEMPO from an N–H bond in the tris iron(II) complex of 2,2′-bi(tetrahydropyrimidine) (1.11), and has shown that the Marcus cross relation accurately models its negative enthalpy of activation [38]. As previously suggested in another context [39], the high point on the enthalpy surface appears to occur before the transition state.

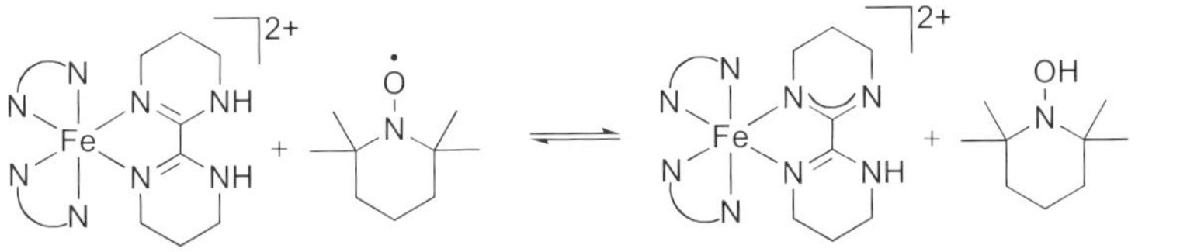

(1.11)

The abstraction of H· from organic substrates by oxo ligands occurs in the chemistry of cytochrome P450 and related biological oxidants [40, 41]. In most cases the resulting hydroxyl ligand is transferred to the organic radical (a "rebound" mechanism), giving an alcohol (Scheme 1.5).

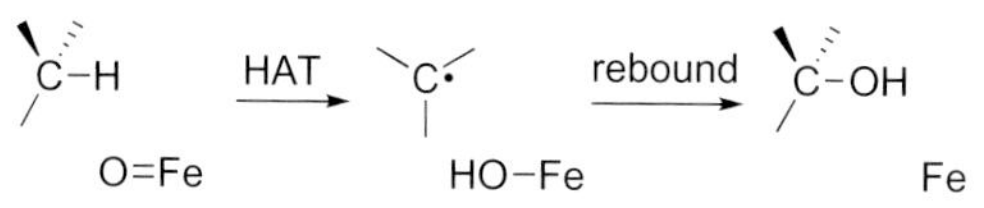

Scheme 1.5 Oxygen rebound mechanism.

Groves has demonstrated that norcarane functions as a mechanistic probe when oxidized by two bacterial enzymes $P450_{cam}$ and $P450_{BM3}$, two mammalian enzymes CYP2B1 and CYP2E1 [42], and the monooxygenase AlkB from the soil organism *Pseudomonas oleovorans* [43]. In all cases (although the yield was lower with CYP2B1) an appreciable amount of the radical rearrangement product 3-(hydroxymethyl)-cyclohexene was formed, and relatively little of the product (3-cyclohepten-1-ol) expected from a cationic rearrangement (Scheme 1.6). In contrast, no radical rearrangement product was observed after the oxidation of spiro[2.5] octane by $P450_{cam}$, $P450_{BM3}$, CYP2B1, and CYP2E1 [42]; the radical generated by H· abstraction rearranges four times more slowly than that from norcarane.

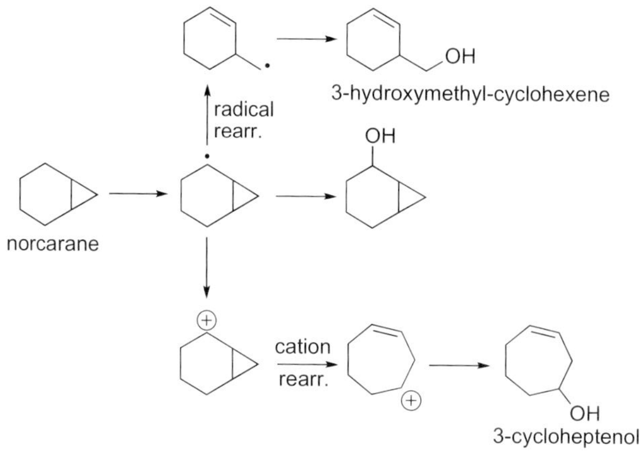

Scheme 1.6 Rearrangements available to norcarane during the oxygen rebound mechanism.

The oxo ligands in a number of model compounds have proven capable of removing H· from C–H bonds [44]. For example, the oxo ligands of [(phen)$_2$Mn(μ-O)$_2$-Mn(phen)$_2$] [4+] remove H· from fluorene (1.12) and xanthene, forming fluorenyl and xanthenyl radicals that dimerize or are oxidized to ketones (by radical addition to the bridging oxo ligand and subsequent oxidation) [45]. The rate of H· transfer in these systems is, to a first approximation, determined by ΔH, that is, by the strength of the O–H bond being made and the C–H bond being broken [44].

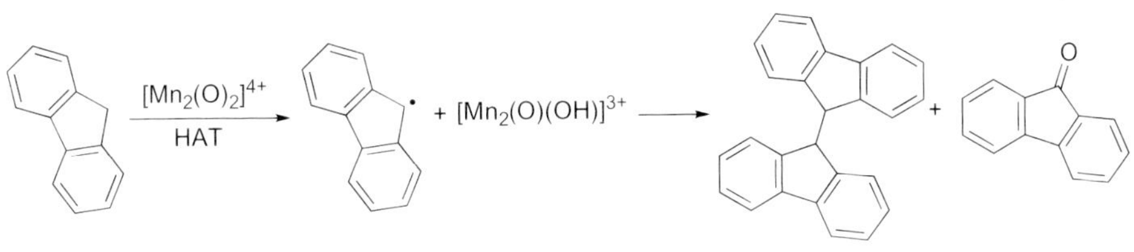

[Structural scheme] (1.12)

The H• transfer in Equation 1.13, to a vanadium oxo ligand, is surprisingly slow. The transfer requires a great deal of structural reorganization, as the V–O distance increases by 0.264 Å when V=O is converted to V–OH. The barrier to transfer of an H• from V–OH to V=O (in Marcus theory, the "intrinsic barrier" to self-exchange) is thus large, and its size is reflected in the size of the barrier to the C→O H• transfer in Equation 1.13 [46].

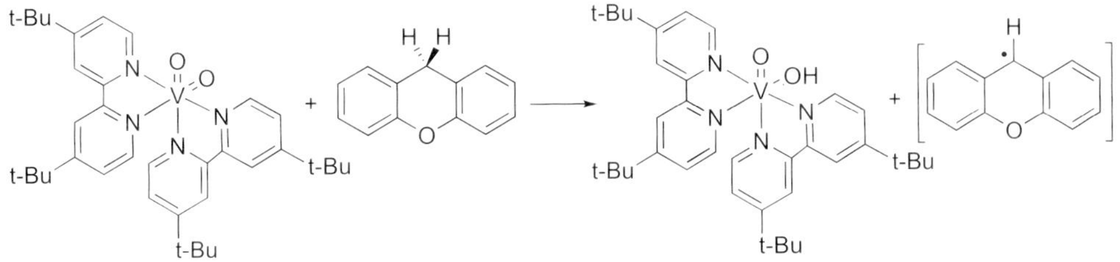

[Structural scheme] (1.13)

1.3
H• Transfer Between M–H and C–C Bonds

In the early 1950s reports appeared of the $Co_2(CO)_8$-catalyzed hydrogenation of polycyclic aromatic hydrocarbons in the presence of synthesis gas (CO/H_2). In 1975 Halpern proposed that these reactions occurred by sequential H• transfer from a cobalt hydride formed *in situ* (Scheme 1.7) [47]. This transfer must be reversible, because treatment of anthracene with $(CO)_4CoD$ leads to deuterated anthracene [48]. (Syn addition of Co–D would not lead to H/D exchange.)

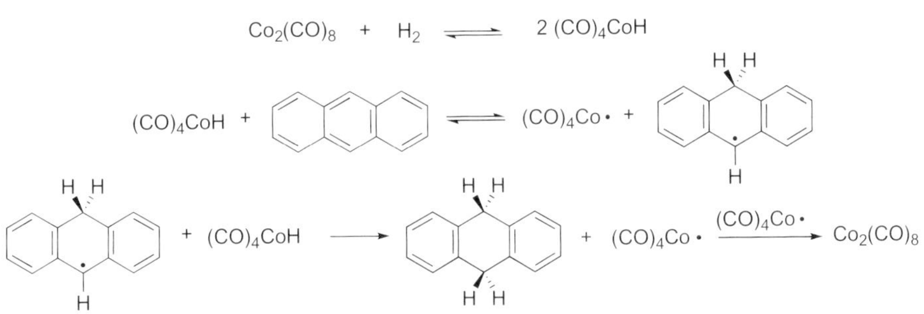

[Scheme]

Scheme 1.7 Reduction of anthracene by $Co_2(CO)_8$ under syngas.

In 1977 Sweany and Halpern reported the hydrogenation of α-methylstyrene by $(CO)_5MnH$ and proposed that it began with a similar step, the reversible transfer of H˙ from Mn–H to the olefin (1.14) [49]. They established reversibility by showing that exchange between $(CO)_5MnH$ and α-methylstyrene was faster than the overall reaction.

$$(CO)_5MnH + \underset{Me}{Ph\diagup\!\!\!=} \;\xrightleftharpoons[k_{-H}]{k_H}\; \underset{Me}{Ph\diagdown\!\!\overset{H}{\underset{\bullet}{\diagup}}} + (CO)_5Mn\bullet \;\xrightleftharpoons{k_{esc}}\; \underset{Me}{Ph\diagdown\!\!\overset{H}{\diagup}} + (CO)_5Mn\bullet \tag{1.14}$$

$$\underset{Me}{Ph\diagdown\!\!\overset{H}{\underset{\bullet}{\diagup}}} + (CO)_5MnH \longrightarrow \underset{Me}{Ph\diagdown\!\!\overset{H}{\diagup}} + Mn_2(CO)_{10} \tag{1.15}$$

Convincing evidence for the radical products in Equation 1.14, and therefore for H˙ transfer, was offered by the observation of enhanced and/or distorted peaks in the ^{1}H NMR known as CIDNP (chemically induced dynamic nuclear polarization) for ^{1}H nuclei possessing significant coupling to unpaired electrons. This phenomenon requires that the pair of radicals formed in Equation 1.14 be contained within a solvent cage, with competition between back transfer (k_{-H}) and escape from the cage (k_{esc}). The changes due to CIDNP are smaller at the higher magnetic fields now in general use, making this phenomenon less commonly observed.

An inverse isotope effect (about 0.4 at 65 °C) was observed for the hydrogenation of α-methylstyrene according to Equations 1.14 and 1.15. (Cage escape, k_{esc} in Equation 1.14, appears to be the rate-determining step.) Sweany and Halpern wrote that an inverse effect in Reaction (1.14) was "not unexpected" in view of the "low initial frequency of the Mn–H bond ($\approx 1800\,cm^{-1}$) relative to that of the C–H bond ($\approx 3000\,cm^{-1}$)" [49]. By assuming 60 kcal mol^{-1} for Mn–H they estimated that $\Delta H°$ for H˙ transfer from $(CO)_5MnH$ to styrene would be +15 kcal mol^{-1} [50], making Equation 1.14 significantly endothermic[3]. A similar $\Delta H°$, +15–16 kcal mol^{-1}, is implied by our BDE (52.4 kcal mol^{-1}) for the α-methylbenzyl radical [50–52] and the Mn–H BDE in Table 1.2 (68 kcal mol^{-1})[4]. Our C–H BDE is consistent with the strengths of C–H bonds next to other radical centers [54]. Inverse isotope effects often result from the presence of pre-equilibria that are less unfavorable for D than for H [55], but it is not clear whether such a pre-equilibrium is established inside the cage in Equation 1.14. It is possible that the inverse isotope effect observed for Equation 1.14 is the result of a late transition state for k_H [56]. It would be useful to know the vibrational frequencies of the α-methylbenzyl radical so that the equilibrium isotope effect could be calculated.

---

3) The fact that such reactions are thermodynamically uphill was pointed out in reference [24].

4) An incorrect estimate of $\Delta H°$ for H˙ transfer from (CO)5MnH to styrene, "+8–10 kcal mol^{-1}", is given in reference [23] of reference [53].

1.3 H· Transfer Between M–H and C–C Bonds

In 1985 Jacobsen and Bergman proposed a similar H· transfer mechanism for the reaction of the dicobalt complex (μ-CCH$_2$)(CpCoCO)$_2$ with Cp(CO)$_2$(L)MoH (L = CO, PMe$_3$, PPh$_3$) (1.16). They reported that H/D exchange was faster than the overall reaction, and that the isotope effect was inverse. (It became more inverse as the solvent viscosity increased, which they explained by suggesting that the greater viscosity made escape slower, increased the ratio of k_H to k_{esc}, and allowed the H· transfer equilibrium within the cage more time to establish itself prior to escape [57].)

$$\text{OC-Co(Cp)-Co(CO)(Cp)(=CH}_2\text{)} \xrightarrow{\text{Cp(CO)}_3\text{MoH}, \ k_H} \text{OC-Co(Cp)-Co(CO)(Cp)(CHCH}_3\text{)} + \text{Cp(CO)}_3\text{Mo·} \quad (1.16)$$

$$\text{CpCo(O)(Cp)(Co)(CO)-Mo(CO)(Cp)(CH}_3\text{)} \xleftarrow{\text{Cp(CO)}_3\text{Mo·}, \ -2\text{CO}} \text{OC-Co(Cp)-Co(CO)(Cp)(CHCH}_3\text{)} \xrightarrow{\text{Cp(CO)}_3\text{MoH}, \ k_{H2}} \text{OC-Co(Cp)-Co(CO)(Cp)(CH}_2\text{CH}_3\text{)}$$

(1.17)

Bullock determined the rate constant k_H for the transfer in Equation 1.18 from the overall rate of hydrogenation, and used the rearrangement of the cyclopropylbenzyl radicals as a clock to determine k_{H2} in Equation 1.19 [58]. The reversibility of Equation 1.18 was demonstrated by the observation that, with Cp(CO)$_3$CrD, H/D exchange was faster than the overall reaction; an inverse isotope effect (0.55 at 100 °C) was obtained with Cp(CO)$_3$WH/Cp(CO)$_3$WD. The relative values of k_H in Equation 1.18 (which is generally endothermic) were largely determined by the strengths of the M–H bonds, while the relative values of k_{H2} in Equation 1.19 (which is relatively exothermic) were largely determined by steric effects.

$$\text{Ph-C(=CH}_2\text{)(cPr)} + \text{M–H} \xrightarrow{k_H} \text{Ph-CH(CH}_3\text{)(cPr)·} + \text{M·} \quad (1.18)$$

$$\text{Ph-CH(CH}_3\text{)(cPr)·} \xleftarrow[k_{H2}]{\text{M–H}} \text{Ph-C·(CH}_3\text{)(cPr)} \xrightarrow{k_R} \text{Ph-CH=CH-CH}_2\text{CH}_2\text{·} \xrightarrow{\text{M–H}} \text{Ph-CH=CH-CH}_2\text{CH}_3 \quad (1.19)$$

Our group has determined the rate constants for H· donation from Cp(CO)$_3$CrH and related Cr hydrides to methyl methacrylate and styrene [51, 52], and compared them to the rates at which the various (P–P)(CO)$_4$VH donate H· to the same substrates [13]. The strength of the Cr–H bond in Cp(CO)$_3$CrH (62.2 kcal mol^{-1} in Table 1.1) is surely one reason why its H· transfer to styrene is relatively slow. The V–H bonds in (P–P)(CO)$_4$VH transfer H· more rapidly, although the difference is

Table 1.3 M–H bond dissociation energies and rates of H· transfer to styrene.

Hydride	BDE, kcal mol^{-1}	$10^{-3} \times k_H$, M^{-1} s^{-1}
dppm(CO)$_4$VH	57.9	≥17
dppe(CO)$_4$VH	57.5	≥9.0
dppp(CO)$_4$VH	56.0	7.0
dppb(CO)$_4$VH	54.9	5.7
Cp(CO)$_3$CrH	62.2	0.85

not as large as would be expected from the decrease in bond strength between these and the Cr hydrides. The vanadium hydride dppe(CO)$_4$VH transfers H• to styrene about 10 times more rapidly than does Cp(CO)$_3$CrH [13].

As Table 1.3 shows, k_H becomes *slower* as the V–H bonds in (P–P)(CO)$_4$VH become *weaker*; presumably the steric effect of the chelating ligand undermines the effect of the weaker bond. It is worth noting a precedent for the fast transfer of H· from vanadium: k_H from the anionic hydride Cp(CO)$_3$VH$^-$ to a hexenyl radical has been reported to be "more than an order of magnitude larger than that measured for Bu$_3$SnH" [59].

The influence of the double bond substituents on k_H has been determined by treating a series of olefins with Cp(CO)$_3$CrH and its D analogue. Comparing the reactivity of the various olefins requires that we estimate k_H (for H· transfer from Cp(CO)$_3$CrH) when we have measured k_D (from Cp(CO)$_3$CrD). As stated previously, inverse isotope effects are generally observed for the transfer of H· from M–H to a carbon–carbon double bond [56, 60]. To date, the only isotope effect reported for Cp(CO)$_3$CrH, by Bullock *et al.* [58], is $k_H/k_D = 0.45 \pm 0.10$ for its reaction with 2-cyclopropylpropene at 68 °C. We have used 0.45 (the effect of the temperature difference is negligible compared to the uncertainty) to estimate the k_H values for hydrogen atom transfer to olefins; we have then calculated the relative rates from these k_H values to deduce the chemoselectivity of this step [61]. Some of the olefins in Table 1.4 undergo hydrogenation, some undergo H/D exchange, but in either case it is straightforward to estimate k_H (Table 1.4)[5].

A substituent on the carbon that receives the H· slows k_H by a factor of about a thousand (compare entries 1 and 2). Substituents that stabilize the resulting radical increase k_H (CH$_3$<CO$_2$Me<Ph) [61].

---

5) Deducing k_H from the rate constant for H/D exchange requires that we estimate the isotope effect for back transfer – the relative rate at which H and D transfer back to the Cr from the organic radical initially formed. In most cases the resulting k_H is not very sensitive to this isotope effect estimate (which we have, from the results quoted after Equation 1.27, taken as 3 in references [62] and [63]).

Table 1.4 Rates of H· transfer from Cp(CO)$_3$CrH to various olefins at 49.8 °C.

Entry	Olefin	$10^{-3} \times k_H$ (M^{-1}s^{-1})	Relative rate
1	Ph₂C=CH(CH₃) (Ph, Ph)	0.59 (2)	1
2	Ph₂C=CPh (Ph, Ph)	460 (60)	780
3	Ph₂C=CH₂ (Ph)	79 (3)	134
4	CH₂=CHCO₂Me	≤3.2 × 10^{-4}	≤5 × 10^{-4}
5	CH₂=CH(CH₂)₅CH₃	≤1.1 × 10^{-4}	≤2 × 10^{-4}
6	CH₂=C(CH₃)CO₂Me	(0.8 to 1.6) × 10^{-2}	≈0.02
7	(CH₃)₂C=C(CH₃)(CH₂)₄	≤3.2 × 10^{-3}	≤0.005
8	PhCH=CH₂	15.8 (6)	27
9	CH₂=C(CH₃)CO₂Me	14 (3)	24

1.4
Chain Transfer Catalysis

H· transfer *from* the α carbon of an organic radical *back* to a metalloradical (the reverse of 1.18) is a key step in CCT (catalytic chain transfer), a process that can be used to control radical polymerizations. Many of the effective CCT catalysts contain inexpensive and abundant transition metals; the first catalysts all contained cobalt.

Suess [64] and Schulz [65] observed that the presence of solvents during the polymerization of styrene lowered the molecular weight. In the case of CCl$_4$, Mayo suggested that the chain-carrying radical abstracted Cl· from CCl$_4$ (1.20), terminating the chain but leaving CCl$_3$·, which added to monomer and initiated a new chain (1.21).

$$\text{(P)-CH}_2\text{-}\overset{\cdot}{\text{CH}}\text{-Ph} + \text{S–X} \longrightarrow \text{(P)-CH}_2\text{-CHX-Ph} + \text{S·} \qquad (1.20)$$

1 Catalysis Involving the H· Transfer Reactions of First-Row Transition Metals

$$S\cdot \;+\; \underset{Ph}{\diagup\!\!\!\diagdown}\;\longrightarrow\; \underset{Ph}{\overset{S}{\diagup\!\!\!\diagdown}}\cdot \tag{1.21}$$

Mayo defined a chain transfer constant that could be empirically determined by Equation 1.22. $\bar{P}$ is the degree of polymerization, S and M are concentrations of solvent and monomer respectively, and $\bar{P}_0$ is the degree of polymerization in the absence of solvent. Thiols are particularly effective (albeit stoichiometrically as in 1.20 and 1.21) reagents for chain transfer [66].

$$\frac{1}{\bar{P}} = C\frac{(S)}{(M)} + \frac{1}{\bar{P}_0} \tag{1.22}$$

It is possibly to carry out chain transfer *catalytically*. The process is related to atom transfer radical polymerization (ATRP) [67, 68] and related "living" polymerizations which keep the concentration of chain-carrying radicals low. ATRP employs a halide complex (often $Ru^{III}X$) that is subject to facile one-electron reduction; that complex reversibly donates $X\cdot$ to the chain-carrying radical (1.23) and thereby decreases the concentration of the latter [69, 70].

$$L_nM \;+\; \underset{P\;\;\;\;R}{\diagup\!\!\!X\!\!\!\diagdown} \;\rightleftharpoons\; L_nM\text{--}X \;+\; \underset{P\;\;\;\;R}{\diagup\!\!\!\cdot\!\!\!\diagdown} \;\xrightarrow{\text{polymerization}} \tag{1.23}$$

In CCT a metalloradical reversibly *abstracts* $H\cdot$ from the chain-carrying radical and starts a new chain. Early work on CCT during radical polymerizations employed cobalt porphyrins during the polymerization of methyl methacrylate, and was carried out in the USSR (Smirnov, Marchenko in 1975; Enikolopyan in 1977). Gridnev discovered in Moscow in 1979 that cobaloximes were effective CCT catalysts, then moved to the US in 1992 (Wayland laboratory, University of Pennsylvania) and joined DuPont in 1994. The basic features of CCT have been described in a series of patents (at first Russian, then largely DuPont) that appeared in the 1980s [71], and in a comprehensive review that appeared in 2001 [72]. The mechanism in Scheme 1.8 has become generally accepted, and CCT has been successfully applied to other monomers (styrene, methacrylonitrile) and comonomers.

In Scheme 1.8, propagation (1.24) occurs by the addition of the chain-carrying radical to the double bond of the monomer. Chain transfer requires that hydrogen atoms shuttle from the growing polymer chain to the same (monomer) double bond. In the first step, Equation 1.25, $H\cdot$ is removed from the chain-carrying radical; in Equation 1.26 its transfer to fresh monomer begins a new chain [72]. The competition of $H\cdot$ removal (1.25) with propagation (1.24) reduces the chain length and molecular weight of the polymer. (During a free radical polymerization the rate constant for propagation is chain length dependent [73].)

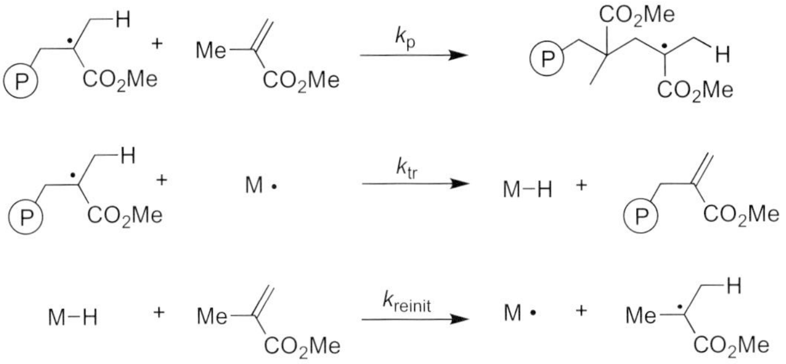

1.4 Chain Transfer Catalysis

Scheme 1.8 Catalytic chain transfer polymerization.

The efficiency of a chain transfer catalyst has traditionally been measured by its chain transfer constant (C_S), which is the ratio of the rate constant for chain transfer (k_{tr}) to that for propagation (k_p). Values of C_S are generally determined from the slope of a plot of $1/DP_n$ vs. [CTA]/[M] (the Mayo method, Equation 1.27 [74]), in which DP_n is the degree of polymerization, CTA is the chain transfer agent, and M is monomer.

$$\frac{1}{DP_n} = \frac{1}{DP_{n0}} + \frac{k_{tr}[CTA]}{k_p[M]} \tag{1.27}$$

By comparing the termination rate constant during a MMA-d_0 polymerization with that during a MMA-d_5 polymerization, the isotope effect $k_{tr(H)}/k_{tr(D)}$ can be estimated as 2.93 at 60 °C [62]. It can also be estimated as 3.5 (between 40 and 80 °C) by comparing the efficiency of a cobalt catalyst for chain transfer during MMA polymerization to the efficiency of the same catalyst during the polymerization of MMA-d_8 [63].

While macrocyclic cobalt(II) complexes are effective catalysts for chain transfer (see above), the hydrides they form during the catalytic cycle are unobservable, presumably because they are so efficient in transferring H˙ back to monomer (1.26) [72]. Chromium and vanadium metalloradicals offer the advantage that their hydrides are stable enough to be observed. For the polymerization of methyl methacrylate, $(C_5Ph_5)(CO)_3Cr˙$ is a good chain transfer catalyst, $Cp(CO)_3Cr˙$ (in equilibrium with its dimer) is a much better one [75], and the vanadium metalloradical $dppe(CO)_4V˙$ shows respectable activity [76]. This effectiveness implies that the corresponding hydrides (($C_5Ph_5)(CO)_3CrH$, $Cp(CO)_3CrH$, and $dppe(CO)_4VH$) transfer H˙ back to methyl methacrylate.

The ratio between the metalloradical and the corresponding hydride during chain transfer catalysis is determined by the balance between H˙ transfer from the chain-carrying radical to M˙ (as in 1.25) and reinitiation by transfer from M–H to the monomer (as in 1.26). We can define a C_S(true) in terms of the apparent C_S (from the slope of a Mayo plot) by Equation 1.28, where $[M˙]_{st}$ is the concentration at steady state. For the various Cr hydrides, with their relatively strong Cr–H bonds, the balance lies slightly on the side of H˙ transfer to Cr˙, so [Cr–H] > [Cr˙], and from Equation 1.28 $C_S(\text{true}) > C_S(\text{app})$. For Co(II) chain transfer catalysts, the balance lies completely on the side of the Co(II) "metalloradical", and $C_S(\text{true})$ is equal to $C_S(\text{app})$ [77].

$$C_S(\text{app}) = C_S(\text{true}) \times \frac{[M\bullet]_{st}}{[M\bullet]_0} \qquad (1.28)$$

The characteristics of an effective chain transfer catalyst are now clear [75]. First, M• must be stable under the conditions of polymerization. Second, the metal center must be crowded enough to discourage (i) the dimerization of M•, (ii) the formation of a bond between M• and the chain-carrying radical R•, and (iii) the hydrogenation of R• by the transfer of H• from M–H. (Dimerization of M• will interfere with its ability to abstract H• from R•; formation of M–R bonds will also deplete M•; transfer of H• to R• will terminate growing chains.) Finally, the BDE of the M–H bond in the corresponding hydride should be as close as possible to the BDE of the C–H bond in the chain-carrying radical. (If the substrate is methyl methacrylate, the C–H BDE formed by the H• transfer in Equation 1.26 will be 50 kcal mol^{-1} [51, 52].)

Cobalt(II) macrocyclic complexes (Figure 1.1, Table 1.5) are the best available chain transfer catalysts [78].

There is a possibility that, during chain transfer catalysis, H• transfer occurs not only to monomer but also to the vinyl-terminated oligomers generated by chain transfer. Wayland and coworkers have reported that the degree of polymerization increases with the extent of conversion during the polymerization of MMA and

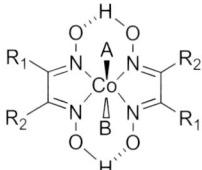

Figure 1.1 Cobaloxime.

Table 1.5 Effect of substituents on cobaloximes: methyl methacrylate polymerization.

R_1	R_2	A	B	C_s
Me	Me	Cl	Py	5 000
Me	Me	PPh$_3$	PPh$_3$	20 000
Ph	Ph	Cl	Py	30 000
Ph	Ph	Cl	PPh$_3$	100 000

methacrylic acid [79] with a Co porphyrin as a chain transfer catalyst. This result suggests that reinitiation can occur by H• transfer to these oligomers.

Treatment of Cp*(CO)$_3$CrD with a large excess of MMA *dimer* allows the measurement of k_{reinit} by monitoring the increase in the hydride resonance due to H/D exchange [80]. This method gives a value of $3 \times 10^{-4}\,M^{-1}\,s^{-1}$, which is much smaller than the k_{reinit} for MMA itself ($1.74 \times 10^{-3}\,M^{-1}\,s^{-1}$) [51].

1.5
Catalysis of Radical Cyclizations

Methods involving carbon-centered radicals are used extensively in synthesis [81–83]. The cyclization of such radicals forms the rings common in natural products, and tolerates functional groups – for example, unprotected hydroxyls – that cannot be present during some heterolytic transformations.

Generally, cyclizations begin with the photo- or initiator-stimulated formation of a tin radical, which abstracts X• from R–X (1.29); where X is Br, I, occasionally Cl, or PhSe or PhS [84]. The resulting R• then undergoes cyclization or rearrangement to R'• before abstracting H• from a tin hydride (typically Bu$_3$SnH), which regenerates the tin radical (1.30) and continues the chain. Such methods are necessarily stoichiometric both in tin and in another heavy element (Br, I, Se, S).

$$R-X + Bu_3Sn• \rightarrow R• + Bu_3SnX \tag{1.29}$$

$$R'• + Bu_3SnH \rightarrow R'-H + Bu_3Sn• \tag{1.30}$$

Related methods begin with the addition of a radical Rad• to the C=S double bond in a xanthate (1.31) or thiohydroxamate (1.32) ester [84, 85]. (Rad• is often R$_3$Sn• but in initial investigations *t*-BuS• and CCl$_3$• were also used [86, 87].) The resulting intermediates give R• by β scission (31 and 32) and decarboxylation (1.32).

(1.31)

(1.32)

Special care is required to handle trialkyltin hydrides and the waste they generate, and standard laboratory purification techniques often leave toxic levels of tin compounds in the product [88]. The industrial application of these methods has been hindered by the need to remove these tin-containing contaminants. Methods catalytic in tin have been developed [89–91], and tin hydride reagents modified to make their removal easier [92–98], but the need for alternatives to tin remains [99–101]. A review entitled "Flight from the Tyranny of Tin" has appeared [102].

Substitutes such as N-ethylpiperidinium hypophosphite [103], $(Me_3Si)_3SiH$ [104], Bu_3GeH [105], $HGaCl_2$ [106], and $HInCl_2$ [107] contain bonds to hydrogen stronger than that of 78 kcal mol^{-1} in Bu_3SnH [29] and are, therefore, likely to be less reactive in H$^\bullet$ transfer. Studer has remarked that "transition metal based hydrides are promising alternatives to the tin hydrides" [108], and $Cp_2Zr(H)Cl$ has been used for the cyclization of halo acetals [109].

Radicals that will undergo cyclization can be generated *without the use of tin* by the transfer of H$^\bullet$ from M–H to one C=C in an appropriate 1,6 diene. The two C=C in the diene must, however, perform different functions: one must accept H$^\bullet$ quickly; the other must undergo facile intramolecular attack by the radical thus generated. Table 1.4 [61] predicts that the "a" double bond in **1** should be kinetically favored for H$^\bullet$ transfer (25 times faster), yielding radical **2** rather than radical **3** (Scheme 1.9).

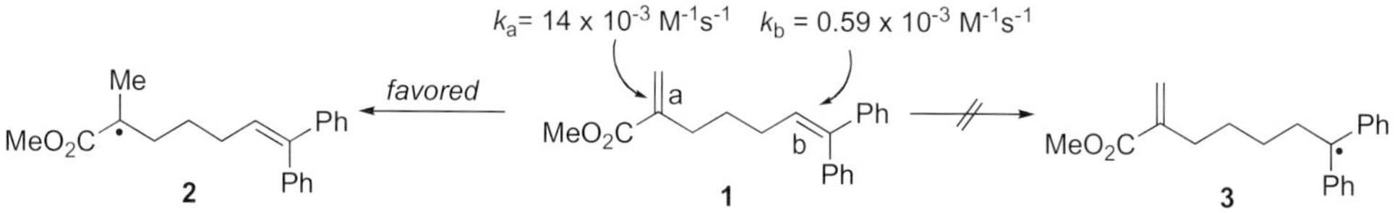

$k_a = 14 \times 10^{-3}$ M^{-1}s^{-1} $k_b = 0.59 \times 10^{-3}$ M^{-1}s^{-1}

Scheme 1.9 H$^\bullet$ transfer to diene **1**.

The ethyl ester analogous to **2** cyclizes rapidly to a five-membered ring with $k_{cyc} > 10^5$ s^{-1} [110]. In general, the cyclization of 5-hexenyl radicals to five-membered rings is favored over their cyclization to six-membered rings. For the 5-hexenyl radical itself the rate constant at 25 °C for 5-exo cyclization is 2×10^5 s^{-1}, and the rate constant for 6-endo cyclization is 4×10^3 s^{-1} (1.33). These reactions have often been used as "radical clocks", to determine the rate constants of competing reactions [111]. The synthetic utility of a particular radical cyclization reaction is often a matter of how successfully its k_{cyc} can compete with other rate constants.

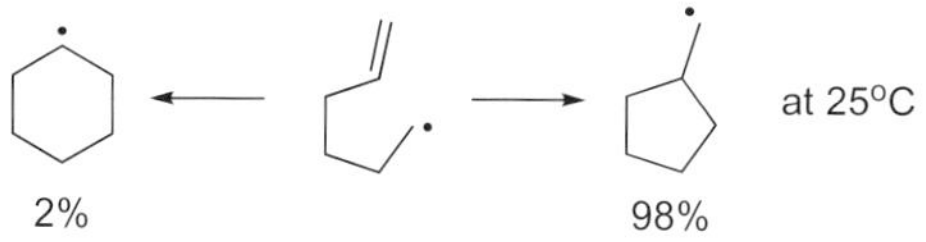

at 25 °C   (1.33)

2%   98%

All the reactions to which **2** is subject are shown in Scheme 1.10. In addition to cyclization it can undergo back transfer (k_{tr}[M$^\bullet$]), hydrogenation (k_{H2}[M–H]), and isomerization by H$^\bullet$ abstraction (k_{tr2}[M$^\bullet$]) from the carbon adjacent to the radical center.

Back transfer and isomerization require M$^\bullet$, and can be avoided if M$^\bullet$ is converted back to M–H rapidly enough – a transformation that has the advantage of making the overall reaction *catalytic*. $Cp(CO)_3Cr^\bullet$ is converted back to $Cp(CO)_3CrH$ with modest hydrogen pressures (3 atm H_2, Equation 1.34). (This transformation

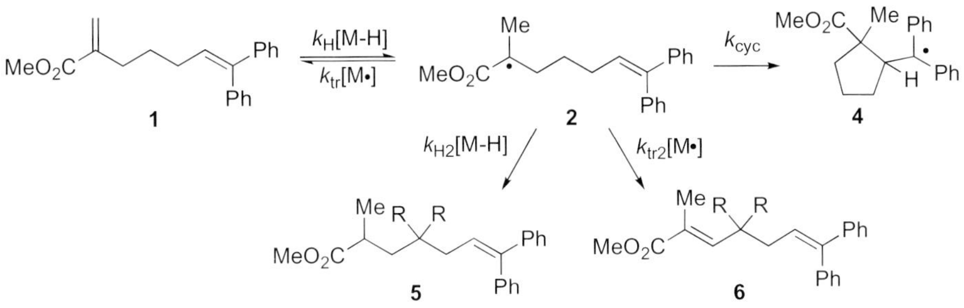

Scheme 1.10 Pathways available to radical **2**.

was reported by Fischer and coworkers in 1955, but under a pressure of 150 atm at 70 °C [112].) Thus Reaction 1.35 is catalytic as well as tin-free, forming **7** in 78% yield together with relatively small amounts of the hydrogenation product **5** and the isomerization product **6**.

$$2\,\mathrm{Cp(CO)_3Cr\bullet} \xrightarrow[\text{3 atm}]{H_2} 2\,\mathrm{Cp(CO)_3CrH} \tag{1.34}$$

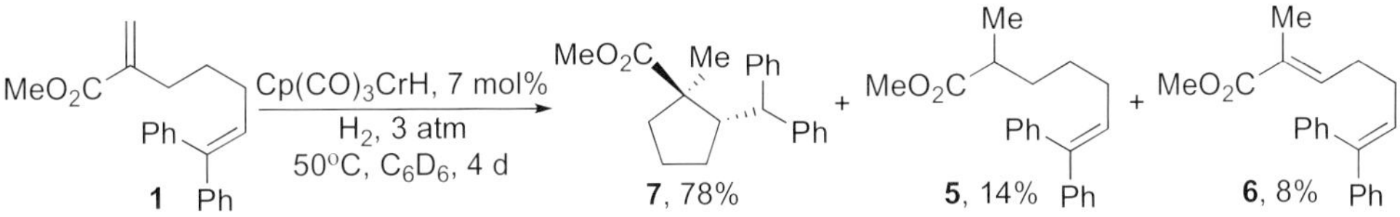

$$\text{(1.35)}$$

Increasing the concentrations of CpCr(CO)$_3$H and **1**, or heating the reaction, shortens the reaction time but increases the extent of hydrogenation to **5**. Increasing the hydrogen pressure increases the rate at which CpCr(CO)$_3\!\!\bullet$ is converted to CpCr(CO)$_3$H and thus increases (as we expect from Scheme 1.10) the ratio of the hydrogenation byproduct **5** to the isomerization byproduct **6**.

When we increase k_{cyc} by placing two substituents on the γ carbon to take advantage of the Thorpe–Ingold effect, the reaction (**8**→**9**) becomes quantitative (1.36) [113].

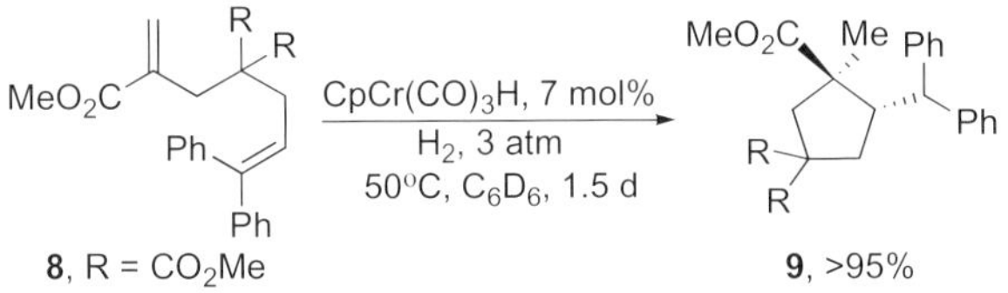

$$\text{(1.36)}$$

In view of the speed with which the weak V–H bonds in (P–P)(CO)$_4$VH transfer H• to styrene (Table 1.3 above), we have tried these hydrides with the cyclization substrates **1** and **8**. Stoichiometrically they give similar results, but *more quickly*, *even at lower temperatures* [13]. For example, treatment of **1** with dppe(CO)$_4$VH

1 Catalysis Involving the H• Transfer Reactions of First-Row Transition Metals

gives **7** in 77 % yield (1.37), while treatment of **8** with any of the vanadium hydrides gives **9** quantitatively. Unfortunately, these vanadium hydrides cannot be regenerated from the corresponding vanadoradicals, at least at 50 °C and 80 psi, so they can only be used as *stoichiometric* reagents.

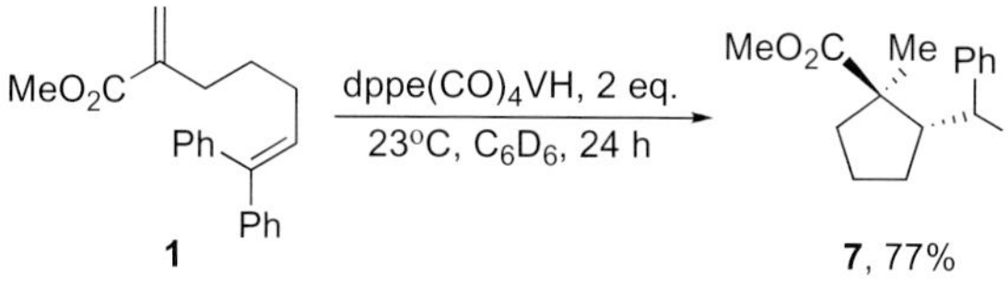

$$\text{MeO}_2\text{C}\diagup\diagdown\text{(Ph)(Ph)} \xrightarrow[\text{23 °C, C}_6\text{D}_6,\ 24\ \text{h}]{\text{dppe(CO)}_4\text{VH, 2 eq.}} \text{MeO}_2\text{C, Me, Ph, Ph cyclopentane} \quad (1.37)$$

1 → **7**, 77%

Additional α-substituted acrylates related to our original substrate **1** are readily assembled by the Morita–Baylis–Hillman reaction [114, 115], but necessarily bear an additional hydroxyl group. However, that substituent appears to have little influence on k_{cyc}. Treatment of diene **10** with a *catalytic* amount (7 mol%) of Cp(CO)$_3$CrH in C$_6$D$_6$ under 3 atm H$_2$ generated the substituted cyclopentanol **11** after three days at 50 °C (1.38), while treatment of **10** with a *stoichiometric* amount of dppe(CO)$_4$VH gave **11** in 77% yield after 3 h at room temperature [53].

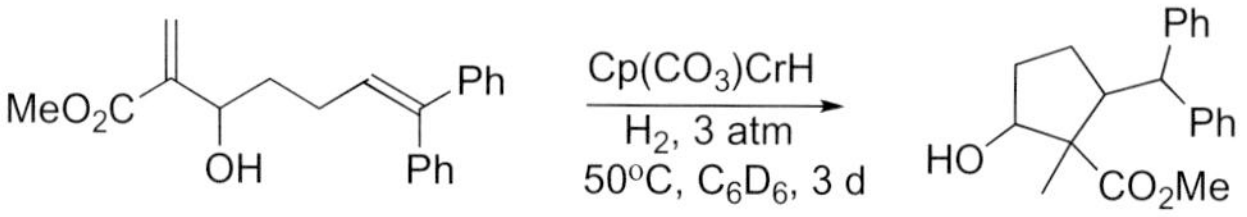

$$\text{MeO}_2\text{C},\ \text{OH},\ \text{Ph},\ \text{Ph} \xrightarrow[\text{H}_2,\ 3\ \text{atm}\ 50\ °\text{C, C}_6\text{D}_6,\ 3\ \text{d}]{\text{Cp(CO}_3\text{)CrH}} \text{HO, Ph, Ph, CO}_2\text{Me} \quad (1.38)$$

10 → **11**, 70%

This chemistry can be used to initiate *sequential* (cascade) cyclizations. From a Morita–Baylis–Hillman reaction and protection of the resulting hydroxyl we have prepared the triene **12**. Treatment of **12** with a catalytic amount of CpCr(CO)$_3$H under H$_2$ for 6 d gave the cyclization product **16** in 23% isolated yield, presumably by the mechanism in Scheme 1.11 [53].

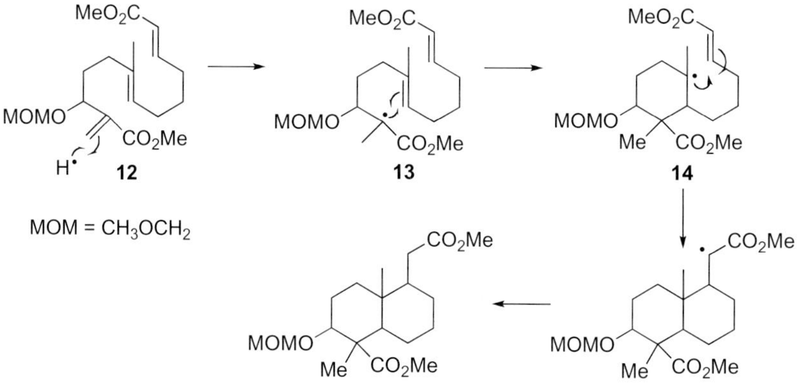

12 (MOMO, H•, CO$_2$Me, MeO$_2$C) → **13** (MOMO, CO$_2$Me, MeO$_2$C) → **14** (MOMO, Me, CO$_2$Me, MeO$_2$C) → **15** (MOMO, Me, CO$_2$Me) → **16** (MOMO, Me, CO$_2$Me)

MOM = CH$_3$OCH$_2$

Scheme 1.11 Proposed 6-endo/6-exo cascade cyclization of **12** to **16**.

1.6
Competing Methods for the Cyclization of Dienes

A number of other organometallic species cyclize 1,5- and 1,6-dienes to five- and six-membered rings. In an early example Molander used 5 mol% Cp*$_2$YMe(THF) under H$_2$ (Scheme 1.12) [116]. Sigma-bond metathesis releases methane and forms a Y–H bond. Sequential insertion of the two double bonds into the Y–H bond gives cyclization; hydrogenolysis of the resulting Y–C bond releases the product and reforms the yttrium hydride.

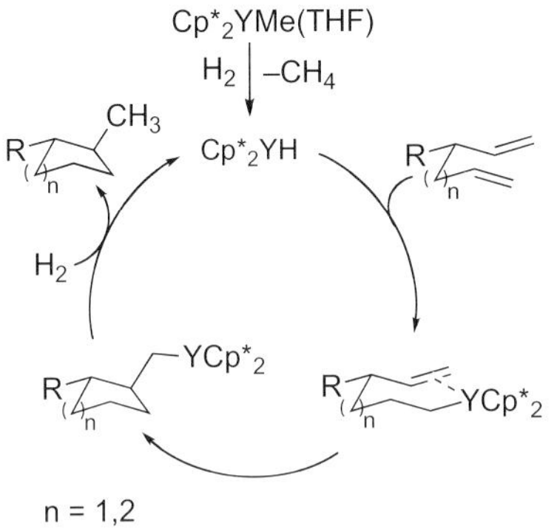

Scheme 1.12 Cyclization of 1,5- and 1,6-dienes by Cp*$_2$YMe(THF) under H$_2$.

Other methodologies capable of cyclizing 1,6 dienes require the use of precious metals. Pd allyls cyclize these dienes to unsaturated products [117], and Ru and Rh catalysts cyclize heterocycle-substituted dienes [118]. Murai's method functions by the mechanism proposed in Scheme 1.13).

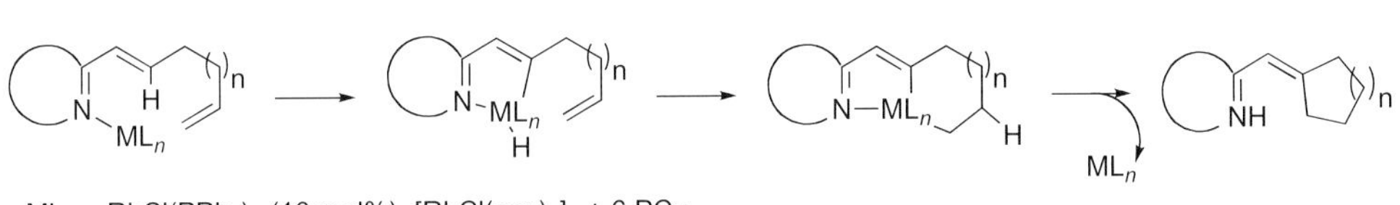

ML$_n$ = RhCl(PPh$_3$)$_3$ (10 mol%), [RhCl(coe)$_2$]$_2$ + 6 PCy$_3$, ...

Scheme 1.13 Cyclization of heterocycle-substituted dienes.

Initial coordination to the nitrogen of the heterocycle guides insertion into the alkenyl C–H bond to form the transition metal hydride. Insertion into the pendant olefin gives the metallobicycle, which reductively eliminates to regenerate the catalytic species and form the cycloalkane.

Rh(I) catalysts under H_2 effect the reductive cyclization of diynes and enynes [119]. In the example below (1.39), $Rh(COD)_2OTf$ (COD = 1,5-cyclooctadiene) is converted by H_2 gas to a Rh hydride, which is thought to undergo oxidative cyclization to produce a rhodacyclopentene intermediate. Cleavage of the Rh–C bonds by reductive elimination to form a vinyl rhodium species, and hydrogenolysis of that intermediate, forms the Rh hydride and the cyclized product.

$$\text{(1.39)}$$

X = CH_2, O, $(CH_3CO_2)_2C$, NTs

1.7
Summary and Conclusions

Steric as well as electronic factors affect the rate at which H· is transferred between a transition metal and the carbon of a double bond. However, the weak M–H bonds of the first-row metals, particularly vanadium, make them uniquely effective in this regard.

Such reactions can be used in the catalysis of *chain transfer* during radical polymerizations. In this process a metalloradical abstracts H· from a chain-carrying radical, transfers it to the double bond of a monomer, and starts a new chain. The resting state of traditional cobalt chain-transfer catalysts is the Co(II) metalloradical, but both the metalloradical and the hydride are present during the operation of newer (Cr) catalysts. Success in catalyzing chain transfer requires (i) that the M–H bond be not too much stronger than the $50 \, \text{kcal} \, \text{mol}^{-1}$ C–H bond in chain-carrying radicals, and (ii) that M· be stable enough to discourage the formation of bonds other than that to hydrogen.

$Cp(CO)_3CrH$ and $(P-P)(CO)_4VH$ can be used to initiate radical *cyclizations* by transferring H· to activated terminal olefins. However, the resulting radicals must cyclize quickly; competing reactions include transfer of a second H· (resulting in hydrogenation) and removal of an H· (resulting in isomerization) (Scheme 1.10). $Cp(CO)_3CrH$ is relatively slow at H· transfer, but can be regenerated with H_2 gas, enabling it to carry out reductive cyclizations catalytically; vanadium hydrides $(P-P)(CO)_4VH$ are faster but operate stoichiometrically.

References

1 Ciancanelli, R., Noll, B.C., DuBois, D.L., and DuBois, M.R. (2002) *J. Am. Chem. Soc.*, **124**, 2984–2992.
2 Curtis, C.J., Miedaner, A., Ellis, W.W., and DuBois, D.L. (2002) *J. Am. Chem. Soc.*, **124**, 1918–1925.
3 Qi, X.J., Fu, Y., Liu, L., Guo, Q.X. (2007) *Organometallics*, **26**, 4197–4203.
4 Tilset, M. (2007) *Comprehensive Organometallic Chemistry III, Introduction: Fundamentals*, vol. 1 (ed. G. Parkin), Elsevier, pp. 279–305.
5 Rautenstrauch, V., Hoang-Cong, X., Churlaud, R., Abdur-Rashid, K., and Morris, R.H. (2003) *Chem. Eur. J.*, **9**, 4954–4967, and references therein.
6 Guan, H., Iimura, M., Magee, M.P., Norton, J.R., and Zhu, G. (2005) *J. Am. Chem. Soc.*, **127**, 7805–7814, and references therein.
7 Casey, C.P., and Guan, H.R. (2007) *J. Am. Chem. Soc.*, **129**, 5816–5817, and references therein.
8 Bullock, R.M. (2007) *Handbook of Homogeneous Hydrogenation*, vol. 1, John Wiley & Sons, Inc., pp. 153–197, and references therein.
9 Samec, J.S.M., Ell, A.H., Aberg, J.B., Privalov, T., Eriksson, L., and Backvall, J.E. (2006) *J. Am. Chem. Soc.*, **128**, 14293–14305.
10 Samec, J.S.M., Backvall, J.E., Andersson, P.G., and Brandt, P. (2006) *Chem. Soc. Rev.*, **35**, 237–248, and references therein.
11 See supporting information in Ellis, W.W., Raebiger, J.W., Curtis, C.J., Bruno, J.W., DuBois, D.L. (2004) *J. Am. Chem. Soc.*, **126**, 2738–2743.
12 Wayner, D.D.M., and Parker, V.D. (1993) *Acc. Chem. Res.*, **26**, 287–294.
13 Choi, J., Pulling, M.E., Smith, D.M., and Norton, J.R. (2008) *J. Am. Chem. Soc.*, **130**, 4250–4252.
14 Zhang, J., Grills, D.C., Huang, K.W., Fujita, E., and Bullock, R.M. (2005) *J. Am. Chem. Soc.*, **127**, 15684–15685.
15 Uddin, J., Morales, C.M., Maynard, J.H., and Landis, C.R. (2006) *Organometallics*, **25**, 5566–5581.
16 Puttfarcken, U., and Rehder, D. (1978) *J. Organomet. Chem.*, **157**, 321–325.
17 Davison, A., and Ellis, J.E. (1972) *J. Organomet. Chem.*, **36**, 131–136.
18 Mader, E.A., Davidson, E.R., and Mayer, J.M. (2007) *J. Am. Chem. Soc.*, **129**, 5153–5166.
19 Manner, V.W., DiPasquale, A.G., and Mayer, J.M. (2008) *J. Am. Chem. Soc.*, **130**, 7210–7211, and references therein.
20 Tishchenko, O., Truhlar, D.G., Ceulemans, A., and Nguyen, M.T. (2008) *J. Am. Chem. Soc.*, **130**, 7000–7010.
21 Litwinienko, G., and Ingold, K.U. (2007) *Acc. Chem. Res.*, **40**, 222–230.
22 Klingler, R.J., and Rathke, J.W. (1994) *J. Am. Chem. Soc.*, **116**, 4772–4785.
23 Calderazzo, F. (1983) *Ann. N. Y. Acad. Sci.*, **415**, 37–46.
24 Eisenberg, D.C., and Norton, J.R. (1991) *Isr. J. Chem.*, **31**, 55–66.
25 Eisenberg, D.C., Lawrie, C.J.C., Moody, A.E., and Norton, J.R. (1991) *J. Am. Chem. Soc.*, **113**, 4888–4895.
26 Rodkin, M.A., Abramo, G.P., Darula, K.E., Ramage, D.L., Santora, B.P., and Norton, J.R. (1999) *Organometallics*, **18**, 1106–1109.
27 Colle, K.S., Glaspie, P.S., and Lewis, E.S. (1975) *J. Chem. Soc., Chem. Commun.*, 266–267.
28 Dunnebacke, D., Neumann, W.P., Penenory, A., and Stewen, U. (1989) *Chem. Ber.*, **122**, 533–535.
29 Laarhoven, L.J.J., Mulder, P., and Wayner, D.D.M. (1999) *Acc. Chem. Res.*, **32**, 342–349.
30 Franz, J.A., Linehan, J.C., Birnbaum, J.C., Hicks, K.W., and Alnajjar, M.S. (1999) *J. Am. Chem. Soc.*, **121**, 9824–9830.
31 O'Connor, J.M., and Friese, S.J. (2008) *Organometallics*, **27**, 4280–4281.
32 Gansäuer, A., Fan, C.-A., and Piestert, F. (2008) *J. Am. Chem. Soc.*, **130**, 6916–6917.
33 Gansäuer, A., Otte, M., Piestert, F., and Fan, C.A. (2009) *Tetrahedron*, **65**, 4984–4991.

34 Jin, J., and Newcomb, M. (2008) *J. Org. Chem.*, **73**, 7901–7905.

35 Cuerva, J.M., Campana, A.G., Justicia, J., Rosales, A., Oller-Lopez, J.L., Robles, R., Cardenas, D.J., Bunuel, E., and Oltra, J.E. (2006) *Angew. Chem. Int. Ed.*, **45**, 5522–5526.

36 Barrero, A.F., Rosales, A., Cuerva, J.M., and Oltra, J.E. (2003) *Org. Lett.*, **5**, 1935–1938.

37 Barrero, A.F., Oltra, J.E., Cuerva, J.M., and Rosales, A. (2002) *J. Org. Chem.*, **67**, 2566–2571.

38 Mader, E.A., Larsen, A.S., and Mayer, J.M. (2004) *J. Am. Chem. Soc.*, **126**, 8066–8067.

39 Houk, K.N., and Rondan, N.G. (1984) *J. Am. Chem. Soc.*, **106**, 4293–4294.

40 Groves, J.T., and Subramanian, D.V. (1984) *J. Am. Chem. Soc.*, **106**, 2177–2181.

41 McLain, J.L., Lee, J., and Groves, J.T. (2000) *Biomimetic Oxidations Catalyzed by Transition Metal Complexes* (ed. B. Meunier), Imperial College Press, London, pp. 91–169.

42 Auclair, K., Hu, Z.B., Little, D.M., de Montellano, P.R.O., and Groves, J.T. (2002) *J. Am. Chem. Soc.*, **124**, 6020–6027.

43 Austin, R.N., Chang, H.K., Zylstra, G.J., and Groves, J.T. (2000) *J. Am. Chem. Soc.*, **122**, 11747–11748.

44 Mayer, J.M. (1998) *Acc. Chem. Res.*, **31**, 441–450.

45 Larsen, A.S., Wang, K., Lockwood, M.A., Rice, G.L., Won, T.J., Lovell, S., Sadilek, M., Turecek, F., and Mayer, J.M. (2002) *J. Am. Chem. Soc.*, **124**, 10112–10123.

46 Waidmann, C.R., Zhou, X., Tsai, E.A., Kaminsky, W., Hrovat, D.A., Borden, W.T., and Mayer, J.M. (2009) *J. Am. Chem. Soc.*, **131**, 4729–4743.

47 Feder, H.M., and Halpern, J. (1975) *J. Am. Chem. Soc.*, **97**, 7186–7188, and references therein.

48 Weil, T.A., Friedman, S., and Wender, I. (1974) *J. Org. Chem.*, **39**, 48–50.

49 Sweany, R.L., and Halpern, J. (1977) *J. Am. Chem. Soc.*, **99**, 8335–8337.

50 Halpern, J. (1979) *Pure Appl. Chem.*, **51**, 2171–2182.

51 Tang, L., Papish, E.T., Abramo, G.P., Norton, J.R., Baik, M.-H., Friesner, R.A., and Rappé, A. (2003) *J. Am. Chem. Soc.*, **125**, 10093–10102.

52 Tang, L., Papish, E.T., Abramo, G.P., Norton, J.R., Baik, M.-H., Friesner, R.A., and Rappé, A. (2006) *J. Am. Chem. Soc.*, **128**, 11314.

53 Hartung, J., Pulling, M.E., Smith, D.M., Yang, D.X., and Norton, J.R. (2008) *Tetrahedron*, **64**, 11822–11830.

54 Blanksby, S.J., and Ellison, G.B. (2003) *J. Am. Chem. Soc.*, **36**, 255–263.

55 See the discussion (and references cited) in reference [23] of Bullock, R.M., Headford, C.E.L., Hennessy, K.M., Kegley, S.E., Norton, J.R. (1989) *J. Am. Chem. Soc.*, **111**, 3897–3908.

56 Bullock, R.M., and Bender, B.R. (2003) *Encyclopedia of Catalysis*, vol. 4 (ed. I.T. Horváth), John Wiley & Sons, Inc., New York.

57 Jacobsen, E.N., and Bergman, R.G. (1985) *J. Am. Chem. Soc.*, **107**, 2023–2032.

58 Bullock, R.M., and Samsel, E.G. (1990) *J. Am. Chem. Soc.*, **112**, 6886–6898.

59 Kinney, R.J., Jones, W.D., and Bergman, R.G. (1978) *J. Am. Chem. Soc.*, **100**, 7902–7915.

60 Bullock, R.M. (1991) *Transition Metal Hydrides* (ed. A. Dedieu), VCH, New York.

61 Choi, J., Tang, L., and Norton, J.R. (2007) *J. Am. Chem. Soc.*, **129**, 234–240.

62 Ayrey, G., and Wong, D.J.D. (1975) *Polymer*, **16**, 623.

63 Gridnev, A.A., Ittel, S.D., Wayland, B.B., and Fryd, M. (1996) *Organometallics*, **15**, 5116–5126.

64 Suess, H., Pilch, K., and Rudorfer, H. (1937) *Z. Phys. Chem. A–Chem. T*, **179**, 361–370.

65 Schulz, G.V., Dinglinger, A., and Husemann, E. (1939) *Z. Phys. Chem. B–Chem. E*, **43**, 385–408.

66 Gregg, R.A., Alderman, D.M., and Mayo, F.R. (1948) *J. Am. Chem. Soc.*, **70**, 3740–3743.

67 Matyjaszewski, K., and Xia, J.H. (2001) *Chem. Rev.*, **101**, 2921–2990.

68 (2002) *Handbook of Radical Polymerization* (eds Matyjaszewski, K. and Davis, T.P.), Wiley–Interscience., New York.

69 Kamigaito, M., Ando, T., and Sawamoto, M. (2001) *Chem. Rev.*, **101**, 3689–3745.

70 Ouchi, M., Terashima, T., and Sawamoto, M. (2008) *Acc. Chem. Res.*, **41**, 1120–1132.

71 The history of CCT is described in Gridnev, A. (2000) *J. Polym. Sci., Part A: Polym. Chem.*, **38**, 1753–1766.

72 Gridnev, A.A., and Ittel, S.D. (2001) *Chem. Rev.*, **101**, 3611–3659.

73 Smith, G.B., Heuts, J.P.A., and Russell, G.T. (2005) *Macromol. Symp.*, **226**, 133–146.

74 Mayo, F.R. (1943) *J. Am. Chem. Soc.*, **65**, 2324–2329.

75 Tang, L., and Norton, J.R. (2004) *Macromolecules*, **37**, 241–243.

76 Choi, J.W., and Norton, J.R. (2008) *Inorg. Chim. Acta*, **361**, 3089–3093.

77 Tang, L.H., and Norton, J.R. (2006) *Macromolecules*, **39**, 8229–8235.

78 Ittel, S.D., and Gridnev, A.A. (2006) U.S. Patent 7022792.

79 Li, Y.Y., and Wayland, B.B. (2003) *Chem. Commun.*, 1594–1595.

80 Tang, L. (2005) Scope and mechanism of chain transfer catalysis with metalloradicals and metal hydrides. PhD Thesis. Columbia University, New York, NY.

81 Zard, S.Z. (2003) *Radical Reactions in Organic Synthesis*, Oxford University Press, New York.

82 Curran, D.P. (1988) *Synthesis*, 417–439.

83 Curran, D.P. (1988) *Synthesis*, 489–513.

84 Chatgilialoglu, C. (2001) *Radicals in Organic Synthesis*, vol. 1 (eds P. Renaud and M.P. Sibi), Wiley-VCH Verlag GmbH, Weinheim, pp. 32–34, 90–93.

85 Quiclet-Sire, B., and Zard, S.Z. (2006) *Chem. Eur. J.*, **12**, 6002–6016.

86 Barton, D.H.R., Crich, D., and Motherwell, W.B. (1983) *J. Chem. Soc., Chem. Commun.*, 939–941.

87 Barton, D.H.R., Crich, D., and Motherwell, W.B. (1985) *Tetrahedron*, **41**, 3901–3924. The authors remarked that "the ideal solution is to completely avoid the use of tri-n-butylstannane".

88 Boyer, I.J. (1989) *Toxicology*, **55**, 253–298.

89 Corey, E.J., and Suggs, J.W. (1975) *J. Org. Chem.*, **40**, 2554–2555.

90 Stork, G., and Sher, P.M. (1986) *J. Am. Chem. Soc.*, **108**, 303–304.

91 Tormo, J., Hays, D.S., and Fu, G.C. (1998) *J. Org. Chem.*, **63**, 5296–5297.

92 Light, J., and Breslow, R. (1990) *Tetrahedron Lett.*, **31**, 2957–2958.

93 Light, J., and Breslow, R. (1995) *Org. Synth.*, **72**, 199–208.

94 Curran, D.P., Hadida, S., Kim, S.-Y., and Luo, Z. (1999) *J. Am. Chem. Soc.*, **121**, 6607–6615.

95 Salomon, C.J., Danelon, G.O., and Mascaretti, O.A. (2000) *J. Org. Chem.*, **65**, 9220–9222.

96 Clive, D.L.J., and Wang, J. (2002) *J. Org. Chem.*, **67**, 1192–1198.

97 Curran, D.P., Yang, F., and Cheong, J. (2002) *J. Am. Chem. Soc.*, **124**, 14993–15000.

98 Stien, D., and Gastaldi, S. (2004) *J. Org. Chem.*, **69**, 4464–4470.

99 Studer, A. (2004) *Chem. Soc. Rev.*, **33**, 267–273.

100 Gilbert, B.C., and Parsons, A.F. (2002) *J. Chem. Soc., Perkin Trans.*, **2**, 367–387.

101 Bowman, W.R., Krintel, S.L., and Schilling, M.B. (2004) *Org. Biomol. Chem.*, **2**, 585–592.

102 Baguley, P.A., and Walton, J.C. (1998) *Angew. Chem. Int. Ed. Engl.*, **37**, 3072–3082.

103 Barton, D.H.R., Jang, D.O., and Jaszberenyi, J.C. (1993) *J. Org. Chem.*, **58**, 6838–6842.

104 Chatgilialoglu, C. (1992) *Acc. Chem. Res.*, **25**, 188–194.

105 Chatgilialoglu, C., Ballestri, M., Escudie, J., and Pailhous, I. (1999) *Organometallics*, **18**, 2395–2397.

106 Mikami, S., Fujita, K., Nakamura, T., Yorimitsu, H., Shinokubo, H., Matsubara, S., and Oshima, K. (2001) *Org. Lett.*, **3**, 1853–1855.

107 Inoue, K., Sawada, A., Shibata, I., and Baba, A. (2002) *J. Am. Chem. Soc.*, **124**, 906–907.

108 Studer, A., and Amrein, S. (2002) *Synthesis*, 835–849.

109 Fujita, K., Nakamura, T., Yorimitsu, H., and Oshima, K. (2001) *J. Am. Chem. Soc.*, **123**, 3137–3138.

110 Newcomb, M., Horner, J.H., Filipkowski, M.A., Ha, C., and Park, S.U. (1995) *J. Am. Chem. Soc.*, **117**, 3674–3684.

111 Griller, D., and Ingold, K.U. (1980) *Acc. Chem. Res.*, **13**, 317–323.

112 Fischer, E.O., Hafner, W., and Stahl, H.O. (1955) *Z. Anorg. Allg. Chem.*, **282**, 47–62.

113 Smith, D.M., Pulling, M.E., and Norton, J.R. (2007) *J. Am. Chem. Soc.*, **129**, 770–771.

114 Baylis, A.B., and Hillman, M.E.D. (1972) U.S. Patent 3,743,669.

115 Morita, K., Suzuki, Z., and Hirose, H. (1968) *Bull. Chem. Soc. Jpn.*, **41**, 2815.

116 Molander, G.A., and Hoberg, J.O. (1992) *J. Am. Chem. Soc.*, **114**, 3123–3125.

117 Radetich, B., and RajanBabu, T.V. (1998) *J. Am. Chem. Soc.*, **120**, 8007–8008.

118 Fujii, N., Kakiuchi, F., Yamada, A., Chatani, N., and Murai, S. (1998) *Bull. Chem. Soc. Jpn.*, **71**, 285–298.

119 Jang, H.Y., and Krische, M.J. (2004) *J. Am. Chem. Soc.*, **126**, 7875–7880.

120 Hartwig, J.F. (2010) *Organotransition Metal Chemistry: From Banding to Catalysis*. University Science Books, Sausalito, USA, pp. 122–136.

2
Catalytic Reduction of Dinitrogen to Ammonia by Molybdenum

Richard R. Schrock

2.1
Introduction

In the 1960s it was first recognized that dinitrogen is reduced to ammonia in the environment by a metalloenzyme, an FeMo nitrogenase [1–5]. Although "alternative" nitrogenases have been discovered that do not contain Mo [6–8], the FeMo nitrogenase is the most accessible. The FeMo nitrogenase has been purified, crystallized, and studied for several decades. It appears to be the most efficient at reducing dinitrogen to ammonia, with only approximately one equivalent of dihydrogen being produced 1–50 atm of dinitrogen. It has also been subjected to X-ray studies that have elicited additional discussion concerning the mechanism of dinitrogen reduction [9–12]. However, no definitive conclusions have yet been reached.

After the discovery of the first dinitrogen complex ($[Ru(NH_3)_5(N_2)]^{2+}$) by Allen and Senoff [13], reduction of dinitrogen to ammonia under mild conditions with an abiological catalyst seemed imminent [14–21]. However, the problem proved to be a great deal more challenging than anticipated. Over a period of several decades, hundreds of man years were invested in attempting to demonstrate that it is possible to reduce dinitrogen abiologically and catalytically under mild conditions (25 °C and 1 atm of N_2). Perhaps it was not fully appreciated how difficult reduction of dinitrogen to ammonia in solution with protons and electrons would be; dinitrogen is an extraordinarily stable molecule, and protons are reduced readily to dihydrogen, a reaction that itself may be catalyzed by transition metals. The only reaction in which dinitrogen is reduced catalytically (to an approximately 10 : 1 mixture of hydrazine and ammonia) under mild conditions was reported by Shilov [22]. That reaction also requires molybdenum and is catalytic with respect to it. The solvent is methanol and a relatively strong reducing agent such as sodium amalgam is required. Hydrazine is the primary product with ammonia being formed through a metal-catalyzed disproportionation of hydrazine to dinitrogen and ammonia, a reaction that is accomplished relatively readily by transition metals. Shilov proposed that dinitrogen is bound and reduced between two metal centers, although direct evidence was not provided.

Catalysis Without Precious Metals. Edited by R. Morris Bullock
© 2010 WILEY-VCH Verlag GmbH & Co. KGaA, Weinheim
ISBN: 978-3-527-32354-8

In the late 1960s research groups led by Chatt [14] and Hidai [16, 17] began to prepare and study Mo(0) and W(0) bisdinitrogen complexes that (usually) contain chelating diphosphines such as 1,2-bis(diphenylphosphino)ethane (diphos), for example, *trans*-W(N$_2$)$_2$(diphos)$_2$. Although many examples of proposed intermediates in a hypothetical catalytic reduction of dinitrogen were prepared and characterized, some of the most important being M(N$_2$), M–N=NH, M=N–NH$_2$, M≡N, M=NH, M–NH$_2$, and M(NH$_3$) species, only stoichiometric reduction of dinitrogen to ammonia was demonstrated, usually with the required six electrons coming from the metal.

The large range of oxidation states for molybdenum (Mo(0) through Mo(VI)) left open the possibility that the higher oxidation states in the right circumstances could be involved in binding and reducing dinitrogen. We demonstrated that high oxidation state dinitrogen complexes of Mo(IV) and W(IV), as well as many that contained relevant dinitrogen-derived ligands, could be prepared [23–28]. Progress quickened after Cummins synthesized complexes that contain tetradentate triamidoamine ligands ([(RNCH$_2$CH$_2$)$_3$N]$^{3-}$) [29–36]. This approach ultimately led to the first system for the reduction of dinitrogen to ammonia with protons and electrons at room temperature and pressure. In this chapter I will trace the development of this catalytic system for the reduction of dinitrogen.

2.2
Some Characteristics of Triamidoamine Complexes

Triamidoamine ligands bind to a transition metal in a tetradentate manner, thereby creating a sterically protected, three-fold-symmetric "pocket" in which three orbitals are available to bond to additional ligands in that pocket, two π orbitals (d_{xz} and d_{yz}) and a σ orbital (approximately d_{z^2}), as shown in Figure 2.1a. The two π orbitals are degenerate in a C$_3$ environment and are especially suitable for

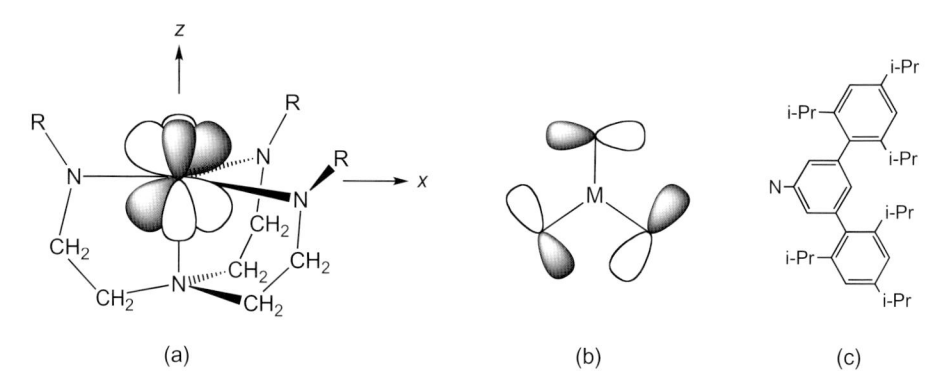

Figure 2.1 (a) A drawing of [(RNCH$_2$CH$_2$)$_3$N]Mo. (b) The ligand-centered non-bonding orbital comprised of in-plane amido p orbitals. (c) the N$_{amido}$-HIPT group.

forming a metal–ligand triple bond. However, combinations of the σ and two π orbitals can be employed to form one double and one single bond, or three single bonds. Only two metal–N_{eq} π bonds are possible in C_{3v} symmetry, since one combination of the three "in-plane" 2p orbitals on the three equatorial nitrogen atoms (A_2 symmetry in C_{3v}; Figure 2.1b) is a ligand-centered nonbonding orbital. Therefore the trianion of a substituted triethylenetetramine can contribute a maximum of 12 electrons to the metal in a C_3-symmetric species. The metal is usually found slightly above the MN_3 plane, for example, by ~0.2 Å for a second or third row metal, which may result in a greater extension of the d_{xz} and d_{yz} orbitals along the z axis in the direction of a ligand in the apical pocket. For a second or third row metal the M–N_{ax} bond distance is that expected from a 2e donor interaction, for example, ~2.2–2.3 Å, while the M–N_{eq} bond distance (~2.0 Å) is consistent with a significant degree of metal–amido π bonding (formally 2/3 of a π bond per N).

$[(Me_3SiNCH_2CH_2)_3N]^{3-}$ ($[TMSN_3N]^{3-}$) was chosen first as a ligand to be explored in transition metal chemistry [36] in view of its facile synthesis, less electron-rich (reducing) character relative to an alkyl-substituted variation such as $[(MeNCH_2CH_2)_3N]^{3-}$ [37], and steric protection of the apical coordination site by the TMS substituents. The Si–N_{eq}–C plane can be "tipped" out of the N_{ax}–M–N_{eq} plane by as much as ~35°, or it can be close to coincident with the N_{ax}–M–N_{eq} plane, depending on which conformation is adopted by the ethylene bridging group and the nature of the ligand or ligands bound in the apical position. Therefore the "depth" of the apical pocket and the nature of the π bonding between the metal and equatorial nitrogens vary to some degree. The M–N_{eq}–Si angles are usually ~125° regardless of the N_{ax}–M–N_{eq}–Si dihedral angle, and N_{eq} is planar. First row metals (Ti–Fe) often form M(III) complexes that do not contain any axial ligand, that is, trigonal monopyramidal species. Trigonal monopyramidal complexes have not yet been isolated for any second or third row metal.

A variety of unusual species have been prepared that contain the $[TMSN_3N]^{3-}$ ligand, among them a tantalum phosphinidene ($[TMSN_3N]Ta=PR$) [38], an alkylidene hydride ($[TMSN_3N]W(H)(cyclopentylidene)$) [39], and a $[TMSN_3N]W(L)$ species, where L is an arsenide [40, 41], phosphide [41, 42], or stibide [43]. Unusual reactions, such as elimination of H_2 from alkyls to give alkylidynes, suggest that the M≡C bond (M = Mo or W) is highly favorable in the triamidoamine setting [39, 44, 45]. However, several problems with TMS substituents were also encountered. Examples include migration reactions (e.g., 2.1 [39]) or CH activation [29, 45] (e.g., 2.2).

(2.1)

$$[N_3N]Ti(H) \underset{+H_2}{\overset{-H_2}{\rightleftharpoons}} \text{[complex]} \qquad (2.2)$$

A reaction between $MoCl_3(THF)_3$ and $Li_3[(t\text{-}BuMe_2SiNCH_2CH_2)_3N]$ under dinitrogen gave paramagnetic, pentane-soluble, purple $\{[(t\text{-}BuMe_2SiNCH_2CH_2)_3N]Mo\}_2(\mu\text{-}N_2)$ in low yield (~10%) [46]. (Low yields of triaminoethylamine (TREN) derivatives that contain silyl substituents on the nitrogens are the norm.) An X-ray study showed that this species can be described as a [Mo(IV)]$_2$[diazenido(2-)] complex (Mo–N=N–Mo) in which Mo–N$_\alpha$ = 1.907(8) Å, N=N = 1.20(2) Å, and Mo–N=N = 178(1)°. Heterobimetallic complexes were prepared that contain the $\{[TMSN_3N]Mo\text{–}N=N\}^-$ ion coordinated to other metals, including the unusual trigonal iron complex, $\{[TMSN_3N]Mo\text{–}N=N\}_3Fe$ [47, 48]. It then was shown that oxidation of the $\{[TMSN_3N]Mo\text{–}N=N\}^-$ ion led to a terminal dinitrogen complex, $[TMSN_3N]Mo(N_2)$ [49]. The reactions shown in Equation 2.3 suggested that chemistry that involves silylated amido nitrogens is likely to be a significant problem in the circumstances that are required for reduction of dinitrogen with protons and electrons.

$$\text{[complex]} \xrightarrow[\text{toluene, -20°C}]{+\ 4\ MeOTf} \text{[complex]}^+ OTf^- \ +\ \text{[complex]}^+ OTf^- \qquad (2.3)$$

NMR techniques provide valuable information for identifying and following the chemistry of paramagnetic Mo triamidoamine complexes, since nuclei that are not directly bound to the metal can often be observed and chemical shifts and line widths are not as large as with first row metal analogs. For example, the number and chemical shifts of the backbone methylene resonances in particular can be diagnostic. In a few species the backbone methylene resonances cannot be observed, for reasons that have not been elucidated.

We then turned to TREN derivatives that contain pentafluorophenyl substituents. Pentafluorophenyl-substituted TREN $(C_6F_5NHCH_2CH_2)_3N$ ($H_3[N_3N_F]$) could be prepared in high yield through nucleophilic attack on hexafluorobenzene [50], and $[N_3N_F]MoCl$ and $[N_3N_F]Mo(OTf)$ could be prepared in high yield. Reduction of $[N_3N_F]Mo(OTf)$ with two equivalents of sodium amalgam produced the sodium salt of $\{[N_3N_F]Mo(N_2)\}^-$ (2.4), while reduction with one equivalent of sodium amalgam gave the dinuclear bridging "diazenido(2-)" species, $[N_3N_F]Mo\text{–}N=N\text{–}Mo[N_3N_F]$ (2.5). Although this chemistry was revealing, we ultimately concluded that pentafluorophenyl was too electron-withdrawing and CF

$$[N_3N_F]Mo(OTf) + 2\,Na + N_2 \xrightarrow{-NaOTf} \{[N_3N_F]Mo(N_2)\}Na \qquad (2.4)$$

$$[N_3N_F]Mo(OTf) + \{[N_3N_F]Mo(N_2)\}Na \xrightarrow{-NaOTf} [N_3N_F]Mo-N=N-Mo[N_3N_F] \qquad (2.5)$$

bonds too reactive, in the presence of strong reducing agents, including a reduced central metal. Therefore, we turned to ordinary aryl substituents. We also anticipated that Mo(μ-N$_2$)Mo species were likely to be relatively stable toward reduction and protonation, and, therefore, that terminal nitrogen complexes would much more readily undergo reactions that would lead to cleavage of the N–N bond, as suggested through studies by Chatt and Hidai (*vide supra*), *if* intermediates can be protected sterically against destructive bimolecular reactions.

The advent of Pd-catalyzed arylation of primary amines allowed aryl-substituted TREN derivatives ([ArN$_3$N]$^{3-}$ species) to be prepared readily [51, 52]. One-electron reductions of [ArN$_3$N]MoCl complexes were found to yield complexes of the type [ArN$_3$N]Mo–N=N–Mo[ArN$_3$N], while two-electron reductions yielded {[ArN$_3$N]Mo–N=N}$^-$ derivatives. Treatment of diazenido complexes (e.g., [ArN$_3$N]Mo–N=N–Na(THF)$_x$) with electrophiles such as Me$_3$SiCl or MeOTf yielded [ArN$_3$N]Mo–N=NR complexes (R = SiMe$_3$ or Me), which reacted further with methylating agents to yield {[ArN$_3$N]Mo–N=NMe$_2$}$^+$ species. Electrochemical studies revealed the expected trends in oxidation and reduction potentials. They also provided evidence for stable neutral dinitrogen complexes of the type [ArN$_3$N]Mo(N$_2$) when Ar was a relatively bulky 3,5-terphenyl substituent, since the bulky terphenyl group blocked formation of an [ArN$_3$N]Mo(μ-N$_2$)Mo[ArN$_3$N] species.

In order to eliminate bimetallic chemistry as much as possible through steric protection of a metal coordination site and to provide increased solubility of Mo complexes in nonpolar solvents, we invested in the [HIPTN$_3$N]$^{3-}$ ligand, where HIPT = 3,5-(2,4,6-i-Pr$_3$C$_6$H$_2$)$_2$C$_6$H$_3$ (hexaisopropylterphenyl; Figure 2.1c) [53–56]. The HIPT derivative was prepared readily through available techniques and on a significant scale. In the HIPT group the two 2,4,6-triisopropylphenyl rings are oriented perpendicular to the central phenyl ring, and no rotation about the C–C bond joining aryl rings is possible. However, rotation about the N–aryl bond is relatively facile. The [HIPTN$_3$N]$^{3-}$ ligand is sterically demanding at some distance from the metal, not near the metal. The [HIPTN$_3$N]$^{3-}$ ligand opened the door to a large amount of new dinitrogen chemistry.

A significant number of theoretical calculations have been carried out on triamidoamine complexes, especially Mo complexes that are relevant to the reduction of dinitrogen. These calculations have been discussed recently and compared with experiment [57]. Perhaps the most relevant calculations are those carried out by Reiher and his group on complexes with the full [HIPTN$_3$N]$^{3-}$ ligand set. Only calculations on complexes with the full [HIPTN$_3$N]$^{3-}$ ligand set will be mentioned here [58–60].

2.3
Possible [HIPTN$_3$N]Mo Intermediates in a Catalytic Reduction of Molecular Nitrogen

A hypothetical catalytic reduction of dinitrogen to ammonia that involves [HIPTN$_3$N]Mo complexes is shown in Figure 2.2. This scheme is based on the proposals by Chatt for reduction of dinitrogen at a single Mo or W center [14]. The oxidation states of Mo are relatively high in this scheme, ranging from Mo(III) to Mo(VI). Eight of these intermediates have been prepared and characterized, while two others have been observed, but not isolated. The eight isolated, crystallographically characterized species are [HIPTN$_3$N]MoN$_2$ (**MoN$_2$**), MoN$_2^-$, Mo–N=NH, Mo–N=NH$_2^+$, Mo≡N, Mo≡NH$^+$, Mo(NH$_3$)$^+$, and Mo(NH$_3$). Each will be discussed here along with some chemistry that is relevant in terms of dinitrogen reduction.

2.3.1
MoN$_2$ and MoN$_2^-$

The [HIPTN$_3$N]Mo system is entered through a reaction between MoCl$_4$(THF)$_2$ and H$_3$[HIPTN$_3$N] in THF followed by 3.1 equivalents of LiN(SiMe$_3$)$_2$. This reaction produces orange [HIPTN$_3$N]MoCl (**MoCl**) in good yield. Reduction of **MoCl** with Mg in THF under a dinitrogen atmosphere produces red, diamagnetic Mo–N=N-MgCl(THF)$_3$. Other salts of the MoN$_2^-$ ion have been isolated, including a tetrabutylammonium salt, and crystal structure determinations have been carried out on Mo–N=N–MgBr(THF)$_3$ [54], {Mo–N=N}{Mg(dme)$_3$}$_{0.5}$ [54], and {Mo–N=N}

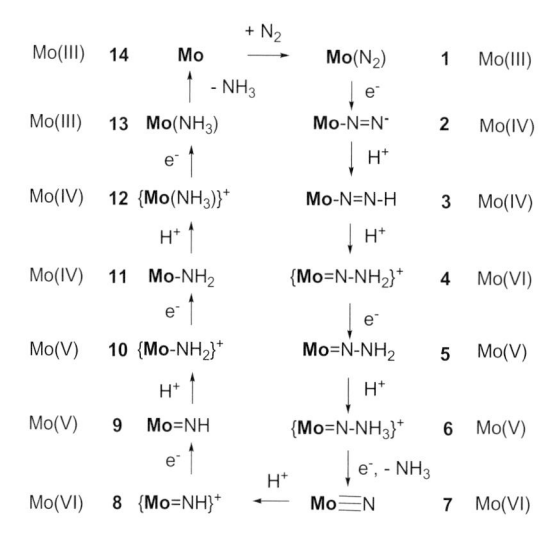

Figure 2.2 Possible intermediates in the reduction of dinitrogen at a Mo center through the stepwise addition of protons and electrons.

2.3 Possible [HIPTN$_3$N]Mo Intermediates in a Catalytic Reduction of Molecular Nitrogen

{K(THF)$_2$(18-crown-6)} [61]. In the last species interaction of the cation with the β nitrogen in the MoN$_2^-$ species is minimal, and v_{NN} therefore is found at a relatively high energy (1858 cm^{-1} in THF), as it is in the tetrabutylammonium salt (1859 cm^{-1} in THF) [54]. The more the cation interacts with N$_\beta$, the more v_{NN} shifts to lower energies. The N–N distance in MoN$_2^-$ salts is ~1.15 Å, while the Mo–N bond to the N$_2$ ligands is ~1.91 Å, all consistent with some reduction of the N$_2$ ligand and a significant degree of Mo–N π bonding. It is believed that reduction of MoCl by Mg in THF under dinitrogen leads first to MoCl$^-$, which is then attacked by dinitrogen to yield Mo(N$_2$)(Cl)$^-$, from which chloride is then lost to yield Mo(N$_2$) (2.6).

$$\text{MoCl} + e^- \rightarrow \text{MoCl}^- \xrightarrow{N_2} \text{Mo(Cl)(N}_2)^- \xrightarrow{-Cl^-} \text{Mo(N}_2) \qquad (2.6)$$

Under the reaction conditions Mo(N$_2$) is then reduced to MoN$_2^-$ by Mg. The slow rate of dissociation of dinitrogen from Mo(N$_2$) ($t_{1/2}$ ~35 h; *vide infra*) suggests that chloride is unlikely to be lost readily from MoCl$^-$ to yield Mo. The one electron added to Mo(N$_2$) to give MoN$_2^-$ is best viewed as being added to the dinitrogen ligand, with the second electron required to yield bound N$_2^{2-}$ being provided through oxidation of Mo from Mo(III) to Mo(IV). In Mo(N$_2$)$^-$ eight electrons are located in two perpendicular π systems made up of d orbitals on Mo (d$_{xz}$ and d$_{yz}$) and p orbitals on each nitrogen (Figure 2.3). The lone pair on the negatively-charged N$_\beta$ is located in an orbital with σ symmetry along the M–N–N axis. The 8 π electron bonding scheme is related to that found in, for example, the azide anion, N$_3^-$.

The reversible Mo(N$_2$)/Mo(N$_2$)$^-$ couple is found at a relatively low potential (E^0) for Mo(N$_2$)$^{0/-}$ is −1.81 V in THF vs. FeCp$_2^{+/0}$) [55]. Therefore Mo(N$_2$)$^-$ is oxidized readily to Mo(N$_2$). ZnCl$_2$ was used routinely as the oxidant until it was discovered that MoCl is formed as an impurity when Mo(N$_2$) is prepared in this fashion [61].

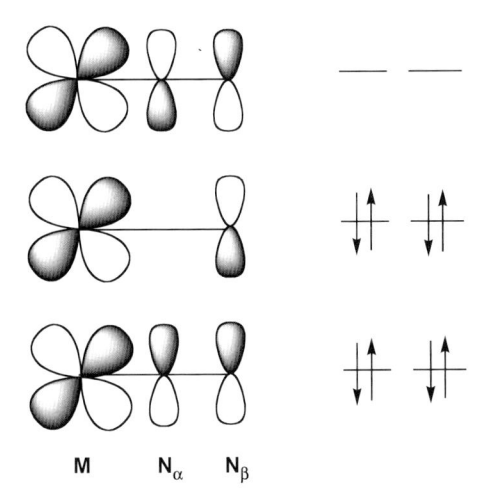

Figure 2.3 The π electron system in Mo(N$_2$)$^-$.

Use of zinc acetate as the oxidant eliminated formation of MoCl and provided Mo(N$_2$) in pure form. The value for v_{NN} (1990 cm^{-1}) in the IR spectrum of Mo(N$_2$) in benzene should be compared with v_{NN} = 1934 cm^{-1} for [TMSN$_3$N]MoN$_2$ in pentane [49]. Stronger backbonding to dinitrogen in [TMSN$_3$N]MoN$_2$ than in Mo(N$_2$) leads to a relatively slow rate of exchange of dinitrogen in [TMSN$_3$N]MoN$_2$ (days at 22 °C). EPR studies [62] confirm that Mo(N$_2$) is a low spin (S = ½) species in which the three d electrons are in the d$_{xz}$ and d$_{yz}$ orbitals.

Mo(N$_2$) can be oxidized relatively easily and reversibly (at −0.66 V in fluorobenzene relative to ferrocene/ferricenium) to Mo(N$_2$)$^+$, in which v_{NN} = 2255 cm^{-1} (in heptane); the difference in v_{NN} between Mo(N$_2$)$^+$ and Mo(N$_2$)$^-$ is 400 cm^{-1}. It is unusual to be able to compare v_{NN} in three complexes (Mo(N$_2$)$^+$, Mo(N$_2$), and Mo(N$_2$)$^-$) that differ by one electron. The high value for v_{NN} in Mo(N$_2$)$^+$ explains why dinitrogen is replaced readily (in minutes) by ^{15}N$_2$. Oxidation of MoN$_2$ with AgBPh$_4$ in THF yielded cationic, paramagnetic [Mo(THF)]BPh$_4$ in which the THF is bound to the metal along the z axis with Mo–O = 2.1811(18) Å and N(4)–Mo–O(1T) = 173.97(7)° [63]. Even 2,6-lutidine can bind to Mo$^+$ to yield [Mo(2,6-Lut)]BPh$_4$ in which the 2,6-lutidine is bound "off-axis" to a significant degree (N$_{amine}$–Mo–N$_{lut}$ = 157°) in a "slot" created by two of the HIPT groups [55]. The THF ligand is considerably less sterically demanding and therefore is bound strictly trans to the amine nitrogen donor. Reductions of [Mo(THF)]BPh$_4$ or [Mo(2,6-Lut)]BPh$_4$ under dinitrogen in CV studies show that Mo(N$_2$) forms in seconds. It is presumed at this stage that THF and 2,6-lutidine (L) are displaced in Mo(L) by dinitrogen, but it has not yet been proven that the reaction is first order in dinitrogen.

A reaction in which Mo(^{15}N$_2$) is converted into Mo(^{14}N$_2$) under 1 atm of ^{14}N$_2$ (2.7) has a half-life of ~35 h at 15 psi (1 atm) [54], 32 h at 30 psi [63], and 30 h at 55 psi [63], which suggests that the exchange of ^{15}N$_2$ for ^{14}N$_2$ is independent of dinitrogen pressure. Therefore *dinitrogen* exchange (2.7) is *dissociative* with the "naked" species, Mo, being the intermediate.

$$[\text{HIPT}_3\text{N}]\text{Mo}(^{15}\text{N}_2) + \text{N}_2 \longrightarrow {}^{15}\text{N}_2 + [\text{HIPT}_3\text{N}]\text{Mo}(\text{N}_2) \qquad (2.7)$$

Calculations suggest that Mo is a low spin doublet, that the quartet is 11.0 kcal mol^{-1} higher in energy, and that binding N$_2$ to Mo is strongly exothermic (ΔE is −37.8 kcal mol^{-1}). The ΔE for dissociation (2.8) seems large in view of the fact that $\Delta G^{\ddagger}$ = 24.8 kcal mol^{-1} for loss of dinitrogen from Mo(N$_2$) to yield intermediate Mo. However, a large and positive ΔS value could lead to a ΔG° that is about equal to $\Delta G^{\ddagger}$, or even smaller. Calculated ΔE values for dinitrogen dissociation in Mo(N$_2$)$^+$ and Mo(N$_2$)$^-$ are 26.5 and 57.0 kcal mol^{-1}, respectively, a trend that one would expect. Mo is calculated to be low spin with an empty d$_{z^2}$ orbital, a circumstance that prepares the metal for efficient backbonding to dinitrogen. Formation of Mo(N$_2$) through displacement of ammonia or dihydrogen from Mo(N$_2$) or Mo(H$_2$), respectively, is considered in a later section.

$$[\text{HIPTN}_3\text{N}]\text{Mo}(\text{N}_2) \rightarrow [\text{HIPTN}_3\text{N}]\text{Mo} + \text{N}_2 \qquad \Delta E = 37.8 \text{ kcal mol}^{-1} \qquad (2.8)$$

2.3.2
Mo–N=NH

Protonation of [Mo–N=N]$^-$ to give red–orange, diamagnetic Mo–N=NH can be carried out with several acids [54, 55]. For example, [Et$_3$NH]BAr$'_4$ (Ar' = 3,5-(CF$_3$)$_2$C$_6$H$_3$) afforded Mo–N=N–H in 50% yield. Some acids oxidize Mo(N$_2$)$^-$ to Mo(N$_2$) and only hydrogen is formed. In proton NMR spectra the diazenido proton resonance in Mo–N=NH is found at 8.57 ppm. Mo–N=NH is remarkably stable, decomposing at 61 °C in C$_6$D$_6$ in a first-order fashion (k_1 = 2.68(5) × 10^{-5} s^{-1}, $t_{1/2}$ ~ 7 h) to yield the paramagnetic hydride derivative MoH (*vide infra*) as the major product, along with small amounts of unidentified impurities (2.9).

$$[HIPTN_3N]Mo–N=NH \xrightarrow{\text{Heat in } C_6D_6} [HIPTN_3N]MoH + N_2 \qquad (2.9)$$

The rate of this decomposition in aliphatic solvents is similar to its rate in C$_6$D$_6$. Traces of acid remaining in Mo–N=NH synthesized through protonation of Mo(N$_2$)$^-$ also appear to catalyze its decomposition. Relatively pure Mo–N=NH is difficult to crystallize from pentane, since MoH, its thermal decomposition product, which is much less soluble, *co-crystallizes* with Mo–N=NH. Traces of MoH are also found upon heating Mo–N=NH in the solid state under vacuum at 60 °C for several hours. However, very pure samples have remained unchanged after being stored at –25 °C for 1.5 years. In an X-ray study the diazenide ligand was found to be linear at N$_\alpha$ (Mo–N5–N6 = 179.5(10)°), and the Mo–N5 and N5–N6 bond lengths of 1.780(9) and 1.302(13) Å are consistent with values found (1.789(2) and 1.229(3) Å, respectively) in a related compound of this general type, [(Me$_3$SiCH$_2$CH$_2$N)$_2$NCH$_2$CH$_2$NMe$_2$]Mo(N=NSiMe$_3$)(Me) [49]. (The N$_\alpha$–N$_\beta$–Si angle in the latter is 170.5°.) Unfortunately, the β proton could not be located in Mo–N=NH, and the N$_\alpha$–N$_\beta$–H angle therefore is not known.

2.3.3
Conversion of Mo(N$_2$) into Mo–N=NH

Treatment of Mo(N$_2$) with CoCp$_2$ and [2,6-LutH]BAr$'_4$ in benzene immediately yields Mo–N=NH, although neither reacts with Mo(N$_2$) in the absence of the other and CoCp$_2$ is too weak a reducing agent to reduce Mo(N$_2$) to Mo(N$_2$)$^-$. Conversion of Mo(N$_2$) to Mo–N=NH takes place with any combination of three acids ([Et$_3$NH]OTf, [Et$_3$NH]BAr$'_4$, [2,6-LutH]BAr$'_4$) and two reductants (CoCp$_2$ and CrCp*$_2$) and is rapid in all cases.

Three possible mechanisms for carrying out this transformation have been proposed [55]. One mechanism involves protonation of N$_\beta$ in the dinitrogen ligand to yield Mo(N$_2$H)$^+$ followed by addition of an electron. The second involves reduction of Mo(N$_2$) to Mo(N$_2$)$^-$ followed by protonation of N$_\beta$ in Mo(N$_2$)$^-$. The third involves a "proton-catalyzed" reduction followed by a protonation. These proposals will be discussed in that order.

Transfer of a proton to N$_\beta$ in Mo(N$_2$) from lutidinium is calculated to have $\Delta E \approx 0$, while $\Delta E = -14.3$ kcal mol^{-1} for the subsequent addition of an electron from CrCp*$_2$ to Mo(N$_2$H)$^+$ [59]. Therefore protonation of N$_\beta$ in Mo(N$_2$) followed by addition of an electron would appear to be a viable pathway for converting Mo(N$_2$) into Mo–N=NH, at least from a thermodynamic perspective, and not accounting for entropic and solvent effects. Attempted electrochemical generation of "[MoN=NH]$^+$" through oxidation of MoN=NH (in PhF with 0.1 M [Bu$_4$N]BAr$'_4$ electrolyte, room temperature, 500 mV s^{-1}) led only to multiple irreversible oxidation waves near the ferrocene/ferrocenium couple, consistent with facile decomposition of [MoN=NH]$^+$ [55].

Addition of an electron from CrCp$^*_2$ to Mo(N$_2$) to give M(N$_2$)$^-$ is calculated to be quite endothermic (ΔE = 71.6 kcal mol^{-1}) [59]. Experimentally it is known that Mo(N$_2$) is reduced reversibly to Mo(N$_2$)$^-$ at −2.01 V in C$_6$H$_5$F or −1.81 V in THF (both versus FeCp$_2$/FeCp$_2^+$) [55]. Since the cobaltocene redox couple is found at −1.33 V in THF, CoCp$_2$ is too weak a reducing agent (by ~500 mV) to reduce Mo(N$_2$) to Mo(N$_2$)$^-$ to any significant degree, even if precipitation of the resulting CoCp$_2^+$ is taken into account; yet Mo(N$_2$) is converted into Mo–N=NH in benzene in the presence of lutidinium and CoCp$_2$.

Addition of one equivalent of [2,6-LutH]BAr$'_4$ to Mo(N$_2$) in fluorobenzene in the absence of a reducing agent reveals that ~20% of a new species is formed with a v_{NN} absorption shifted by 67 cm^{-1} to higher energy compared to that for Mo(N$_2$), consistent with decreased backbonding into the dinitrogen ligand. No new *low* energy absorption is observed, as would be expected if N$_\beta$ of the dinitrogen ligand were protonated. We also have found that Mo(CO) (v_{CO} = 1885 cm^{-1}) is ~40% protonated by one equivalent of [2,6-LutH]BAr$'_4$ in fluorobenzene to yield "{[HIPTN$_3$N]Mo(CO)H}$^+$" in which v_{CO} = 1932 cm^{-1} (Δ = 47 cm^{-1}) [63], a result that appears to be analogous to protonation of Mo(N$_2$) and one that further suggests that neither N$_2$ nor CO is protonated directly. Unfortunately, although the amount of protonated Mo(N$_2$) increases with the amount of [2,6-LutH]BAr$'_4$ added, attempts to measure the equilibrium between Mo(N$_2$) and protonated Mo(N$_2$) have failed as a consequence of slow decomposition of both with time and as more acid is added. We propose that the ligand is at least partially removed at high acid concentrations to give other species, and "free" ligand is formed with time. We have also not yet been able to obtain any useful information concerning the precise nature of protonated Mo(N$_2$) from IR or NMR studies.

In a calculation in which the full [HIPTN$_3$N]$^{3-}$ ligand is employed, protonation of an amido nitrogen by lutidinium has been found to take place with ΔE = −9.0 kcal mol^{-1}, and no evidence could be obtained for formation of [Mo(N$_2$)(H)]$^+$ (in which H is bound to the metal) on the potential energy hypersurface [59]. Reiher therefore concluded that protonation of an amido nitrogen is more favored thermodynamically (by ~9 kcal mol^{-1}) than protonation of the dinitrogen ligand. Therefore, of the three proposed pathways for conversion of Mo(N$_2$) into Mo–N=NH the "proton-catalyzed reductive protonation" (2.10) appears to be the most favored mechanism for conversion of Mo(N$_2$) into Mo–N=NH. Reiher believes that kinetic conclusions based solely on the thermodynamic aspects of protonation reactions are justified, that is, protonation reactions of Mo(N$_2$) and Mo(N) with lutidinium have barriers of only ~5 kcal mol^{-1}.

2.3 Possible [HIPTN₃N]Mo Intermediates in a Catalytic Reduction of Molecular Nitrogen

$$(2.10)$$

A fourth possible mechanism that has not yet been explored theoretically consists of reduction of the protonated species to a neutral species followed by migration of the H on N to dinitrogen, either intramolecularly or intermolecularly (if the NHHIPT arm dissociates). If the dinitrogen were coordinated side-on (2.11), a 1,3 migration would complete formation of Mo–N=NH upon addition of an electron to protonated Mo(N₂).

$$(2.11)$$

2.3.4
[Mo=N–NH₂]⁺

Further protonation of Mo–N=NH with [H(OEt$_2$)$_2$]BAr$_4'$ afforded dark red, diamagnetic [Mo=N–NH$_2$]BAr$_4'$ in 68% yield. Protonation of MoN=NH with one equivalent of [2,6-LutH]BAr$_4'$ in C$_6$D$_6$ yields a 44:56 equilibrium mixture of MoN=NH and Mo=NNH$_2^+$ (K_{eq} = 1.6) after 15 min at 22 °C [55]. Since resonances for both MoN=NH and Mo=NNH$_2^+$ can be observed, the proton exchange rate between MoN=NH and Mo=NNH$_2^+$ cannot be fast on the NMR time scale under these conditions. An X-ray study of [Mo=N–NH$_2$]BAr$_4'$ revealed bond distances and angles consistent with a hydrazido(2-) species.

Attempts to observe Mo=NNH$_2$ in electrochemical (cyclic voltammetry) studies through reduction of Mo=NNH$_2^+$ were not successful. Reduction of Mo=NNH$_2^+$ is irreversible in THF at all scan rates employed. However, in fluorobenzene a quasi-reversible reduction could be observed at −1.56 V at scan rates as slow as 100 mV s⁻¹. Bulk reductions with three different metallocenes (CoCp$_2$, CrCp$^*_2$, or CoCp$^*_2$) in C$_6$D$_6$ resulted in complete consumption of Mo=NNH$_2^+$ at room temperature over a period of 10 min to give mixtures that contain complexes in which dinitrogen is further reduced (i.e., Mo≡N, Mo(NH$_3$)⁺ or Mo(NH$_3$)) along with molecular hydrogen and ammonia. It was proposed that Mo=NNH$_2$ is protonated by Mo=NNH$_2^+$ to yield Mo=NNH$_3^+$ and Mo≡N–NH, and that Mo=NNH$_3^+$ is reduced irreversibly to give Mo≡N and ammonia. The instability of Mo=NNH$_2$ suggests that the steric protection of the apical Mo pocket may be compromised in some circumstances, perhaps through protonation of an amido nitrogen and dissociation of that protonated arm (*vide infra*).

The NNH$_2$ ligand in Mo=NNH$_2$ ($S = \frac{1}{2}$) is calculated to be bent at N$_\alpha$ in Mo=NNH$_2$ [59]. The SOMO is a d$_{xz}$ orbital that is antibonding with respect to the lone pair on N$_\alpha$. Bending the Mo–N–N system at N$_\alpha$ creates the possibility that N$_\alpha$ could be protonated or that the NNH$_2$ unit could bind in an η^2 fashion. Either (or both) may be the source of the relatively dramatic disproportionations that are observed when Mo=NNH$_2^+$ is reduced. Details of the disproportions are not known.

Perhaps the most elusive species is Mo=NNH$_3^+$, the presumed immediate precursor to Mo≡N and ammonia upon addition of an electron. There is no experimental evidence for Mo=NNH$_3^+$. Calculations suggest that if Mo=NNH$_3^+$ is formed, addition of an electron would lead spontaneously to Mo≡N and NH$_3$ [59]. Reiher considers a number of other possibilities, among them protonation of an amido nitrogen followed by addition of an electron.

2.3.5
Mo≡N and [Mo=NH]$^+$

The Mo(VI) nitride, Mo≡N, was prepared by heating a C$_6$D$_6$ solution of MoCl with 1.2 equiv of Me$_3$SiN$_3$ at 90 °C for three days. Mo≡N could be isolated in 77% yield as a bright yellow crystalline solid. Mo≡N is by far the most stable of any of the Mo compounds toward air.

Protonation of Mo≡N with [H(OEt$_2$)$_2$]BAr$'_4$ in C$_6$D$_6$ eventually afforded [Mo=NH]BAr$'_4$ [55]. Upon mixing the reagents in an NMR tube in C$_6$D$_6$, a variety of unidentified diamagnetic intermediates were formed, but the system evolved to give [Mo=NH]$^+$ in high yield after 24 h. The unidentified diamagnetic intermediates are proposed to contain partially protonated ligands, but no further information has been forthcoming. In diethyl ether the [Mo=NH]BAr$'_4$ salt was isolated in 93% yield 1 h after mixing the reagents. An X-ray structure determination of [Mo=NH]$^+$ revealed no surprises [55]. Protonation of Mo≡N with 2,6-lutidinium is calculated to be favorable ($\Delta E = -20.3$ kcal mol^{-1}) [59].

Reaction of Mo=NH$^+$ with CrCp*$_2$ in C$_6$D$_6$ instantly yields a paramagnetic, yellow–brown product whose ^{1}H NMR spectrum features a relatively sharp peak at −5.7 ppm and a very broad resonance at +32 ppm [55]. The product has been assigned tentatively as the neutral imido species "Mo=NH". Mo=NH can be generated in solution and that solution can be stored for a period of time at room temperature, as long as a small amount of CrCp*$_2$ is present. In the absence of CrCp*$_2$, Mo=NH is slowly converted into a mixture of Mo≡N and "Mo(NH$_2$)" (*vide infra*). Resonances for "Mo(NH$_2$)" then disappeared over a period of hours. The kinetic stability of "Mo=NH" allows electrochemical reduction of Mo=NH$^+$ to be quasi-reversible in THF (at −1.25 V) and PhF (at −1.38 V).

Mo(NH$_2$)$^+$ should be preparable through protonation of Mo=NH, as suggested by the calculated ΔE (−33.6 kcal mol^{-1}). However, no experimental results are yet available. Protonation of the imido ligand in Mo=NH by lutidinium ($\Delta E = -33.6$ kcal mol^{-1}, 2.12) is favored over protonation of a ligand amido nitrogen ($\Delta E = -20.5$ kcal mol^{-1}) or reduction of Mo=NH ($\Delta E = 93.9$ kcal mol^{-1}). Subsequent reduction of Mo(NH$_2$)$^+$ with CrCp$^*_2$ is calculated to take place with $\Delta E = 8.7$ kcal mol^{-1}.

$$\mathbf{Mo=NH} \xrightarrow[-33.6\text{ kcal/mol}]{H^+} [\mathbf{Mo=NH_2}]^+ \xrightarrow[8.7\text{ kcal/mol}]{e^-} \mathbf{Mo(NH_2)} \quad (2.12)$$

Mo(NH$_2$) can be observed in solution upon deprotonation of Mo(NH$_3$)$^+$ (*vide infra*) with LiN(SiMe$_3$)$_2$ in THF; Mo(NH$_3$)$^+$ is not deprotonated by Et$_3$N. ^{1}H NMR spectra of Mo(NH$_2$) suggest that it has the $S = 0$ ground state configuration, with the high spin state being thermally accessible. This circumstance has been encountered in related compounds, [RN$_3$N]MoNMe$_2$ (R = Me$_3$Si, C$_6$F$_5$) [41]. Calculations suggest that high spin Mo(NH$_2$) is only 1.8 kcal mol^{-1} above the ground state, a circumstance that would give rise to the proposed spin equilibrium detectable in temperature-dependent proton NMR spectra.

2.3.6
Mo(NH$_3$) and [Mo(NH$_3$)]$^+$

A combination of MoCl, NaBAr$'_4$, and ammonia in CD$_2$Cl$_2$ produced the cationic ammonia complex, [Mo(NH$_3$)]BAr$'_4$, while use of NaBPh$_4$ led to [Mo(NH$_3$)]BPh$_4$ in high yield. Since reactions of MoCl with ammonia in the presence of NaBAr$'_4$ or NaBPh$_4$ are complete at room temperature in CD$_2$Cl$_2$ within 30 min, it seems plausible that ammonia displaces chloride in MoCl to yield [Mo(NH$_3$)]Cl in a pre-equilibrium, and the reaction is then driven to completion through formation of NaCl. An X-ray study of [Mo(NH$_3$)]BAr$'_4$ reveals a Mo–NH$_3$ distance of 2.236(10) Å; the ammonia protons could not be located.

Electrochemical reduction of Mo(NH$_3$)$^+$ is fully reversible in both THF and PhF. Interestingly, at slow scan rates (~10 mV s^{-1}), waves corresponding to the MoN$_2^{0/-}$ couple become discernible (near −1.8 V in THF). Therefore exchange of ammonia in Mo(NH$_3$) (*vide infra*) with dinitrogen takes place to an observable extent after the beginning of the Mo(NH$_3$)$^+$ reduction sweep (10 mV s^{-1}), a period of 30–40 s. Since the potentials of CrCp$^{*+/0}_2$ and Mo(NH$_3$)$^{+/0}$ in fluorobenzene are both −1.63 V, CrCp$^*_2$ is a viable reductant of Mo(NH$_3$)$^+$.

The energy difference between the ground ($S = 1$) and $S = 0$ state in Mo(NH$_3$)$^+$ is calculated to be 10.7 kcal mol^{-1}, while the $S = \frac{1}{2}$ state of Mo(NH$_3$) is calculated to be 23.7 kcal mol^{-1} below the $S = 3/2$ state [59]; high spin Mo(NH$_3$) therefore is unlikely to be relevant in the chemistry of Mo(NH$_3$). Both protonation of Mo(NH$_2$) and reduction of Mo(NH$_3$)$^+$ are calculated to be favorable (2.13).

$$\mathbf{Mo(NH_2)} \xrightarrow[-21.2\text{ kcal/mol}]{H^+} [\mathbf{Mo(NH_3)}]^+ \xrightarrow[4.5\text{ kcal/mol}]{e^-} \mathbf{Mo(NH_3)} \quad (2.13)$$

Mo(NH$_3$) was prepared by treating MoN$_2$ with ammonia (~4 equiv) in C$_6$D$_6$ solution in the absence of dinitrogen for 16 h at 22 °C. An X-ray study of Mo(NH$_3$) showed that the Mo–N$_{amide}$ bond distance is somewhat longer (0.05 Å) than it is in Mo(NH$_3$)$^+$, the Mo–NH$_3$ bond is shorter (by 0.07 Å), and the trans Mo–N(amine) bond is longer (by 0.06 Å). These (small) changes in bond distances are consistent with an electron being added to a d$_{xz}$ or d$_{yz}$ orbital on Mo, although it is not clear why the Mo–NH$_3$ bond shortens upon reduction of Mo(NH$_3$)$^+$ to Mo(NH$_3$). Loss of ammonia from Mo(NH$_3$) is calculated to require 28.2 kcal mol^{-1} [59].

2.4
Interconversion of Mo(NH$_3$) and Mo(N$_2$)

It has been established that [Mo(NH$_3$)]BAr$_4'$ is formed upon treatment of Mo(N$_2$) in benzene in the presence of 1 equiv each of CoCp$_2$ and [2,6-LutH]BAr$_4'$ and that a stronger reducing agent such as CrCp*$_2$ will reduce [Mo(NH$_3$)]BAr$_4'$ to Mo(NH$_3$) in benzene. Therefore, conversion of Mo(NH$_3$) into Mo(N$_2$) (2.14) would complete a hypothetical catalytic cycle. The equilibrium constant for conversion of Mo(NH$_3$) in the presence of dinitrogen into a mixture of Mo(N$_2$), Mo(NH$_3$), and ammonia has been measured in C$_6$D$_6$ at 50 °C and found to be 1.2 [55]. It also has been shown that an equilibrium between Mo(NH$_3$), dinitrogen, Mo(N$_2$), and ammonia is established in 1–2 h under an atmosphere of ammonia (0.28 atm, ~21 equiv vs. Mo)

(2.14)

and dinitrogen [63]. At room temperature K_{eq} for the displacement of N$_2$ in Mo(N$_2$) by ammonia ([Mo(NH$_3$)][N$_2$]/[Mo(N$_2$)][NH$_3$]) is ~10. Since the exchange of dinitrogen in Mo(^{15}N$_2$) for N$_2$ has been shown to be independent of pressure and to have a half-life of 30–35 h, the mechanism of replacement of dinitrogen in Mo(N$_2$) by ammonia cannot consist of rate-limiting loss of dinitrogen from Mo(N$_2$) to yield Mo.

The rate of conversion of Mo(NH$_3$) into Mo(N$_2$) under dinitrogen in the absence of [2,6-LutH]BAr$_4'$ has been studied by IR spectroscopy and Differential Pulse Voltammetry and found to depend on dinitrogen pressure to the first order [63]. The conversion of Mo(NH$_3$) into Mo(N$_2$) has also been found to be accelerated dramatically at 1 atm in the presence of BPh$_3$ (4 or 10 equiv), most likely because BPh$_3$ removes much or all of the ammonia formed in solution. From these data it can be estimated that the half-life for conversion of Mo(NH$_3$) into Mo(N$_2$) is 10–15 min [63]. Although conversion of Mo(NH$_3$) into Mo(N$_2$) is not a simple reaction (ammonia that is generated in solution not only back reacts with Mo(N$_2$) to give Mo(NH$_3$), but also begins to enter the head space above solvents such as benzene or heptane), the conversion of Mo(NH$_3$) into Mo(N$_2$) follows what is close to first order kinetics in Mo with a half-life for the conversion at 22 °C and 1 atm of ~115 min. At 2 atm the half-life is ~45 min, which suggests that the conversion depends on the concentration of dinitrogen to the first power. An inescapable consequence is that the rate of conversion of Mo(N$_2$) into Mo(NH$_3$) is necessarily dependent upon ammonia and that ammonia will be an inhibitor of further dinitrogen reduction as a consequence of limiting formation of Mo(N$_2$) from Mo(NH$_3$).

Reiher has recently carried out calculations concerned with the reaction between Mo(NH$_3$) and dinitrogen to yield Mo(N$_2$) [60]. As outlined above, this reaction is relatively fast from either direction, reaching equilibrium in 1–2 h. He finds that a dissociative mechanism in which Mo forms through dissociation of NH$_3$ or N$_2$ is unlikely in view of the energy required to dissociate nitrogen from Mo(N$_2$) (37.8 kcal mol^{-1}) or ammonia from Mo(NH$_3$) (28.2 kcal mol^{-1}). However, calculations show that nitrogen can attack the metal approximately in the plane of the three amido nitrogens to form an intermediate six-coordinate species, Mo(N$_2$)$_{eq}$(NH$_3$)$_{ax}$ (2.15). The analogous six-coordinate species in which ammonia and dinitrogen have switched places (Mo(N$_2$)$_{ax}$(NH$_3$)$_{eq}$) is only 5.7 kcal mol^{-1} lower in energy than Mo(N$_2$)$_{eq}$(NH$_3$)$_{ax}$, so Mo(N$_2$)$_{ax}$(NH$_3$)$_{eq}$ could be accessed readily if ammonia must be lost from the equatorial position. Several other possible mechanisms for loss of ammonia were considered. The main point is that the [HIPTN$_3$N]$^{3-}$ ligand system is relatively flexible and allows pseudooctahedral six-coordinate [HIPTN$_3$N]Mo(NH$_3$)(N$_2$) to form, as was proposed on the basis of experimental studies, and possibly also for Mo(N$_2$)$_{eq}$(NH$_3$)$_{ax}$ and Mo(N$_2$)$_{ax}$(NH$_3$)$_{eq}$ to interconvert through an intermediate that contains a trigonal Mo core.

(2.15)

2.5
Catalytic Reduction of Dinitrogen

Substoichiometric formation of ammonia could be demonstrated readily. For example, addition of 7.0 equiv of [2,6-LutH]BAr$'_4$ and 8.2 equiv of CoCp$_2$ in benzene to Mo(N$_2$) produced (by NMR) ~60% [Mo(NH$_3$)]BAr$'_4$ and ~10% of free ligand (H$_3$[HIPTN$_3$N]). When Bu$_4$NCl and NEt$_3$ were added to this mixture 1.09(2) equiv of ammonia were produced. Upon treating Mo≡N with 3.5 equiv of [2,6-LutH]BAr$'_4$ and 4.2 equiv of CoCp$_2$ in benzene, [Mo(NH$_3$)]BAr$'_4$ was formed in ~80% yield (by NMR); further treatment of the reaction mixture with Bu$_4$NCl and NEt$_3$ afforded 0.88(2) equiv of ammonia. However, *catalytic* reduction of a molecule as stable as dinitrogen with protons and electrons is significantly more demanding. Not only must all steps that involve protonation and subsequent reduction of a cationic species (Figure 2.2) be fast relative to reduction of protons, but the dozen or so intermediates must not undergo any side reactions that compete with the desired reactions and the reaction must of course "turn over."

Heptane was chosen as the solvent for catalytic reduction in order to minimize the concentration of relatively insoluble [2,6-LutH]BAr$'_4$. [2,6-LutH]BAr$'_4$ was chosen because it is a relatively weak acid (pK_a in water = 6.75), because we felt that the 2,6-lutidine that is formed after delivery of the proton is unlikely to bind strongly to a neutral (Mo(III)) center in competition with dinitrogen, and because BAr$'^{-}_4$ is large and unlikely to interact with any neutral Mo center. We had also found that BAr$'^{-}_4$ salts that we had prepared were soluble in heptane. Decamethylchromocene was chosen as the reducing agent since it was a sufficiently good reducing agent for all species examined except Mo(N$_2$) itself, even though conversion of Mo(N$_2$) to Mo–N=NH by CoCp$_2$ and [2,6-LutH]BAr$'_4$ in benzene appeared to point to reduction of Mo(N$_2$) to Mo(N$_2$)$^-$ not being required. For example, decamethylchromocene was found to reduce [Mo(NH$_3$)]BAr$'_4$ to Mo(NH$_3$) in benzene ($E°$ for Mo(NH$_3$)$^{+/0}$ = −1.51 V relative to FeCp$_2^{+/0}$ while $E°$ for CrCp$^*_2$ [0/+] = −1.47 V in THF), while CoCp$_2$ cannot; CrCp$^*_2$ cannot fully reduce Mo(N$_2$) to Mo(N$_2$)$^-$ ($E°$ for Mo(N$_2$)$^{0/-}$ is −1.81 V in THF vs. FeCp$_2^{+/0}$). The reducing agent was added slowly over a period of 6 h in order to allow the ammonia in Mo(NH$_3$) to be displaced by dinitrogen. A standard reaction involved addition of 36 equiv of CrCp$_2^*$ to a stirred solution containing some molybdenum species thought to be a plausible intermediate in a catalytic reduction in heptane in the presence of 48 equiv of [2,6-LutH]BAr$'_4$ under dinitrogen. The sensitivity of most molybdenum species described above to air, water, and possible impurities in CrCp$_2^*$ and [2,6-LutH]BAr$'_4$ required very high standards of solvent and reagent purity; for example, CrCp$_2^*$ had to be purified by multiple sublimations or crystallizations. The containment and analysis of any ammonia produced (by the indophenol method) required the apparatus to be all glass, self-contained, and vacuum tight. Compounds **1**, **3**, **7**, **12**, or **13** (Figure 2.2) were usually employed as "catalysts." [2,4,6-trimethylpyridinium]$^+$ (collidinium) eventually replaced [2,6-LutH]$^+$ in order to eliminate any possible reductive coupling of 2,6-dimethylpyridyl radicals in the para position of the pyridine ring.

In a typical run a total of 7–8 equiv of ammonia are formed out of ~12 possible (the theoretical total varies with the Mo derivative employed), which suggests an efficiency of ~65% based on the reducing equivalents available. If the amount of ammonia formed from the N$_x$H$_y$ fragment on the metal is subtracted from the total ammonia formed, then the efficiency of formation of ammonia from gaseous dinitrogen is 55–60%. A run employing Mo–^{15}N=^{15}NH under ^{15}N$_2$ yielded entirely ^{15}N-labeled ammonia. No hydrazine (estimated to be <0.01 equiv) was formed. Reducing equivalents that were not employed to reduce dinitrogen were used to form dihydrogen; in a typical run dihydrogen (33%) and ammonia (61%) account for 94% of the electrons. CoCp$_2$ can also be employed as the reducing agent for catalytic dinitrogen reduction, although it is approximately half as efficient as CrCp$^*_2$. In the absence of any molybdenum but in the presence of [collidinium]BAr′₄, CrCp*₂ yields >95% of the expected hydrogen over a period of ~1 h. Reduction is more efficient when the amount of reducing agent and acid is halved (67% ammonia from dinitrogen). Reduction is less efficient (24% ammonia from dinitrogen) when the reducing agent is added over a period of ~30 s.

The yield of ammonia increases only marginally (~1 equiv) when all volatiles are removed from a completed run and the reactor is recharged with CrCp₂* and [2,6-LutH]BAr′₄. Evidently little molybdenum remains in a catalytically viable form. Since formation of free ligand is observed in most reactions in which some Mo species is treated with CrCp₂* and [2,6-LutH]BAr′₄, since large amounts of free ligand is present in the residue from a catalytic reaction, and since the ligand in Mo(N₂) (and Mo(CO)) is known to be protonated at the amido nitrogen, turnover is believed to be limited primarily as a consequence of loss of ligand from the metal.

The nature of the acid is crucial [63]. The amount of ammonia formed from dinitrogen was found to be 34% when [2,4-Me₂C₆H₃NH]BAr′₄ was employed, 23% for [2,6-Et₂C₆H₃NH]BAr′₄, and 0% for [2,6-Ph₂C₆H₃NH]BAr′₄, [3,5-Me₂C₆H₃NH]BAr′₄, and [Et₃NH]BAr′₄. Possible explanations of these results include pK_a, the nature of the conjugate base, steric factors, solubility in heptane, and impurities in the acid employed. Only 8% ammonia was formed from dinitrogen employing [2,6-LutH]BAr′₄ in the presence of 145 equivalents of 2,6-lutidine and 28% was formed employing [2,6-LutH]BAr′₄ in the presence of 150 equivalents of THF.

Finally, the yield of ammonia increased at 15 psi overpressure of dinitrogen, from 63% total ammonia to 71% total ammonia. Therefore, an important question is whether the total ammonia will continue to increase at higher pressures, or whether formation of 1 equiv of hydrogen is required for reduction of dinitrogen. This question has still not been answered.

There are several other results that could be put in the category of variations of standard conditions, including other ligands and other metals. These are covered in Section 2.7. Before we turn to these variations we want to explore the issues surrounding hydride and dihydrogen complexes.

2.6
MoH and Mo(H₂)

At one point we considered the possibility that MoH, which had been shown to be the product of thermal decomposition of Mo–N=NH, could be formed during catalytic dinitrogen reduction and that it might be resistant to being incorporated into the catalytic cycle. It turns out that catalytic runs in which MoH is the "precatalyst" reveal that it is as competent (65–66% efficiency in electrons) as any other molybdenum species that has been employed as the "catalyst." MoH might be drawn into the catalytic scheme through protonation and reduction to give MoH₂. Therefore we also became interested in preparing MoH₂. Another reason for exploring MoH₂ is that dihydrogen, the other product formed during catalytic reduction of dinitrogen, might behave as an inhibitor, as it does in the biological

FeMo system. Hydrides and hydrogen complexes have been considered as part of dinitrogen reduction by the natural FeMo system, either in an integral fashion, or as part of a side reaction that leads to formation of hydrogen.

A convenient, one-pot synthesis of MoH directly from MoCl is shown in Equation 2.16. [Mo(NH$_3$)]BPh$_4$ need not be isolated, but can be treated with the hydride reagent immediately. The IR spectrum of MoH features a relatively strong ν(Mo–H) absorption at 1748 cm^{-1} (in C$_6$D$_6$, 1751 cm^{-1} in Nujol), shifted by 70 cm^{-1} to higher energies relative to that observed in [TMSN$_3$N]MoH [45].

$$\text{MoCl} \xrightarrow[\text{1.1 NaBPh}_4]{\text{4 NH}_3} [\text{Mo(NH}_3)\text{BPh}_4] \xrightarrow[\text{THF}]{\text{LiBHEt}_3} \text{MoH} \qquad (2.16)$$

Exposure of a C$_6$D$_6$ solution of MoH to 1 atm of H$_2$ resulted in formation of a mixture of MoH and ~20% of a diamagnetic product. This process is reversed upon removal of hydrogen. The precise nature of this diamagnetic product is not yet known. It should be noted that classical tungsten trihydride species, [RN$_3$N]WH$_3$, have been identified in which R = Me$_3$Si [64] or HIPT [65], and that [HIPTN$_3$N]WH$_3$ will lose H$_2$ when heated *in vacuo* to form mixtures of [HIPTN$_3$N]WH$_3$ and [HIPTN$_3$N]WH. Exposure of a solution of MoH to D$_2$ led to formation of an observable quantity of HD after 50 min (according to proton NMR spectroscopy) and yielded largely MoD after 23 h, as determined by IR, consistent with reversible formation of MoH$_3$. MoH reacts with [2,6-LutH]BAr$_4'$ in C$_6$D$_6$ to yield [Mo(2,6-Lut)]BAr$_4'$, as noted earlier.

[HIPTN$_3$N]Mo(N$_2$) (MoN$_2$) reacts with dihydrogen slowly (days at 22 °C in C$_6$D$_6$) and reversibly in solution to yield a compound with $S = \frac{1}{2}$ that has the stoichiometry [HIPTN$_3$N]Mo(H$_2$) (MoH$_2$). The ability to prepare MoD$_2$ rules out any H/D scrambling that might result from CH activation in the ligand and subsequent exchange of D for H. Attempts to prepare MoHD resulted in formation of MoH$_2$, MoHD, and MoD$_2$ in a 1 : 2 : 1 ratio. Interconversion of MoN$_2$ and MoH$_2$ consists of (slow) rate-limiting loss of dinitrogen or dihydrogen, respectively, to yield unobservable trigonal monopyramidal Mo. All evidence is consistent with MoH$_2$ being a Mo(III) dihydrogen complex at room temperature.

In contrast to the slow reaction between MoN$_2$ and H$_2$ through rate-limiting loss of dinitrogen, MoNH$_3$ reacts with H$_2$ within minutes to yield MoH$_2$, consistent with a reaction that is first order in dihydrogen. K_{eq} (K_{eq} = [MoH$_2$][NH$_3$]/[MoNH$_3$][H$_2$]) for this reaction was found to be ~12 in C$_6$D$_6$ at 22 °C. Since [MoNH$_3$][N$_2$]/[MoN$_2$][NH$_3$] is ~10 (*vide supra*), the equilibrium constant for the reaction between MoN$_2$ and H$_2$ mentioned above ([MoH$_2$][N$_2$]/[MoN$_2$][H$_2$]) is ~120. Reactions that involve ammonia are second order overall, while the one that involves only hydrogen and nitrogen is first order overall.

Catalytic runs with MoH$_2$ as a catalyst suggest that MoH$_2$ is competent for reduction of N$_2$ with protons and electrons under standard conditions. Ammonia is formed in 52% yield (from N$_2$) relative to reducing equivalents (cf. 60–65% yield from N$_2$ with other catalysts such as MoN$_2$.) All evidence suggests that dihydrogen inhibits reduction of dinitrogen. Both ammonia and dihydrogen shift MoN$_2$ toward MoNH$_3$ and MoH$_2$, respectively, and therefore turnover slows as ammonia and dihydrogen are formed.

2.6 MoH and Mo(H₂)

Reactions involving H$_2$, D$_2$, and HD proved to be relatively complex and revealing. The slowest reaction (several days) is formation of MoHD from MoH$_2$ and MoD$_2$ under argon. It is proposed that MoH$_2$ is in equilibrium with a species in which one H has migrated to an amido nitrogen (MoH$_{Mo}$H$_N$; 2.17), and that MoH$_{Mo}$H$_N$ is in slow equilibrium with a small amount of a species in which one arm is dissociated ("MoH$_{Mo}$H$_{Noff}$"; 2.17). One way for H to exchange with D is to form a dimer that contains bridging hydrides (2.18).

$$(2.17)$$

$$(2.18)$$

When a solution that contains a small quantity of MoH$_2$ is placed under a 1:1 mixture of H$_2$ and D$_2$, a 1:2:1 ratio of H$_2$, HD, and D$_2$ is formed rapidly. This "fast" H/D exchange process is the reason why it is not possible to obtain the MoHD complex; only a mixture of MoH$_2$, MoHD, and MoD$_2$ is formed, as noted earlier. It is proposed that this fast H/D exchange involves oxidative addition to Mo(III) in MoH$_{Mo}$H$_{Noff}$ to give a Mo(V) trihydride intermediate, MoH$_3$H$_{Noff}$ (2.19), or a related dihydrogen/hydride in which Mo is still Mo(III).

$$(2.19)$$

The picture that has emerged provides strong evidence for formation of MoH$_{Mo}$H$_N$ and MoH$_{Mo}$H$_{Noff}$. Labilization of the ligand in a similar manner whenever an H is present on an amido nitrogen atom is a plausible way to allow bimetallic reactions to take place that would be excluded if the ligand were to remain bound to the metal in a tetradentate fashion. We now suspect that "arm-off" species are formed under catalytic conditions, since an amido nitrogen in MoN$_2$ is known to be protonated and the ligand ultimately stripped from the metal. Can intermediate MoN$_x$H$_y$ complexes rearrange to yield a species that contains a protonated amido nitrogen and ultimately an "arm-off" species? Many puzzling results suggest that the answer may be yes. For example, (i) Mo–N=NH is known to decompose slowly to MoH. (ii) Mo=NH is known to decompose to yield hydrogen and Mo≡N. (iii) [Hybrid]Mo–N=NH species, in which the "hybrid" ligand is significantly smaller than [HIPTN$_3$N]$^{3-}$ (vide infra), are relatively unstable toward 2,6-lutidine, the conjugate base of the acid employed in a typical catalytic reduction of dinitrogen; hydrogen is evolved and [Hybrid]MoN$_2$ species are formed. (iv) [Hybrid]Mo–^{15}N=^{15}NH species have been observed to exchange with N$_2$ to yield [Hybrid]Mo–N=NH species at a rate slightly faster than the rate at which they decompose to yield [Hybrid]MoN$_2$.

2.7
Ligand and Metal Variations

Three "symmetric" variations of the [HIPTN$_3$N]$^{3-}$ ligand have been explored, a hexa-t-butylterphenyl-substituted ([HTBTN$_3$N]$^{3-}$) ligand, a hexamethylterphenyl-substituted ([HMTN$_3$N]$^{3-}$) ligand, and a [pBrHIPTN$_3$N]$^{3-}$ ligand, which is a [HIPTN$_3$N]$^{3-}$ ligand in which the para position of the central phenyl ring is substituted with a bromide [66]. One would expect [HTBTN$_3$N]Mo complexes to be significantly more crowded sterically than [HIPTN$_3$N]Mo complexes. (An X-ray study of [HTBTN$_3$N]MoCl shows this to be the case.) [HMTN$_3$N]Mo complexes are expected to be significantly less crowded than [HIPTN$_3$N]Mo complexes, while [pBrHIPTN$_3$N]Mo complexes are sterically similar to [HIPTN$_3$N]Mo complexes. Tansformations of [HTBTN$_3$N]$^{3-}$ derivatives that involve electron and proton transfer (e.g., conversion of [HTBTN$_3$N]Mo(N$_2$) into [HTBTN$_3$N]Mo–N=NH) are slower by perhaps an order of magnitude compared to analogous conversions of [HIPTN$_3$N]$^{3-}$ derivatives, consistent with a high degree of steric crowding in [HTBTN$_3$N]$^{3-}$ derivatives.

[pBrHIPTN$_3$N]Mo≡N was found to be a catalyst for the formation of ammonia in yields only slightly less than those observed employing [HIPTN$_3$N]$^{3-}$ derivatives. Generation of [pBrHIPTN$_3$N]Mo(NH$_3$) and observation of its conversion to [pBrHIPTN$_3$N]Mo(N$_2$) under 1 atm of dinitrogen in heptane at 22 °C showed that the half-life for formation of [pBrHIPTN$_3$N]Mo(N$_2$) is ~2 h, the same as for conversion of [HIPTN$_3$N]Mo(NH$_3$) into [HIPTN$_3$N]Mo(N$_2$).

[HTBTN$_3$N]Mo≡N was found to be a poor catalyst for the reduction of dinitrogen, with only ~0.1 equiv being formed from gaseous dinitrogen. One measurement of the rate of conversion of [HTBTN$_3$N]Mo(NH$_3$) into [HTBTN$_3$N]Mo(N$_2$) under conditions analogous to those employed for Mo(NH$_3$) reveals that the "half-life" for conversion is approximately 30 h instead of 2 h, as it is for conversion of Mo(NH$_3$) into Mo(N$_2$). The slower coupled reduction of {[HTBTN$_3$N]Mo(NH$_3$)}$^+$ to [HTBTN$_3$N]Mo(NH$_3$) coupled with slow conversion of [HTBTN$_3$N]Mo(NH$_3$) to [HTBTN$_3$N]Mo(N$_2$) apparently cannot compete with formation of dihydrogen via "direct" reduction of protons.

One puzzling result is that exchange of ^{15}N$_2$ for N$_2$ in [HTBTN$_3$N]Mo(N$_2$) takes place approximately 20 times more slowly ($t_{1/2}$ ~ 750 h at 22 °C or $k = 2.6 \times 10^{-7}$ s^{-1}) than in Mo(N$_2$) ($k = 5.5 \times 10^{-6}$ s^{-1} at 5 atm); it is also independent of pressure [66]. Since v_{NN} in [HTBTN$_3$N]Mo(N$_2$) and [HIPTN$_3$N]Mo(N$_2$) are identical (v_{NN} = 1990 cm^{-1}), the Mo–N$_2$ bond strength must be the same in the two species. The difference in the unimolecular rate constant for loss of dinitrogen in [HTBTN$_3$N]Mo(N$_2$) and [HIPTN$_3$N]Mo(N$_2$) currently is ascribed to a trapping of the dissociated dinitrogen near the metal in the more crowded [HTBTN$_3$N]Mo(N$_2$) system where it re-coordinates more often than it escapes through the ligand "canopy" compared to the [HIPTN$_3$N]Mo(N$_2$) system. We do not know whether N$_2$ loss in [HIPTN$_3$N]Mo(N$_2$) is also "slow" relative to some less crowded analog, only that the rate of N$_2$ loss from [HTBTN$_3$N]Mo(N$_2$) is slow relative to the rate of loss in [HIPTN$_3$N]Mo(N$_2$).

Only 0.47 equiv of ammonia were formed from dinitrogen when [HMTN$_3$N]Mo≡N was employed as a catalyst under standard conditions. Although the low solubility of [HMTN$_3$N]$^{3-}$ intermediates (in heptane) is one possible problem, studies of other "hybrid" alternatives to the [HIPTN$_3$N]$^{3-}$ ligand (*vide infra*) suggest that a lack of sufficient steric protection may also be problematic.

"Hybrid" ligands of the type [(HIPTNCH$_2$CH$_2$)$_2$NCH$_2$CH$_2$NAr]$^{3-}$, could be prepared where Ar, for example, is 3,5-dimethylphenyl, 3,5-dimethoxyphenyl, or 3,5-bistrifluoromethylphenyl [67]. When [(HIPTNCH$_2$CH$_2$)$_2$NCH$_2$CH$_2$NAr]Mo≡N species are employed in a standard attempted catalytic reaction, *no ammonia is produced from dinitrogen using any of the three Ar variations*. The half-life for conversion of [(HIPTNCH$_2$CH$_2$)$_2$NCH$_2$CH$_2$NAr]Mo(NH$_3$) into [(HIPTNCH$_2$CH$_2$)$_2$NCH$_2$CH$_2$NAr]Mo(N$_2$) was shown to be ~170 min when Ar = 3,5-(CF$_3$)$_2$C$_6$H$_3$ and < 30 min when Ar = 3,5-(OMe)$_2$C$_6$H$_3$. Since the half-life for conversion of Mo(NH$_3$) into Mo(N$_2$) is ~120 min, reduction must fail in the "hybrid" cases for some reason other than a slow conversion of [Hybrid]Mo(NH$_3$) into [Hybrid]Mo(N$_2$). [(HIPTNCH$_2$CH$_2$)$_2$NCH$_2$CH$_2$NAr]Mo≡N complexes may also be prepared where Ar is mesityl or triisopropylphenyl. These species produce 0.2 and 2.0 equiv of ammonia from dinitrogen, respectively. [(HIPTNCH$_2$CH$_2$)$_2$NCH$_2$CH$_2$N-3,5-(CF$_3$)$_2$C$_6$H$_3$]Mo–N=NH was shown to be decomposed by 2,6-lutidine. Decomposition of [(HIPTNCH$_2$CH$_2$)$_2$NCH$_2$CH$_2$N-3,5-(CF$_3$)$_2$C$_6$H$_3$]Mo–N=NH by lutidine constitutes a "shunt" in the catalytic cycle that consumes protons and electrons to yield hydrogen. Formation of dihydrogen is proposed to be bimolecular in Mo, although that has not been proven.

[HIPTN$_3$N]$^{3-}$ species can be prepared for Cr [68]. Attempted catalytic reduction of dinitrogen under standard conditions with [HIPTN$_3$N]Cr≡N yields only 0.8 equiv of ammonia, that is, none from dinitrogen. The fact that $[HIPTN_3N]Cr(N_2)$ and $\{[HIPTN_3N]Cr-N=N\}^-$ could not be observed suggest that $[HIPTN_3N]Cr$ simply does not bind dinitrogen strongly. Even the reaction between high spin $[HIPTN_3N]$ Cr and CO is slow as a consequence (it was proposed) of a restricted access to a low spin form that binds CO readily.

Most $[HIPTN_3N]^{3-}$ species that are known for Mo can also be prepared for W [65], the one exception being $[HIPTN_3N]W(NH_3)$. CV studies of $W(NH_3)^+$ in fluorobenzene revealed that W is much more difficult to reduce than Mo ($<-2\,V$ vs. $-1.63\,V$ for Mo) and the reduction is irreversible. All attempted reductions of dinitrogen with $W(N_2)$ as a catalyst yielded no ammonia from dinitrogen. The difficulty of forming $W(NH_3)$ and its apparent instability is one of the reasons why W fails as a catalyst for reducing dinitrogen.

$[HIPTN_3N]V(THF)$ could be prepared [69], but no ammonia was formed in a standard catalytic reduction. If a series of intermediates similar to that shown in Figure 2.2 is anticipated, then all neutral species in the Mo series will be anions, for example, $\{[HIPTN_3N]V\equiv N\}^-$, and all cationic species will be neutral, for example, $[HIPTN_3N]V=NH$. It is not known whether this fact alone limits the options for vanadium. In order to preserve the charges on all V species analogous to those in Figure 2.2, the ligand on V would have to be a dianion instead of a trianion.

There are several possible alternatives to the $[(HIPTNCH_2CH_2)_3N]^{3-}$ ligand, one being increasing the length of carbon chain by one methylene group. The synthesis of $(HIPTNHCH_2CH_2CH_2)_3N$ is straightforward [70]. $[pHIPTN_3N]$ MoCl ($[pHIPTN_3N] = [(HIPTNHCH_2CH_2CH_2)_3N]$) can be prepared, as can $\{[pHIPTN_3N]Mo(NH_3)\}BAr'_4$. However, the yields and solubilities of $[pHIPTN_3N]^{3-}$ derivatives are in general much poorer than the already difficult (soluble and sensitive) $[HIPTN_3N]^{3-}$ derivatives, and few species can be isolated. For example, $[pHIPTN_3N]Mo-N=NK$ ($v_{NN} = 1810\,cm^{-1}$) and $[pHIPTN_3N]Mo(N_2)$ ($v_{NN} = 2030\,cm^{-1}$) can only be observed *in situ*. The lower value for v_{NN} in $[pHIPTN_3N]Mo(N_2)$ ($2030\,cm^{-1}$) compared to MoN_2 ($1990\,cm^{-1}$) suggests that the π backbonding in $[pHIPTN_3N]Mo(N_2)$ is less efficient than in MoN_2. An X-ray structure of $\{[pHIPTN_3N]Mo(NH_3)\}BAr'_4$ reveals that the metal is in the $N_{amido}-N_{amido}-N_{amido}$ plane in $\{[pHIPTN_3N]Mo(NH_3)\}BAr'_4$ (cf. $0.6\,\text{Å}$ above the plane in $\{[HIPTN_3N]Mo(NH_3)\}BAr'_4$) and that the $C-N_{amido}-C$ planes are tipped by $\sim35°$, compared to $\sim10°$ in $\{[HIPTN_3N]Mo(NH_3)\}BAr'_4$. Both may compromise the efficiency of π backbonding into dinitrogen. All indications are the the $[pHIPTN_3N]^{3-}$ ligand is simply not bound as securely as is the $[HIPTN_3N]^{3-}$ ligand and, therefore, is more susceptible to being lost from the metal. Attempted catalytic dinitrogen reductions lead to no ammonia from gaseous dinitrogen.

In an effort to prepared a suitable trianionic ligand in which the "amido" nitrogens will not be protonated readily when bound as a trianion to a metal, three substituted tris(pyrrolyl-α-methyl)amines ($H_3[Aryl_3TPA]$) (Aryl = $2,4,6-C_6H_2Me_3$, $2,4,6-C_6H_2(i\text{-}Pr)_3$ (Trip), or $3,5-C_6H_3(CF_3)_2$) were prepared [71]. An X-ray study of $[Trip_3TPA]MoCl$ shows it to be a distorted trigonal bipyramidal species in which the Mo–N and Mo–Cl bond lengths are in the expected range, and in which the

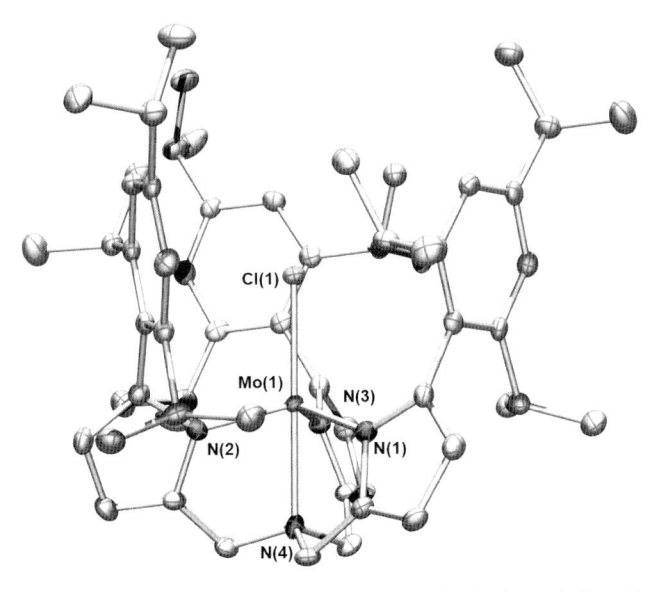

Figure 2.4 POV-Ray rendering of [Trip₃TPA]MoCl with thermal ellipsoids at 50% probability; (Å) Mo–N(1) = 2.004(3), Mo–N(2) = 2.009(3), Mo–N(3) = 2.009(3), Mo–N(4) = 2.213(3), Mo–Cl(1) = 2.3095(13).

2,4,6-triisopropylphenyl substituents surround and protect the apical chloride (Figure 2.4); other [Aryl₃TPA]MoCl derivatives could not be prepared. So far preliminary studies suggest that derivatives relevant to dinitrogen reduction will be difficult to prepare, if they can be prepared at all; the steric protection in fact may be too extreme and the ligand too inflexible in tris(pyrrolyl-α-methyl)amine complexes.

2.8 Comments

Given the degree to which catalytic reduction of dinitrogen is fraught with difficulties it is remarkable that it can be observed in an abiological system. One of the most important, and perhaps most difficult steps, the reaction of MoN$_2$ with one proton and one electron (H$^+$/e) to give Mo–N=NH, is actually relatively efficient. Mo–N=NH itself is unusual since M–N=NH species are inherently rare, Mo–N=NH is thermally stable, and Mo–N=NH is protonated on N$_\beta$ cleanly and reduced to "Mo–N=NH$_2$." The main problem limiting turnover appears to be loss of ligand. In theory, loss of ligand might be limited if the three substituted amido arms could be connected in some fashion, although that is synthetically challenging without altering the electronic or steric nature of the ligand in the process. It is also likely that one or more essential reactions (e.g., displacement of ammonia in Mo(NH$_3$)

via formation of Mo(NH$_3$)(N$_2$)) requires some ligand flexibility and room near the metal to make a six-coordinate intermediate. Recent evidence that amido nitrogens that bear one H can dissociate to give "arm-off" species is consistent with several apparent disproportionations and/or bimetallic reactions that would be untenable if the ligand actually remained fully bound to the metal. An interesting question, therefore, is whether any "arm-off" species are of any *direct relevance* to reduction of dinitrogen, that is, *required* for reduction. Since all evidence suggests that reduction is efficient only when the ligand remains bound to the metal, the answer is likely to be no. However, it must be admitted that this question cannot be answered at this stage. How to keep the ligand bound to the metal without damaging other features of the reduction process will be one of the most important challenges of the future. The second will be to find a way to make reduction practical, that is, ultimately to employ dihydrogen as the reducing agent, either directly, or after conversion into protons and electrons.

Acknowledgements

R.R.S. is grateful to the National Institutes of Health (GM 31978) for research support.

References

1 Burgess, B.K., and Lowe, D.J. (1996) *Chem. Rev.*, **96**, 2983.
2 Hardy, R.W.F., Bottomley, F., and Burns, R.C. (1979) *A Treatise on Dinitrogen Fixation*, Wiley-Interscience, New York.
3 Veeger, C., and Newton, W.E. (1984) *Advances in Nitrogen Fixation Research*, Dr. W. Junk/Martinus Nijhoff, Boston.
4 (1980) *Molybdenum and Molybdenum-Containing Enzymes* (ed. M.P. Coughlan), Pergamon, New York.
5 Noodleman, L., Lovell, T., Han, W.-G., Li, J., and Himo, F. (2004) *Chem. Rev.*, **104**, 459.
6 Rehder, D. (1999) *Coord. Chem. Rev.*, **182**, 297.
7 Smith, B.E. (1999) *Adv. Inorg. Chem.*, **47**, 159.
8 Eady, R.R. (1996) *Chem. Rev.*, **96**, 3013.
9 Einsle, O., Tezcan, F.A., Andrade, S.L.A., Schmid, B., Yoshida, M., Howard, J.B., and Rees, D.C. (2002) *Science*, **297**, 1696.
10 Rees, D.C., and Howard, J.B. (2000) *Curr. Opin. Chem. Biol.*, **4**, 559.
11 Bolin, J.T., Ronco, A.E., Morgan, T.V., Mortenson, L.E., and Xuong, L.E. (1993) *Proc. Natl. Acad. Sci. USA*, **90**, 1078.
12 Kim, J., and Rees, D.C. (1992) *Science*, **257**, 1677.
13 Allen, A.D., and Senoff, C.V. (1965) *J. Chem. Soc., Chem. Commun.*, 621.
14 Chatt, J., Dilworth, J.R., and Richards, R.L. (1978) *Chem. Rev.*, **78**, 589.
15 Fryzuk, M.D., and Johnson, S.A. (2000) *Coord. Chem. Rev.*, **200–202**, 379.
16 Hidai, M., and Mizobe, Y. (1995) *Chem. Rev.*, **95**, 1115.
17 Hidai, M. (1999) *Coord. Chem. Rev.*, **185–186**, 99.
18 Barriere, F. (2003) *Coord. Chem. Rev.*, **236**, 71.
19 Henderson, R.A., Leigh, G.J., and Pickett, C.J. (1983) *Adv. Inorg. Chem. Radiochem.*, **27**, 197.
20 Allen, A.D., Harris, R.O., Loescher, B.R., Stevens, J.R., and Whiteley, R.N. (1973) *Chem. Rev.*, **73**, 11.
21 Richards, R.L. (1996) *Coord. Chem. Rev.*, **154**, 83.

22 Shilov, A.E. (2003) *Russ. Chem. Bull. Int. Ed.*, **52**, 2555.
23 Glassman, T.E., Liu, A.H., and Schrock, R.R. (1991) *Inorg. Chem.*, **30**, 4723.
24 Glassman, T.E., Vale, M.G., and Schrock, R.R. (1991) *Organometallics*, **10**, 4046.
25 Schrock, R.R., Glassman, T.E., and Vale, M.G. (1991) *J. Am. Chem. Soc.*, **113**, 725.
26 Glassman, T.E., Vale, M.G., and Schrock, R.R. (1992) *Inorg. Chem.*, **31**, 1985.
27 Glassman, T.E., Vale, M.G., and Schrock, R.R. (1992) *J. Am. Chem. Soc.*, **114**, 8098.
28 Glassman, T.E., Vale, M.G., Schrock, R.R., and Kol, M. (1993) *J. Am. Chem. Soc.*, **115**, 1760.
29 Cummins, C.C., Schrock, R.R., and Davis, W.M. (1992) *Organometallics*, **11**, 1452.
30 Cummins, C.C., Lee, J., Schrock, R.R., and Davis, W.M. (1992) *Angew. Chem. Int. Ed. Engl.*, **31**, 1501.
31 Cummins, C.C., Schrock, R.R., and Davis, W.M. (1993) *Angew. Chem. Int. Ed. Engl.*, **32**, 756.
32 Cummins, C.C., and Schrock, R.R. (1994) *Inorg. Chem.*, **33**, 395.
33 Cummins, C.C., Schrock, R.R., and Davis, W.M. (1994) *Inorg. Chem.*, **33**, 1448.
34 Freundlich, J., Schrock, R.R., Cummins, C.C., and Davis, W.M. (1994) *J. Am. Chem. Soc.*, **116**, 6476.
35 Schrock, R.R., Cummins, C.C., Wilhelm, T., Kol, M., Lin, S., Reid, S., and Davis, W.M. (1996) *Organometallics*, **15**, 1470.
36 Schrock, R.R. (1997) *Acc. Chem. Res.*, **30**, 9.
37 Verkade, J.G. (1993) *Acc. Chem. Res.*, **26**, 483.
38 Freundlich, J.S., Schrock, R.R., and Davis, W.M. (1996) *J. Am. Chem. Soc.*, **118**, 3643.
39 Schrock, R.R., Seidel, S.W., Mösch-Zanetti, N.C., Dobbs, D.A., Shih, K.-Y., and Davis, W.M. (1997) *Organometallics*, **16**, 5195.
40 Johnson-Carr, J.A., Zanetti, N.C., Schrock, R.R., and Hopkins, M.D. (1996) *J. Am. Chem. Soc*, **118**, 11305.
41 Mösch-Zanetti, N.C., Schrock, R.R., Davis, W.M., Wanninger, K., Seidel, S.W., and O'Donoghue, M.B. (1997) *J. Am. Chem. Soc.*, **119**, 11037.

42 Zanetti, N.C., Schrock, R.R., and Davis, W.M. (1995) *Angew. Chem. Int. Ed. Engl.*, **34**, 2044.
43 Balzs, G., Gregoriades, L.J., and Scheer, M. (2007) *Organometallics*, **26**, 3058.
44 Shih, K.-Y., Totland, K., Seidel, S.W., and Schrock, R.R. (1994) *J. Am. Chem. Soc.*, **116**, 12103.
45 Schrock, R.R., Seidel, S.W., Mösch-Zanetti, N.C., Shih, K.-Y., O'Donoghue, M.B., Davis, W.M., and Reiff, W.M. (1997) *J. Am. Chem. Soc.*, **119**, 11876.
46 Shih, K.-Y., Schrock, R.R., and Kempe, R. (1994) *J. Am. Chem. Soc.*, **116**, 8804.
47 O'Donoghue, M.B., Zanetti, N.C., Schrock, R.R., and Davis, W.M. (1997) *J. Am. Chem. Soc.*, **119**, 2753.
48 O'Donoghue, M.B., Davis, W.M., Schrock, R.R., and Reiff, W.M. (1999) *Inorg. Chem.*, **38**, 243.
49 O'Donoghue, M.B., Davis, W.M., and Schrock, R.R. (1998) *Inorg. Chem.*, **37**, 5149.
50 Kol, M., Schrock, R.R., Kempe, R., and Davis, W.M. (1994) *J. Am. Chem. Soc.*, **116**, 4382.
51 Greco, G.E., and Schrock, R.R. (2001) *Inorg. Chem.*, **40**, 3850.
52 Greco, G.E., and Schrock, R.R. (2001) *Inorg. Chem.*, **40**, 3861.
53 Yandulov, D.V., and Schrock, R.R. (2002) *J. Am. Chem. Soc.*, **124**, 6252.
54 Yandulov, D.V., Schrock, R.R., Rheingold, A.L., Ceccarelli, C., and Davis, W.M. (2003) *Inorg. Chem.*, **42**, 796.
55 Yandulov, D., and Schrock, R.R. (2005) *Inorg. Chem.*, **44**, 1103.
56 Yandulov, D.V., and Schrock, R.R. (2003) *Science*, **301**, 76.
57 Schrock, R.R. (2008) *Angew. Chem. Int. Ed.*, **47**, 5512.
58 Reiher, M., Le Guennic, B., and Kirchner, B. (2005) *Inorg. Chem.*, **44**, 9640.
59 Schenk, S., Le Guennic, B., Kirchner, B., and Reiher, M. (2008) *Inorg. Chem.*, **47**, 3634; erratum 7934.
60 Schenk, S., Kirchner, B., and Reiher, M. (2009) *Chem. Eur. J.*, **15**, 5073.
61 Hetterscheid, D.G.H., Hanna, B.S., and Schrock, R.R. (2009) *Inorg. Chem.*, **45**, 8569.
62 McNaughton, R., Chin, J., Weare, W.W., Schrock, R.R., and Hoffman,

B.M. (2007) *J. Am. Chem. Soc.*, **129**, 3480.
63 Weare, W.W., Dai, X., Byrnes, M.J., Chin, J.M., Schrock, R.R., and Mueller, P. (2006) *Proc. Natl. Acad. Sci. USA*, **103**, 17099.
64 Dobbs, D.A., Schrock, R.R., and Davis, W.M. (1997) *Inorg. Chim. Acta*, **36**, 171.
65 Yandulov, D.V., and Schrock, R.R. (2005) *Can. J. Chem.*, **83**, 341.
66 Ritleng, V., Yandulov, D.V., Weare, W.W., Schrock, R.R., Hock, A.R., and Davis, W.M. (2004) *J. Am. Chem. Soc.*, **126**, 6150.
67 Weare, W.W., Schrock, R.R., Hock, A.S., and Müller, P. (2006) *Inorg. Chem.*, **45**, 9185.
68 Smythe, N.C., Schrock, R.R., Müller, P., and Weare, W.W. (2006) *Inorg. Chem.*, **45**, 7111.
69 Smythe, N.C., Schrock, R.R., Müller, P., and Weare, W.W. (2006) *Inorg. Chem.*, **45**, 9197.
70 Chin, J., Weare, W.W., Schrock, R.R., and Müller, P. (2010) *Inorg. Chem.*, submitted.
71 Wampler, K.M., and Schrock, R.R. (2007) *Inorg. Chem.*, **46**, 8463.

3
Molybdenum and Tungsten Catalysts for Hydrogenation, Hydrosilylation and Hydrolysis

R. Morris Bullock

3.1
Introduction

Hydrogenations occupy a central role in the development of a detailed understanding of the reactivity and mechanism of molecular catalysts [1]. Addition of H_2 to C=C and C=O bonds is required in many processes that are widely used in industry, particularly in the synthesis of pharmaceutical and agricultural chemicals. Noyori has called catalytic hydrogenation a "Core Technology in Synthesis" [2]. The field of homogeneous hydrogenation has become large and diverse, but is dominated by the use of precious metal catalysts, particularly ruthenium [3] and rhodium [4]. Detailed studies of the kinetics have revealed many variants on the mechanism, but Equation 3.1 shows in generalized form a key step in the traditional mechanism, involving insertion of a bound ketone into an M–H bond.

$$\text{M-H, O=CR}_2 \longrightarrow \text{M-O-CHR}_2 \xrightarrow{H_2} \text{R}_2\text{CH-OH} \quad (3.1)$$

Devising alternative mechanisms that do not require an insertion reaction requires finding abundant, inexpensive metals that exhibit different reactivity patterns, and then combining those steps into a new catalytic pathway to accomplish the hydrogenation. The use of "Cheap Metals for Noble Tasks" is appealing not only due to the lower cost of base metals, but also for the potentially reduced environmental and toxicological impact that their complexes can have. It is recognized, however, that the cost of the metal, while contributing a large amount to the overall cost of a catalytic cycle, is not the only consideration, as specialized phosphines or other ligands can contribute substantially to the cost of a catalyst.

New catalysts using precious metals have been developed by altering the ligand, but that approach is less applicable to catalysts based on non-traditional metals,

Catalysis Without Precious Metals. Edited by R. Morris Bullock
© 2010 WILEY-VCH Verlag GmbH & Co. KGaA, Weinheim
ISBN: 978-3-527-32354-8

where far less is known about catalysis with any type of ligand framework. Many of the catalytic reactions to be described in this chapter on Mo and W complexes were preceded by fundamental studies of individual reactions under stoichiometric conditions. These studies of kinetics, thermochemical properties, and reactivity then formed a base of knowledge upon which new classes of catalysts that use Mo or W were developed.

The widespread utility of metal hydrides in many types of reactions is due in part to the versatility of M–H cleavage modes available (Scheme 3.1) [5]. The M–H bond of a metal hydride can be homolytically cleaved, resulting in hydrogen atom transfer reactions [6]. Examples of the utility of hydrogen atom transfer reactions are given in Chapter 1. Heterolytic rupture of M–H bonds leads to either proton transfer reactions or hydride transfer reactions, and both of these reactions are involved in the hydrogenations described in this chapter.

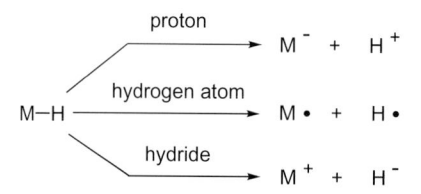

$$\text{M-H} \begin{cases} \xrightarrow{\text{proton}} \text{M}^- + \text{H}^+ \\ \xrightarrow{\text{hydrogen atom}} \text{M}\bullet + \text{H}\bullet \\ \xrightarrow{\text{hydride}} \text{M}^+ + \text{H}^- \end{cases}$$

Scheme 3.1

Studies of the kinetics and thermodynamics of proton transfer and hydride transfer reactions have led to a better fundamental understanding of the range of reactivity available, and how it is influenced by different metals and ligands. This information is also central to the rational development of molecular catalysts for oxidation of H_2 and production of H_2 described in Chapter 7, and in the broad context of other reactions pertinent to energy production and energy utilization that require control of multi-proton and multi-electron reactivity.

3.2
Proton Transfer Reactions of Metal Hydrides

Even though conventions of nomenclature designate all metal complexes with a bond to hydrogen as metal hydrides, many metal hydrides can exhibit acidic as well as hydridic behavior. Norton and coworkers studied extensively the thermodynamic and kinetic acidities of a wide series of metal carbonyl hydrides in acetonitrile [7]. Table 3.1 shows a few of the pK_a values reported to document the range of pK_a values and to illustrate some of the trends; more extensive compilations are available [7]. The acidity of $(CO)_4CoH$, the metal hydride involved in catalytic hydroformylations, is 8.3, which is similar to the acidity of HCl in CH_3CN. Clearly, some metal hydrides can exhibit quite acidic behavior; Morris

Table 3.1 pK_a values of metal carbonyl hydrides in CH$_3$CN.

Metal hydride	pK_a	Ref.
(CO)$_4$CoH	8.3	[9]
Cp(CO)$_3$MoH	13.9	[9]
(CO)$_5$MnH	14.2	[7]
(CO)$_3$(PPh$_3$)CoH	15.4	[9]
Cp(CO)$_3$WH	16.1	[9]
(CO)$_5$ReH	21.1	[9]
Cp(CO)$_2$(PMe$_3$)WH	26.6	[9]

Table 3.2 pK_a values of bis(diphosphine) metal hydrides in CH$_3$CN.

Metal hydride[a]	pK_a	Ref.
HCo(dppe)$_2$	38.1	[10]
[HCo(dppe)$_2$]$^+$	23.6	[10]
[(H)$_2$Co(dppe)$_2$]$^+$	22.8	[10]
[HCo(dppe)$_2$(NCCH$_3$)]$^{+2}$	11.3	[10]
[HNi(dppe)$_2$]$^+$	14.2	[11]
[HPt(dppe)$_2$]$^+$	22.0	[11]
[HNi(dmpe)$_2$]$^+$	24.3[b]	[11]
[HPt(dmpe)$_2$]$^+$	28.5[b]	[11]

a) dmpe = 1,2-bis(dimethylphosphino)ethane;
dppe = 1,2-bis(diphenylphosphino)ethane.
b) Value determined in PhCN, though these values are usually similar to those found in CH$_3$CN, typically differing by less than 1 pK_a unit between the two solvents.

and coworkers have discovered some remarkable dicationic dihydrogen complexes that have acidities comparable to or greater than that of HOTf [8].

Replacement of an electron-withdrawing CO ligand by an electron-donating phosphine ligand diminishes the acidity of the metal hydride considerably, by about 7 pK_a units in (CO)$_4$CoH versus (CO)$_3$(PPh$_3$)CoH, and by about 10 pK_a units for the more electron-donating PMe$_3$ ligand in Cp(CO)$_3$WH versus Cp(CO)$_2$(PMe$_3$)WH. The first-row metal hydride (CO)$_5$MnH is more acidic than the third row congener (CO)$_5$ReH.

Studies of the acidity of a series of bis(diphosphine) metal hydrides in CH$_3$CN show that they exhibit a wide range of acidity; as shown in Table 3.2, cobalt bis(diphosphine) hydrides alone span about 27 pK_a units in acidity. One-electron oxidation of the diamagnetic cobalt hydride HCo(dppe)$_2$ gives the paramagnetic Co(II) hydride [HCo(dppe)$_2$]$^+$, which is more acidic by about 14 pK_a units. The dicationic cobalt hydride [HCo(dppe)$_2$(NCCH$_3$)]$^{+2}$ is very acidic, even with two electron-donating diphosphine ligands, demonstrating that the metal, ligands and charge all play a role in determining the acidity. The effect of ligands on acidity is shown in the lower half of Table 3.2, where Ni and Pt hydrides with dmpe ligands are 6–10 pK_a units less acidic than analogous hydrides with the less electron-donating dppe ligands.

3.3
Hydride Transfer Reactions of Metal Hydrides

DuBois and coworkers carried out extensive studies of the thermodynamic hydricity of metal hydrides (3.2); in many cases the metal product resulting following hydride transfer will bind a CH_3CN ligand (not shown in 3.2), so comparisons of hydricity will be solvent-dependent.

$$M-H \xrightarrow{\Delta G^0_{H^-}} M^+ + H^- \quad (3.2)$$

The data in Table 3.3 show an extraordinary range of hydride donating ability of metal hydrides, spanning over 40 kcal mol^{-1}! Note that the smaller numbers are

Table 3.3 $\Delta G^\circ_{H^-}$ (kcal mol^{-1}) for hydride transfer in CH_3CN.

Hydride donor[a]	$\Delta G^\circ_{H^-}$ (kcal mol^{-1})[b]	Ref.
HRh(dppb)$_2$	34	[12]
HW(CO)$_4$(PPh$_3$)$^-$	36	[13]
[HPt(dmpe)$_2$]$^+$	39	[11]
HW(CO)$_5^-$	40	[13]
HCo(dppe)$_2$	49	[10]
HCo(dppb)$_2$	48	[12]
[HNi(dmpe)$_2$]$^+$	51	[11]
[HPt(dppe)$_2$]$^+$	52	[14]
HMo(CO)$_2$(PMe$_3$)Cp	55	[15]
[HCo(dppe)$_2$]$^+$	60	[10]
[HNi(dppe)$_2$]$^+$	63	[11]
[(H)$_2$Co(dppe)$_2$]$^+$	65	[10]
[HPt(EtXantphos)$_2$]$^+$	76	[16]
H$_2$	76	[14]
HCPh$_3$	99	[17]

a) dmpe = 1,2-bis(dimethylphosphino)ethane;
depe = 1,2-bis(diethylphosphino)ethane;
dppb = 1,2-bis(diphenylphosphino)benzene;
dppe = 1,2-bis(diphenylphosphino)ethane;
EtXantphos = 9,9-dimethyl-4,5-bis(diethylphosphino)xanthene.
b) Estimated uncertainties of ΔG° values are typically ±2 kcal mol^{-1}.

associated with stronger hydride donors, just as lower pK_a values indicate higher acidity. The strongest hydride donor listed in the table, HRh(dppb)$_2$, is exceptionally hydridic; HRh(dmpe)$_2$ will transfer a hydride to BEt$_3$ [18]. Included in this table are bis(diphosphine) hydrides whose reactivity is described in Chapter 7, as well as anionic and neutral tungsten and molybdenum carbonyl hydrides used as hydride donors in the hydrogenation reactions described in this chapter. Third row metal hydrides are more hydridic than their first row analogs, as shown by the hydricity of [HPt(dmpe)$_2$]$^+$ exceeding that of [HNi(dmpe)$_2$]$^+$ by 12 kcal mol^{-1}. As was found in the acidity studies, changing the ligands on the same metal system has a large effect, as shown by the hydricity of [HNi(dmpe)$_2$]$^+$ exceeding that of [HNi(dppe)$_2$]$^+$ by 12 kcal mol^{-1}. Studies of the hydricity of Pd hydrides revealed that the hydricity of a series of HPd(diphosphine)$_2^+$ complexes varied by 27 kcal mol^{-1} as the bite angle (P–Pd–P) changed by 27° [19]. For the five-coordinate metal hydrides HM(diphosphine)$_2^+$ the order of hydricity is PdH > PtH > NiH [20]. Further studies on hydrides with other geometries and ligands will be needed to show whether this trend of hydricity (second row > third row > first row) is more generally true.

Thermochemical studies of acidity and hydricity are extremely valuable, since they can help determine the energies of potential intermediates in catalytic cycles, and can thus guide the choice of complexes proposed as catalysts. But, since catalysis is a kinetic phenomenon, the kinetics of delivery of a proton or hydride are also important. The kinetics of proton transfer from metal hydrides to amines [21] or metal alkynyl complexes [22], as well as degenerate proton transfers between metal hydrides and metal anions [21] led to the conclusion that proton transfers from metal hydrides have a high intrinsic barrier.

The kinetics of hydride transfer from metal carbonyl hydrides to trityl cation were studied using stopped-flow methods (3.3) [23, 24].

$$\text{M–H} + \text{Ph}_3\text{C}^+\text{BF}_4^- \xrightarrow{k_{H^-}} \text{M–F–BF}_3 + \text{Ph}_3\text{C–H} \tag{3.3}$$

The 16-electron metal cation that results following hydride transfer is captured by the BF$_4^-$ anion, leading to the formation of complexes with weakly bound FBF$_3$ ligands. As shown in Table 3.4, the rate constants for kinetic hydricity of this series of metal hydrides span about six orders of magnitude. Second row metal hydrides are faster hydride donors than their third row congeners, as shown by HMo > HW and HRu > HOs. As was found for the *thermodynamic* hydricity, changes of ligands can have a large influence on *kinetic* hydricity, with the rate constant for hydride transfer from *trans*-HMo(CO)$_2$(PMe$_3$)Cp exceeding that for HMo(CO)$_3$Cp by a factor of about 10^4. Electronic effects dominate over steric effects, as indicated by the faster rates of hydride transfer for HMo(CO)$_3$Cp* compared to HMo(CO)$_3$Cp. The Cp* ligand is much larger sterically than Cp, yet the higher electron-donating ability of this ligand compared to Cp makes its hydride complexes more hydridic. Even one methyl group on a Cp ligand can have a measurable effect, with the hydricity of HW(CO)$_3$(C$_5$H$_4$Me) being about three times that of HW(CO)$_3$Cp. The silicon hydride HSiEt$_3$, which is used as a hydride donor in the stoichiometric ionic hydrogenations described below, was

Table 3.4 Rate constants for hydride transfer from metal hydrides to $Ph_3C^+BF_4^-$ (CH_2Cl_2 solvent; 25 °C) [23, 24].

Metal hydride	k_{H^-} ($M^{-1}s^{-1}$)
$HRu(CO)_2Cp^*$	$>5 \times 10^6$
trans-$HMo(CO)_2(PMe_3)Cp$	4.6×10^6
$HFe(CO)_2Cp^*$	1.1×10^6
trans-$HMo(CO)_2(PCy_3)Cp$	4.3×10^5
$HOs(CO)_2Cp^*$	3.2×10^5
cis-$HRe(CO)_4(PPh_3)$	1.2×10^4
$HMo(CO)_3Cp^*$	6.5×10^3
$HRe(CO)_5$	2.0×10^3
$HW(CO)_3Cp^*$	1.9×10^3
$HMo(CO)_3Cp$	3.8×10^2
$HW(CO)_3(C_5H_4Me)$	2.5×10^2
cis-$HMn(CO)_4(PPh_3)$	2.3×10^2
$HSiEt_3$	1.5×10^2
$HW(CO)_3Cp$	7.6×10^1
$HMn(CO)_5$	5.0×10^1
$HW(CO)_3(C_5H_4CO_2Me)$	7.2×10^{-1}

shown to be among the slower hydride donors compared to transition metal hydrides.

3.4
Stoichiometric Hydride Transfer Reactivity of Anionic Metal Hydride Complexes

Anionic metal carbonyl hydrides, such as $(CO)_5MH^-$ (M = W or Cr) have been studied in detail by Darensbourg and coworkers [25]. Alkyl halides are readily converted to alkanes through reaction with anionic metal carbonyl hydrides (3.4) [26].

$$MH^- + R-Br \rightarrow MBr^- + R-H \qquad (3.4)$$

The kinetics of these reactions were found to be second-order overall, first order in metal carbonyl hydride anion and first order in alkyl halide. For the conversion of n-BuBr to n-butane, the second-order rate constant at 26 °C in THF was $1.79 \times 10^{-3} M^{-1}s^{-1}$ for $(CO)_5CrH^-$; the third-row analog, $(CO)_5WH^-$, had a larger rate constant ($3.31 \times 10^{-3} M^{-1}s^{-1}$). The reactivity of the $[(CO)_4P(OCH_3)_3MH]^-$ was about one order of magnitude higher, indicating a significant increase in hydride transfer rate for the hydrides with stronger phosphite donors compared to those with only electron-accepting CO ligands. Surprisingly, the rate constants for reaction of t-BuBr with $(CO)_5MH^-$ (M = W or Cr) were similar to those for the reaction with the primary bromide, n-BuBr. Kinetic studies using radical clocks provided mechanistic insight that helped to explain this observation [27]. While the reaction of the metal hydrides with the primary alkyl halide proceeded mostly by nucleophilic displacement of the halide by the hydride, an alternate mechanism was found to be operative in the reactions with the hindered alkyl halides. Evidence was presented favoring a radical chain reaction for the hindered alkyl halides, involving formation of a carbon-centered radical that abstracted a hydrogen *atom* from the metal hydride.

These anionic metal hydrides also readily react with acyl chlorides at 25 °C, converting them to aldehydes [28]. An advantage is that in many cases further reduction of the aldehyde to alcohols, which is a problem with some borohydride reagents, is avoided with these metal hydrides. The hydride in $(CO)_5CrH^-$ or $(CO)_5WH^-$ readily exchanges with deuterated alcohols, so use of the hydride in the presence of CH_3OD provides a means of converting acyl chlorides or alkyl halides to deuterated products, as shown in Equation 3.5 [29].

$$R-C(=O)-Cl + [(CO)_5CrH]^- \xrightarrow{CH_3OD} R-C(=O)-D + [(CO)_5CrCl]^- \quad (3.5)$$

Reaction of paraformaldehyde with $(CO)_5CrH^-$ leads to the alkoxide complex $[(CO)_5CrOCH_3]^-$, indicating that insertion of the C=O bond into the Cr–H bond occurred readily (3.6) [30]. Addition of acetic acid (3.7) or water produced methanol.

$$[(CO)_5CrH]^- + (CH_2O)_n \rightarrow [(CO)_5Cr-OCH_3]^- \quad (3.6)$$

$$[(CO)_5Cr-OCH_3]^- + HOAc \rightarrow [(CO)_5Cr-OAc]^- + CH_3OH \quad (3.7)$$

In contrast, reaction of either $(CO)_5CrH^-$ PPN^+ or $(CO)_5WH^-$ PPN^+ $\{PPN^+ = N(PPh_3)_2^+\}$ with propionaldehyde or cyclohexanone was very slow, suggesting that the insertion reaction of these substrates is much slower than that of paraformaldehyde. Addition of acetic acid rapidly gave the alcohol product, *n*-propanol or cyclohexanol. This divergent reactivity suggests that the hydrogenation of propionaldehyde and cyclohexanone proceeds by an ionic hydrogenation mechanism, in which protonation of the aldehyde activates it towards hydride transfer, as shown in Equation 3.8.

$$R-C(=OH^+)-H \xrightarrow{[H-W(CO)_5]^-} RCH_2OH \quad (3.8)$$

The results discussed above show that insertion of the C=O into the M–H bond can occur, but that an ionic hydrogenation pathway can occur when insertion is disfavored. Brunet and coworkers found that the reactivity of $(CO)_5CrH^-$ can also depend on the counterion [31]. In contrast to the lack of reactivity of cyclohexanone with $HCr(CO)_5^-$ PPN^+, the potassium salt $(CO)_5CrH^-$ K^+ reacts with cyclohexanone at 25 °C in the *absence* of added acid. Subsequent hydrolysis gave a 50% yield of cyclohexanol. The striking difference in reactivity is attributed to activation of the C=O by K^+, accelerating its reaction with $(CO)_5CrH^-$. Effects of ion-pairing between metal carbonyl anions and various counterions have been summarized in a review [32].

3.5
Catalytic Hydrogenation of Ketones with Anionic Metal Hydrides

The understanding of the reactivity of anionic metal carbonyl hydrides that resulted from the studies discussed above provided the basis for development of hydrogenation catalysts using Cr, Mo and W. Darensbourg and coworkers found that cyclohexanone could be hydrogenated to cyclohexanol by $(CO)_5M(OAc)^-$ (5 mol%) at 125 °C under H_2 (600 psi) [33].

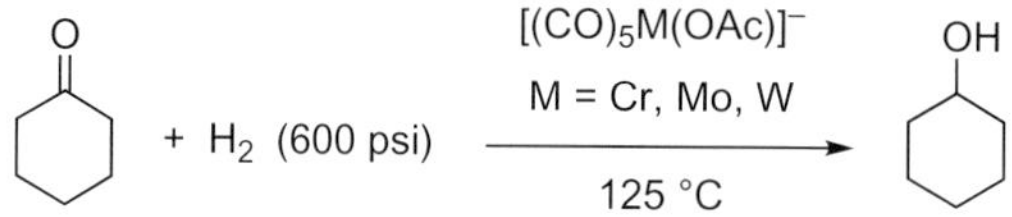

$$\text{cyclohexanone} + H_2 \text{ (600 psi)} \xrightarrow[125 \,°C]{[(CO)_5M(OAc)]^-,\ M = Cr, Mo, W} \text{cyclohexanol} \quad (3.9)$$

After 24 h, turnover numbers were 18 for Cr, 3.5 for Mo, and 10 for W. The proposed mechanism is shown in Scheme 3.2. Loss of CO from $(CO)_5M(OAc)^-$ followed by heterolytic cleavage of H_2 would give the anionic hydride $(CO)_5MH^-$ and HOAc. Insertion of the C=O bond into the M–H bond would give an anionic alkoxide complex that would produce the alcohol upon reaction with HOAc, regenerating $(CO)_5M(OAc)^-$. Monitoring of the reaction of $(CO)_5W(OAc)^-$ using high-

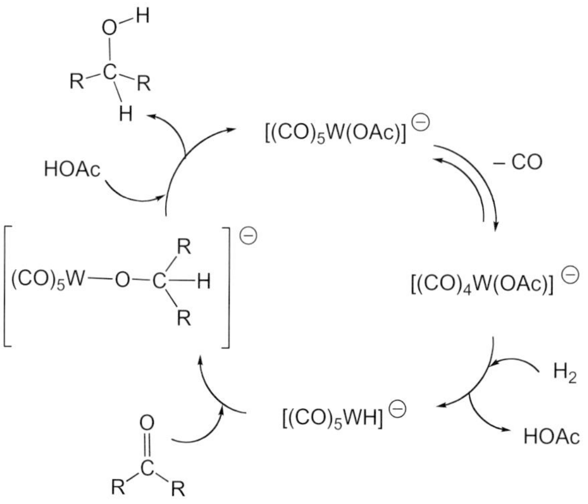

Scheme 3.2

pressure IR spectroscopy gave evidence for $(CO)_4W(THF)(OAc)^-$, $[(\mu\text{-}H)W_2(CO)_{10}]^-$, and the catalyst precursor, $(CO)_5W(OAc)^-$. The bimetallic complex with a bridging hydride, $[(\mu\text{-}H)W_2(CO)_{10}]^-$, was shown to result from reaction of $(CO)_5W(OAc)^-$ with H_2, and the proposed intermediate $(CO)_4W(THF)(OAc)^-$ was independently synthesized to confirm the assignment of its IR bands. Use of $[(\mu\text{-}H)W_2(CO)_{10}]^-$ as the hydrogenation catalyst precursor under the same reaction conditions produced similar results.

Markó and coworkers found that hydrogenation of ketones (acetone, acetophenone, cyclohexanone, etc.) or aldehydes (benzaldehyde, n-butyraldehyde) were catalytically hydrogenated by Cr, Mo or W carbonyl complexes in methanol under 100 bar H_2 [34]. Their experiments used $M(CO)_6$ as the catalyst precursor, with $NaOCH_3$ added; under these conditions $(CO)_5MH^-$ and $[(\mu\text{-}H)M_2(CO)_{10}]^-$ are formed, so these reactions are similar to those discussed above. Most of their experiments were carried out at 100–160 °C, but with the Mo complex hydrogenations could be carried out at temperatures as low as 70 °C.

Brunet and coworkers found that catalytic transfer hydrogenation of ketones could be carried out using 20 mol% $(CO)_5CrH^-\ K^+$. These experiments were carried out in THF solution with 5 equiv each (relative to Cr) of formic acid and NEt_3. The purpose of the NEt_3 is to moderate the strength of the formic acid, since in the absence of NEt_3, formic acid was shown to convert $(CO)_5CrH^-$ to $[(\mu\text{-}H)Cr_2(CO)_{10}]^-$. Catalytic transfer hydrogenation of cyclohexanone at room temperature gave 95% conversion, with lower conversions being found for methyl isopropyl ketone (45%), 2-hexanone (38%), or acetophenone (25%). The mechanism is similar to that shown in Scheme 3.2, with the product alcohol being released by reaction of formic acid with the anionic chromium alkoxide. Regeneration of $(CO)_5CrH^-$ would occur by decarboxylation of the formate complex (3.10), a reaction that had been previously shown to proceed by a mechanism involving initial dissociation of CO [35].

$$\left[(CO)_5Cr-O-\underset{\underset{O}{\|}}{C}-H\right]^{\ominus} \longrightarrow \left[(CO)_5Cr-H\right]^{\ominus} + CO_2 \qquad (3.10)$$

3.6
Ionic Hydrogenation of Ketones Using Metal Hydrides and Added Acid

The use of stoichiometric ionic hydrogenations in organic synthetic applications was established prior to the use of transition metal hydride complexes for ketone hydrogenations [36, 37]. In traditional stoichiometric ionic hydrogenations, CF_3CO_2H is used as the acid and $HSiEt_3$ serves as the hydride donor, though other acids and other hydride sources have been used as well. The C=O bonds of ketones can be hydrogenated by this method, as well as certain types of C=C bonds.

Using transition metal hydrides, rather than the H–Si bond of silanes, as hydride donors offers several significant advantages. The steric and electronic properties of silane hydride donors can be altered by changing the organic groups (e.g., $HSiEt_3$ vs. $HSiPh_3$) and by changing the number of H–Si bonds (e.g., $HSiPh_3$ vs. H_2SiPh_2). These changes are relatively modest, however, in comparison to the huge range of hydricity that can be obtained from a series of transition metal hydrides, as discussed above. (Tables 3.3 and 3.4). The versatility of reaction pathways available to metal hydrides upon reaction with acid provides another contrast with the reactivity of hydridosilanes. Addition of acids to silanes readily leads to formation of H_2, so this has to be considered in designing ionic hydrogenations using H–Si bonds. The same reaction can occur with metal hydrides, but metal hydrides can form dihydrogen complexes [38] or dihydrides when reacted with acids (3.11).

$$M-H + H^{\oplus} \longrightarrow \underset{\text{Dihydrogen Complex}}{M^{\oplus}\!\!-\!\!\genfrac{}{}{0pt}{}{H}{H}|} \rightleftharpoons \underset{\text{Dihydride Complex}}{M^{\oplus}\genfrac{}{}{0pt}{}{H}{H}} \rightleftharpoons M^{\oplus} + H_2 \quad (3.11)$$

The initial site of protonation is often at the M–H bond, converting a neutral metal hydride into a cationic dihydrogen complex. Such dihydrogen complexes are often stable and can be isolated. In many cases, dihydrogen complexes have been shown to be in equilibrium with the corresponding dihydride complex, in which the metal has undergone a formal oxidative addition of the H–H bond to rupture the H–H bond to generate two M–H bonds. For purposes of using these types of complexes in hydrogenations, it makes little difference whether the complex exists as a dihydrogen complex, a dihydride complex, or an interconverting mixture of the two. The important requirement is that this species has sufficient thermodynamic and kinetic acidity to transfer a proton the ketone or other substrate. It must also have sufficient thermal stability to carry out the proton transfer rather than losing H_2, though the loss of H_2 may be reversible.

In addition to the differences highlighted above, a particularly important difference in the reactions of silanes versus metal complexes is the ability of metal complexes to react with hydrogen, thus enabling pathways involving metal complexes to become catalytic. In the case of ionic hydrogenations, the ability to elicit such pathways relies on the ability of the metal complex to bind H_2, then heterolytically cleave it into H^+ and H^-. Precious metals like ruthenium have been found to employ heterolytic cleavage of H_2, and such reaction pathways have been successfully exploited in many catalytic hydrogenation reactions, often proceeding through metal–ligand bifunctional catalysis [39].

Stoichiometric ionic hydrogenations using acids as the H^+ source and transition metal hydrides as the H^- donor were studied prior to the development of catalytic hydrogenations. Hydrogenation of C=C bonds through the reaction with triflic acid (HOTf) and metal hydrides can be accomplished within minutes at low temperature (3.12).

3.6 Ionic Hydrogenation of Ketones Using Metal Hydrides and Added Acid

$$\text{(CH}_3)_2\text{C}=\text{C(CH}_3)_2 + \text{HOTf} + \text{Cp(CO)}_3\text{W–H} \xrightarrow[5 \text{ min}]{-50\,°\text{C}}$$

$$(\text{CH}_3)_2\text{CH–CH(CH}_3)_2 + \text{Cp(CO)}_3\text{W–OTf}$$

(3.12)

This reaction was shown to proceed cleanly for several metal hydrides, including Cp(CO)$_3$WH, Cp*(CO)$_3$WH, (Cp* = η^5-C$_5$Me$_5$), Cp(CO)$_3$MoH, (CO)$_5$MnH, (CO)$_5$ReH and Cp*(CO)$_2$OsH, thus including first, second and third-row metal hydrides from Groups 6, 7 and 8. Excellent yields (>90%) were found for the stoichiometric hydrogenations of C=C bonds for 1,1-disubstituted, trisubstituted and tetrasubstituted alkenes. The common feature of these alkenes is that all of them generate a tertiary carbenium ion upon protonation. This is the same limitation encountered in hydrogenations of C=C bonds by the traditional CF$_3$CO$_2$H and HSiEt$_3$ reagents [36, 37]. Styrene, stilbene and related C=C compounds were hydrogenated by HOTf and metal hydrides but at lower yields (around 50%).

Stoichiometric hydrogenations of C=C bonds using traditional ionic hydrogenation reagents, CF$_3$CO$_2$H and HSiEt$_3$, are typically carried out for a few hours at 50 °C [36, 37]. It was surprising that the very stong acid (HOTf) could be used, since it had been stated in a review [36] of ionic hydrogenations using CF$_3$CO$_2$H and HSiEt$_3$, that "stronger acids cannot be used in conjunction with silanes because they react with the latter". Indeed, addition of HOTf to HSiEt$_3$ (with no alkenes present) does result in prompt evolution of H$_2$. This high reactivity does not preclude the use of HOTf/HSiEt$_3$ for ionic hydrogenations, however, as long as the reagents are added in the correct order. When HOTf is added to a solution containing the alkene and HSiEt$_3$, good yields of hydrogenation products can be obtained. Another distinction between the use of HOTf and CF$_3$CO$_2$H as the acid in ionic hydrogenations concerns the trapping of the carbenium ion by the counterion. Trifluoroacetate esters (CF$_3$CO$_2$R) have been observed in hydrogenations of C=C bonds by CF$_3$CO$_2$H/HSiEt$_3$, with subsequent reversion of the trifluoroacetate ester to the carbenium ion occurring upon reaction with acid. In contrast, alkyl triflates (ROTf) were not observed as intermediates in C=C hydrogenations using HOTf as the acid, except in the case of hydrogenation of allylbenzene [40].

Successful ionic hydrogenations require the ability to accomplish a hydride transfer *in the presence of acid*, avoiding significant formation of H$_2$ from the reaction of the hydride and acid. All of the metal hydrides used for ionic hydrogenations do react with acids. The extent of protonation, and the stability of the protonated products, influence the suitability of different hydrides for use in ionic hydrogenations. The tungsten hydride Cp(CO)$_3$WH is partially protonated by HOTf, leading to the dihydride [Cp(CO)$_3$W(H)$_2$]$^+$OTf$^-$, but the release of H$_2$ from this dihydride, forming Cp(CO)$_3$WOTf, is very slow, taking weeks at room temperature [41]. In contrast, reaction of HOTf with either (CO)$_5$MnH or (CO)$_5$ReH is very fast, so more than one equivalent of these metal hydrides is needed to get good yields in hydrogenations of C=C bonds. Some metal hydrides (e.g., $Cp(CO)_2(PPh_3)MoH$, $Cp^*(CO)_3MoH$ and $Cp^*(CO)_2FeH$) evolve H_2 so quickly upon reaction with HOTf that they are not effective hydride donors for ionic hydrogenation of C=C bonds.

The C=O bond of ketones can also be hydrogenated through reactions with HOTf and metal hydrides [42]. Reaction of acetone with HOTf and $Cp(CO)_3WH$ gives stoichiometric hydrogenation of the C=O bond, initially giving a product with the alcohol bound to the metal as a kinetically stabilized complex (3.13).

$$\text{(3.13)}$$

In the ^{1}H NMR spectrum, the OH of the alcohol appeared as a doublet at δ7.34. This chemical shift is at least 5 ppm downfield of the corresponding OH resonance of the free alcohol, suggesting that the alcohol was hydrogen bonded in solution to the triflate counterion. A crystal structure of the complex of isopropyl alcohol showed that the OH was strongly hydrogen-bonded in the solid state as well, with an O•••O distance of 2.63(1) Å. While alcohol complexes can be isolated, solutions of these complexes release free alcohol, forming the metal triflate; an approximate half-life of 14 h at room temperature was found for the alcohol complex shown in Equation 3.13.

The kinetics of hydrogenation of isobutyraldehyde were studied at −30 °C in CD_2Cl_2, using CF_3CO_2H as the acid and $CpMo(CO)_3H$ as the hydride. The reaction becomes slower as it proceeds, since the acid concentration decreases. However, in the presence of a buffer to hold the acid concentration constant, the rate of disappearance of the aldehyde follows pseudo-first order kinetics. The rate was also first order in metal hydride, and the second-order rate constant was $k = 3.5 \times 10^{-4} \, M^{-1} s^{-1}$ at −30 °C. The mechanism proposed in Scheme 3.3 is consistent with these observations. Pre-equilibrium protonation of the aldehyde is followed by rate-determining hydride transfer.

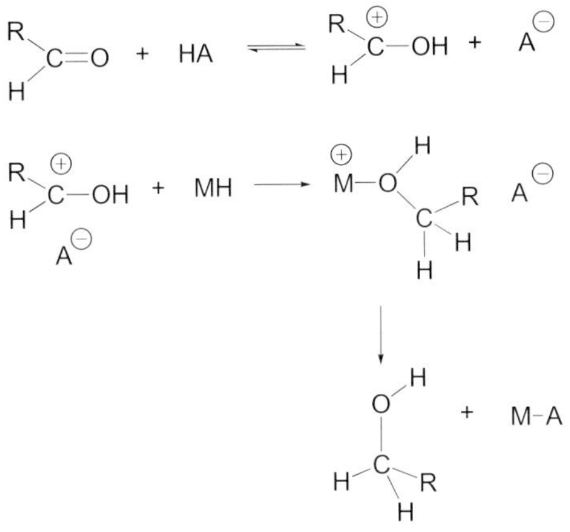

3.6 Ionic Hydrogenation of Ketones Using Metal Hydrides and Added Acid

$$\underset{H}{\overset{R}{>}}C=O + HA \rightleftharpoons \underset{H}{\overset{R}{>}}C-OH^+ + A^-$$

$$\underset{H}{\overset{R}{>}}C-OH^+ + MH \longrightarrow M-O\underset{H}{\overset{H}{>}}C\underset{H}{\overset{R}{<}} + A^-$$

$$\downarrow$$

$$\underset{H}{\overset{H}{>}}O\underset{H}{\overset{C}{<}}R + M-A$$

Scheme 3.3

Ethylbenzene results from reaction of acetophenone with HOTf and Cp(CO)$_3$WH (3.14).

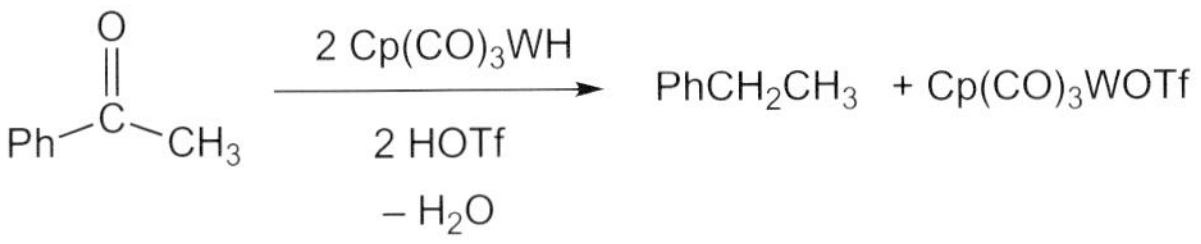

$$\text{Ph-C(=O)-CH}_3 \xrightarrow[\substack{2\ \text{HOTf} \\ -\text{H}_2\text{O}}]{2\ \text{Cp(CO)}_3\text{WH}} \text{PhCH}_2\text{CH}_3 + \text{Cp(CO)}_3\text{WOTf} \qquad (3.14)$$

With two equivalents each of HOTf and Cp(CO)$_3$WH, ethylbenzene is formed in good yield, though when only one equivalent each of HOTf and Cp(CO)$_3$WH are used, half of the acetophenone remains and half is converted to ethylbenzene. This suggests that the *sec*-phenethyl alcohol formed by hydrogenation of the C=O bond is consumed more rapidly than it is formed. Separate experiments confirmed that it is converted to ethylbenzene under the same reaction conditions, releasing water (3.15).

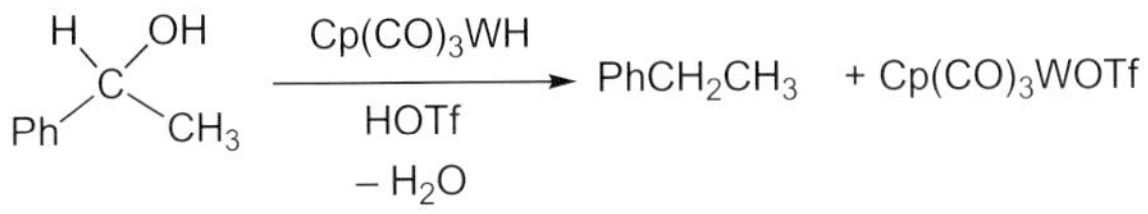

$$\text{Ph-CH(OH)-CH}_3 \xrightarrow[\substack{\text{HOTf} \\ -\text{H}_2\text{O}}]{\text{Cp(CO)}_3\text{WH}} \text{PhCH}_2\text{CH}_3 + \text{Cp(CO)}_3\text{WOTf} \qquad (3.15)$$

Intermolecular and intramolecular competition experiments gave information about the selectivity of ionic hydrogenations of two different C=O bonds. Addition of HOTf to a solution containing Cp(CO)$_3$WH, pivaldehyde and acetone led to hydrogenation only of the aldehyde, not the ketone. A similar competition experiment with acetophenone and acetone resulted in hydrogenation of the acetone. The relative thermodynamics of basicity of the two ketones would influence these reactions, as well as the rate of hydride transfer following the protonation of the ketone. In this example, protonation of acetophenone produces a carbenium ion that is stabilized by the Ph group, thus making it a weaker hydride acceptor compared to protonated acetone. An intramolecular competition between two ketone C=O bonds was carried out (3.16), leading to a product in which the dialkyl ketone moiety was preferentially hydrogenated over the ketone stabilized by an aryl group [43].

$$Ph-C(O)-CH_2-C(O)-CH_3 \xrightarrow[HOTf]{Cp(CO)_3WH} [Cp(CO)_3W\cdots O=C(CH_3)-CH_2-C(O)-Ph\cdots H]^+ OTf^- \quad (3.16)$$

Similar stoichiometric ionic hydrogenations with $Cp(CO)_3WH$ as the hydride donor and HOTf as the acid lead to the formation of aldehyde or ketone complexes [44]. Reaction of $Ph(C=O)Cl$ with $Cp(CO)_3WH/HOTf$ leads to the formation of $[Cp(CO)_3W(\eta^1\text{-PhCHO})]^+OTf^-$, and reactions of α,β-unsaturated ketones result in hydrogenation of the C=C bond, producing $[Cp(CO)_3W(\eta^1\text{-ketone})]^+OTf^-$ complexes [44].

3.7
Ionic Hydrogenations from Dihydrides: Delivery of the Proton and Hydride from One Metal

The hydrogenations discussed above show that hydride transfer from metal hydrides constitutes the C–H bond-forming step in ionic hydrogenations using an added acid as the proton source. Catalytic hydrogenations require that *both* of the added hydrogens come from H_2, delivered by the metal catalyst. Finding examples where a proton *and* a hydride can both be delivered from a metal center to accomplish hydrogenations is an important step in establishing the viability of this step in catalysis.

Moïse and coworkers had shown that a cationic dihydride of tantalum, $Cp_2(CO)Ta(H)_2^+$, reacts with acetone to give the alcohol complex $Cp_2(CO)Ta(HO^iPr)^+$ [45]. This reaction was proposed to occur by proton transfer from the dihydride, followed by hydride transfer to the protonated ketone. Unfortunately, attempts to make this reaction catalytic under H_2 were not successful. Early transition metals like Ta form strong bonds to oxygen, with no alcohol being observed at 60 °C in attempted catalytic reactions.

The cationic dihydride $[Cp(CO)_2(PMe_3)W(H)_2]^+OTf^-$ can be isolated, and has been fully characterized, including a crystal structure [41]. It reacts readily at room temperature with acetone, giving cis- and trans-isomers of the isopropyl alcohol complex (3.17); an analogous reaction occurs with propionaldehyde.

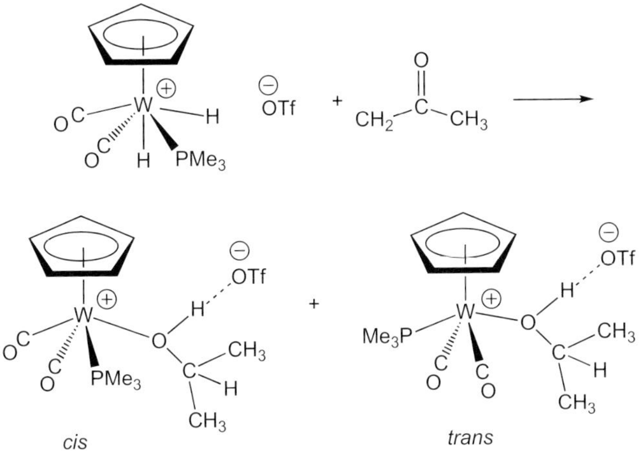

(3.17)

3.8
Catalytic Ionic Hydrogenations With Mo and W Catalysts

The reactivity studies described above documented that the cationic dihydride $[Cp(CO)_2(PMe_3)W(H)_2]^+$ hydrogenated ketones, but closing the catalytic cycle requires a reaction with H_2 to regenerate the metal hydride bonds. The tungsten and molybdenum ketone complexes $[Cp(CO)_2(PR_3)M(O=CEt_2)]^+ BAr'_4$ [Ar′ = 3,5-bis(trifluoromethyl)phenyl; R = Me, Ph, Cy] catalyze the hydrogenation of $Et_2C=O$ at 23 °C under 4 atm of H_2 (3.18) [46, 47].

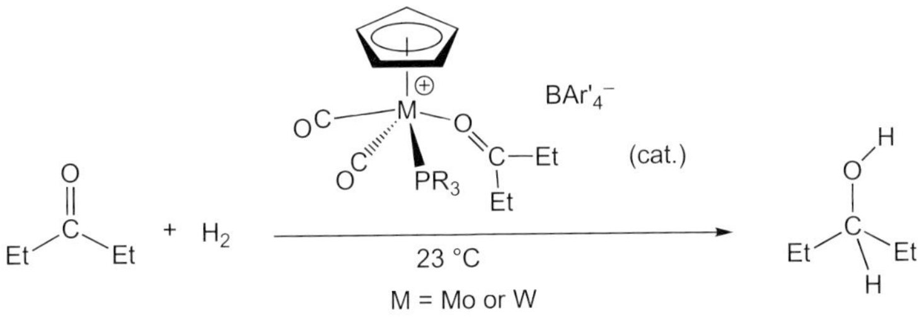

(3.18)

^{1}H NMR spectroscopic monitoring of the hydrogenation of $Et_2C=O$ catalyzed by $[Cp(CO)_2(PPh_3)W(O=CEt_2)]^+$ showed that the decline in concentration of the ketone complex was accompanied by the formation of the alcohol complex trans-$[CpW(CO)_2(PPh_3)(Et_2CHOH)]^+$. At later reaction times the concentration of the alcohol complex exceeds that of the ketone complex. The formation of alcohol complexes under catalytic conditions is anticipated by their observation in stoichiometric hydrogenations as shown in Equations 3.13 and 3.16. When a solution of $[Cp(CO)_2(PPh_3)W(O=CEt_2)]^+$ was heated under H_2 (950 psi) for 17 h, the trans-isomer of $[CpW(CO)_2(PPh_3)(Et_2CHOH)]^+$ was produced. The formation of the trans-isomer suggests that the alcohol ligand occupies the site from which hydride transfer occurred. Studies of the kinetics of hydride transfer from $Cp(CO)_2(PCy_3)MoH$ to Ph_3C^+ had shown that hydride transfer from trans-$Cp(CO)_2(PCy_3)MoH$ is much faster than that from cis-$Cp(CO)_2(PCy_3)MoH$ [23].

In contrast to the formation of the trans-isomer under catalytic conditions, hydride transfer from $Cp(CO)_2(PPh_3)WH$ to Ph_3C^+, followed by addition of Et_2CHOH, led to the isolation of cis-$[CpW(CO)_2(PPh_3)(Et_2CHOH)]^+$ (3.19).

$$Cp(CO)_2(PPh_3)W-H \;+\; Ph_3\overset{\oplus}{C}\;\overset{\ominus}{BAr'_4} \;+\; Et_2CHOH \longrightarrow$$

(3.19)

[structure of cis-$[CpW(CO)_2(PPh_3)(Et_2CHOH)]^+$] $\overset{\ominus}{BAr'_4}$ + Ph_3C-H

Conversion of the kinetically favored trans-isomer to the thermodynamically favored cis-isomer has been observed in several cases of Mo and W complexes with similar structures [23, 48].

The rate of catalysis of hydrogenation of $Et_2C=O$ was faster for Mo than for W using the phosphine complexes $[Cp(CO)_2(PR_3)M(O=CEt_2)]^+$ as catalyst precursors [46, 47]. For the PPh_3 complexes, the initial rate with the Mo complex was about eight times faster than with the W complex. A larger difference was observed for the complexes with PCy_3 ligands, with the Mo complex being about two orders of magnitude faster in initial rate than the W analog. The rates are modest, however, with about 2 turnovers per hour being found initially for the Mo-PCy_3 complex at 23 °C under 4 atm H_2. For both Mo and W complexes, the relative rates of hydrogenation as a function of phosphine were $PCy_3 > PPh_3 > PMe_3$. The catalysts with the very bulky PCy_3 ligand are substantially faster than those with PMe_3, despite the similar electronic properties of these two trialkylphosphine ligands. This indicates that steric factors predominate over electronic factors, with more sterically demanding phosphines leading to faster rates. The increased rate with larger phosphines is thought to be due to destabilization of ketone binding with the larger steric bulk of the phosphine ligand, such that promoting dissociation of the ketone will lead to more facile formation of the dihydride.

The stoichiometric and catalytic studies provide strong evidence for the proposed mechanism shown in Scheme 3.4. Ketone complexes $[Cp(CO)_2(PR_3)M(O=CEt_2)]^+$ (M = Mo or W) can be isolated and used as catalyst precursors for both R = Ph and R = Me. Formation of the cationic dihydride through displacement of the ketone ligand by H_2 was independently verified: in the absence of excess ketone, the reaction of H_2 with $[Cp(CO)_2(PPh_3)W(O=CEt_2)]^+$ gives the dihydride $[Cp(CO)_2(PPh_3)W(H)_2]^+$ (3.20) [47]. In contrast, for the PCy_3 complexes, the corresponding ketone complex, $[Cp(CO)_2(PCy_3)M(O=CEt_2)]^+$, was not isolated, but catalysis was obtained by hydride abstraction in situ from $Cp(CO)_2(PCy_3)MH$ by Ph_3C^+.

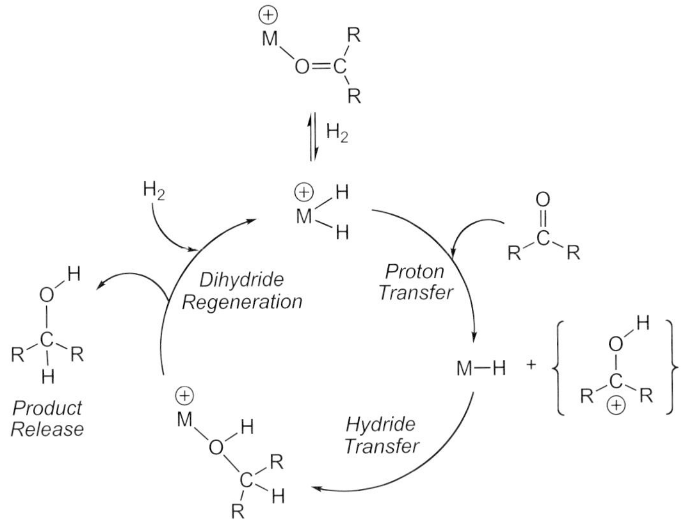

3.8 Catalytic Ionic Hydrogenations With Mo and W Catalysts

Scheme 3.4

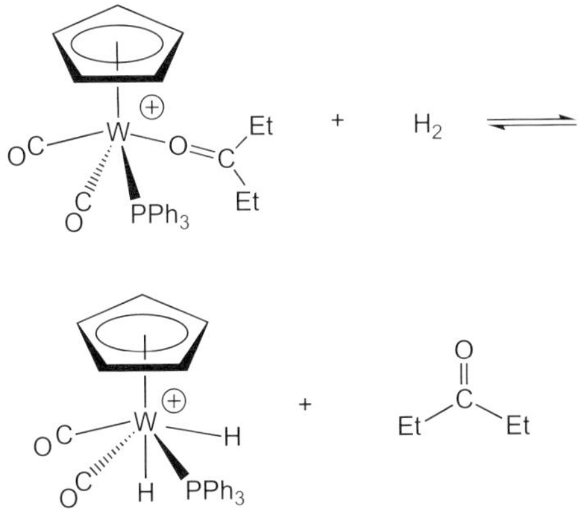

(3.20)

The next step in the catalytic cycle, ionic hydrogenation of ketones from cationic dihydrides, was separately shown to occur under stoichiometric conditions, as shown in Equation 3.17. Alcohol complexes have been synthesized and isolated, and some were characterized by crystallography [43]. Analogous alcohol complexes were observed spectroscopically during the catalytic reaction, so displacement of the alcohol by H_2 regenerates the dihydride and completes the catalytic cycle.

Evidence for tungsten dihydride complexes was obtained directly, as discussed above, but analogous complexes of molybdenum have not been directly observed. Indirect evidence for molybdenum dihydrides (or dihydrogen complexes) comes from the reaction shown in Equation 3.21, in which the molybdenum ketone complex heterolytically cleaves H_2 in the reaction with a hindered base.

$$\text{Cp(CO)}_2\text{(PPh}_3\text{)(Et}_2\text{C=O)Mo}^+ + \text{H}_2 + \text{2,6-di-tert-butyl-4-methylpyridine} \longrightarrow$$

$$\text{Cp(CO)}_2\text{(PPh}_3\text{)Mo–H} + \text{Et}_2\text{C=O} \cdot \cdot \cdot + \text{[2,6-di-tert-butyl-4-methylpyridinium]}^+ \tag{3.21}$$

This reaction proceeds in 5 min at room temperature. As a second row metal, Mo may be more likely to form a dihydrogen complex, in contrast to the dihydride found for the third row W analog.

Norton and coworkers reported a pK_a of 5.6 in CH$_3$CN for the cationic dihydride complex [CpW(CO)$_2$(PMe$_3$)(H)$_2$]$^+$ [49]. The pK_a of protonated acetone in CH$_3$CN is about −0.1 [50, 51]. The actual pK_a values will be different in CD$_2$Cl$_2$, the solvent in which these hydrogenations were carried out, compared to MeCN, the solvent in which the pK_a values were measured. However, making the assumption [7] that the *relative* pK_a values will be similar in different solvents shows that the proton transfer step, in which the cationic dihydride protonates a ketone, is significantly uphill thermodynamically in all of these cases. The tungsten dihydride, [CpW(CO)$_2$(PMe$_3$)(H)$_2$]$^+$, for which the pK_a value was reported, is the slowest catalyst of the series studied, so this specific example may be the least favorable case thermodynamically. Based on trends in acidity discussed above (Table 3.1) the acidity of [CpW(CO)$_2$(PPh$_3$)(H)$_2$]$^+$ will likely be greater than that of [CpW(CO)$_2$(PMe$_3$)(H)$_2$]$^+$, though perhaps by only 2 or 3 pK_a units, so that the proton transfer step in all these tungsten complexes is likely thermodynamically unfavorable. The Mo analogs, whether they form dihydrides or dihydrogen complexes, are likely to be significantly more acidic than the W congeners. Despite the proton transfer step being uphill, the catalytic ionic hydrogenations still occur, driven in part by the very favorable hydride transfer step that follows the proton transfer. The rate constants for hydride transfer from neutral metal hydrides to protonated ketones have not been studied as extensively as those for hydride transfer to Ph$_3$C$^+$ [23, 24], but the rate constant of 1.2×10^4 M^{-1}s^{-1} at 25 °C for hydride transfer from Cp(CO)$_2$(PPh$_3$)MoH to protonated acetone shows that such reactions can occur rapidly [51].

Martins and coworkers recently reported experimental and computational studies on Mo and W complexes with a pentabenzylcyclopentadienyl ligand [52]. The crystal structures of both (C$_5$Bz$_5$)Mo(CO)$_3$H (Bz = CH$_2$Ph) and (C$_5$Bz$_5$)W(CO)$_3$H were reported [53]. Catalytic hydrogenation of Et$_2$C=O was studied in CD$_2$Cl$_2$, with Ph$_3$C$^+$BAr$_4^-$ being used to abstract hydride from the neutral metal hydride. The catalytic activity found for the PMe$_3$-substituted complex (C$_5$Bz$_5$)Mo(CO)$_2$(PMe$_3$)H was similar to that reported earlier [46] for the related complex with an unsubstituted Cp ligand, CpMo(CO)$_2$(PMe$_3$)H. In contrast, Martins and coworkers found that the rate of catalytic hydrogenation activity using (C$_5$Bz$_5$)Mo(CO)$_3$H was higher than that obtained with CpMo(CO)$_3$H. This trend, with the more sterically bulky (C$_5$Bz$_5$)Mo(CO)$_3$H system being faster than that derived from CpMo(CO)$_3$H, parallels that found in the [Cp(CO)$_2$(PR$_3$)M(O=CEt$_2$)]$^+$ system, where rates were faster for more sterically demanding phosphines. DFT calculations provided further detailed insight. The overall calculated mechanism is consistent with that shown in Scheme 3.4, and the calculations identified the displacement of the bound ketone by H$_2$ as the step having the highest thermodynamic barrier. The results of the calculations corroborate experimental observations for the [Cp(CO)$_2$(PPh$_3$)W(O=CEt$_2$)]$^+$ system, where the ketone complex was observed during the hydrogenation reaction.

3.9
Mo Phosphine Catalysts With Improved Lifetimes

The observation of phosphonium cations (HPR$_3^+$) in the catalytic reactions (3.18, Scheme 3.4) showed that dissociation of a phosphine leads to catalyst deactivation. Under the reaction conditions, the free phosphine is protonated, presumably by a cationic dihydride complex, thus removing another metal from the catalytic cycle. Identification of this mode of catalyst deactivation prompted efforts to synthesize a second generation of catalysts that would be less prone to loss of a phosphine. Molybdenum complexes with a two-carbon linkage between the phosphine and cyclopentadienyl ring were synthesized [54], and were characterized by X-ray crystallography for R = Ph, ^{t}Bu and Cy. Hydride transfer from HMo(CO)$_2$[η^5:η^1-C$_5$H$_4$(CH$_2$)$_2$PR$_2$] (R = Cy, ^{t}Bu, and Ph) to Ph$_3$C$^+$BAr$'_4^-$ in Et$_2$C=O solvent gave ketone hydrogenation catalysts (3.22) that offered several advantages over the first-generation catalysts shown in Equation 3.18.

$$\text{[structure]} + Ph_3C^+ \ BAr'_4^- + H_2 \xrightarrow[\text{(solvent-free)}]{\text{catalytic hydrogenation of Et}_2\text{C=O}} \quad (3.22)$$

Significant increases in catalyst lifetimes were found for these C$_2$-bridged catalysts, with up to 500 turnovers being observed. These catalytic reactions were carried out in neat ketone (solvent-free conditions) [55], and the longer catalyst lifetime is due in part to an increased stability in ketones, suggesting that earlier reactions with CD$_2$Cl$_2$ contributed to the decomposition of the [Cp(CO)$_2$(PR$_3$)M(O=CEt$_2$)]$^+$ catalyst precursors. In addition to the increased lifetime, the C$_2$-bridged catalysts exhibited higher thermal stability compared to the unbridged catalyst precursors. The maximum turnover frequency found for HMo(CO)$_2$[η^5:η^1-C$_5$H$_4$(CH$_2$)$_2$PCy$_2$]

was about 90 turnovers per day (measured over three days) at 50 °C, under 800 psi H_2 pressure. These C_2-bridged Mo catalysts were used at low catalyst loadings, typically 0.2–0.4 mol%, but loadings as low as 0.1 mol% were shown to be effective. As was found with the unbridged catalysts, faster rates were found with R = Cy than with R = Ph.

Most of the studies were carried out with cationic complexes with the $BAr'_4{}^-$ anion. Catalysis was also observed with BF_4^- and PF_6^- anions, though with lower turnover numbers. When $HMo(CO)_2[\eta^5{:}\eta^1{-}C_5H_4(CH_2)_2PCy_2]$ is protonated by HOTf, H_2 is evolved, presumably from an unobserved dihydrogen or dihydride complex, and $TfOMo(CO)_2[\eta^5{:}\eta^1{-}C_5H_4(CH_2)_2PCy_2]$ can be isolated. This triflate complex was also used in catalytic hydrogenations since the triflate counterion is much less expensive than the $BAr'_4{}^-$ anion. In addition, the triflate complexes can be used directly as catalyst precursors, rather than having to conduct the hydride abstraction from the metal hydride as shown in Equation 3.22.

3.10
Tungsten Hydrogenation Catalysts with N-Heterocyclic Carbene Ligands

The use of N-heterocyclic carbene (NHC) ligands in catalytic reactions has become widespread in recent years [56], and many complexes with NHC ligands offer better performance than analogous complexes with phosphine ligands. Calorimetric studies show that NHC ligands are usually significantly more strongly bound to the metal than are phosphines [57]. Hydride transfer from the tungsten hydride complex $Cp(CO)_2(IMes)WH$ (IMes = the carbene ligand 1,3-bis(2,4,6-trimethylphenyl)-imidazol-2-ylidene) to $Ph_3C^+B(C_6F_5)_4^-$ gave an unusual complex shown in Equation 3.23.

(3.23)

Crystallographic and computational evidence showed that one C=C of a mesityl ring forms a weak bond to the tungsten [58]. This complex serves as a catalyst precursor for the hydrogenation of ketones, but only 2 turnovers per day were found for catalytic hydrogenation of solvent-free $Et_2C=O$ at 23 °C under 4 atm H_2, using 0.34% catalyst loading. At higher pressure (800 psi H_2), 86 turnovers were found after 10 days at 23 °C. Higher turnover rates were found at 50 °C (15 turnovers at 4 atm H_2 in 1 day), but the higher activity comes at the expense of catalyst decomposition. Protonated imidazolium cation, $[H\text{-}IMes]^+$, was identified as a decomposition product. This highlights a vulnerability of this type of NHC ligand to decomposition. A decomposition product thought to contain a protonated imidazolium, $[H\text{-}IMes]^+[Co(CO)_4]^-$, was also observed from attempted catalytic hydroformylation using $Co_2(CO)_6(IMes)_2$ [59], but the details of the proton transfer steps involved in the formation of $[H\text{-}IMes]^+$ have not been determined. In other cases, NHC complexes have been shown to be stable in the presence of acid [60], so the generality of this type of vulnerability to decomposition will depend on the specific catalyst system.

3.11
Catalysts for Hydrosilylation of Ketones

The hydrosilylation reaction [61], in which a Si–H bond is added across a C=C or C=O bond, shares many similarities to catalytic hydrogenations. Hydrosilylation of ketones is catalyzed by $[Cp(CO)_2(IMes)W(ketone)]^+$ complexes; these reactions are carried out under solvent-free conditions, as the ketones and $HSiEt_3$ are liquids (3.24) [62].

$$R\text{-}C(=O)\text{-}R' + HSiEt_3 \xrightarrow{[Cp(CO)_2(IMes)W(ketone)]^+ \text{ (cat.)}} R\text{-}C(OSiEt_3)(H)\text{-}R'$$

(3.24)

These reactions produce alkoxysilanes in high yields at 23 °C using 0.2 mol% catalyst. The approximate initial turnover frequency for hydrosilylation of $Et_2C=O$ at 23 °C is 370 h^{-1}, which is faster than that of 110 h^{-1} found for the hydrosilylation of acetophenone under the same conditions. The most unusual feature of these reactions is that when aliphatic ketones are hydrosilylated using $HSiEt_3$, the tungsten catalyst precipitates at the end of the reaction, so it is readily recycled. This reaction thus offers an advantage – ease of separation – that is normally associated with heterogeneous catalysts. The catalyst still maintains activity after five cycles of recycling and recovery.

The proposed mechanism for the catalytic hydrosilylation (Scheme 3.5) shares some similarities to that proposed for catalytic ionic hydrogenations (Scheme 3.4). Displacement of the bound ketone by $HSiEt_3$ gives $[CpW(CO)_2(IMes)(SiEt_3)H]^+$, which has been independently synthesized and characterized. Transfer of the

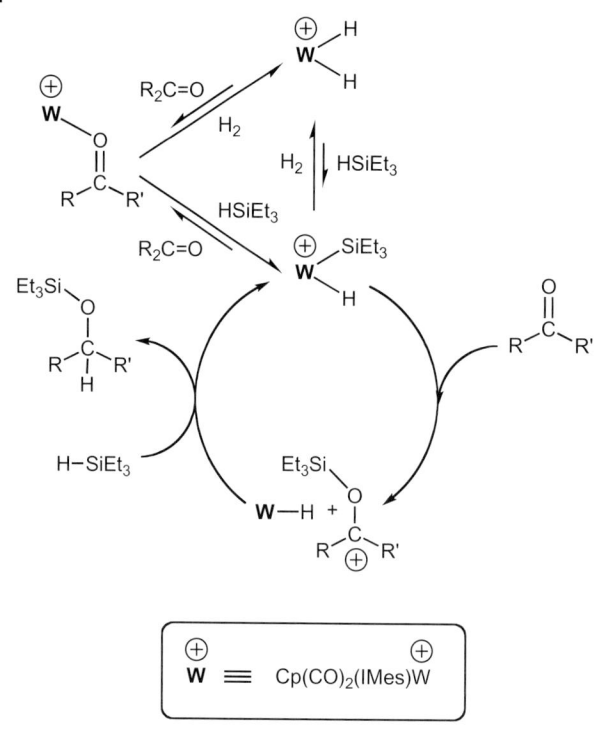

Scheme 3.5

"SiEt$_3^+$" to the oxygen of the ketone forms a strong Si–O bond. Hydride transfer to produce the alkoxysilane might occur from the W–H bond of CpW(CO)$_2$(IMes)H, but more likely occurs by hydride transfer from HSiEt$_3$, which is present in much higher concentrations than the metal. While these proposed steps account for the formation of the product and closure of the catalytic cycle, they do not account for all the observations made in the experiments. In experiments where the catalyst is recycled, the resting state of the catalyst was found to be a mixture of [CpW(CO)$_2$(IMes)(SiEt$_3$)H]$^+$ and the dihydride [CpW(CO)$_2$(IMes)(H)$_2$]$^+$. Small amounts of H$_2$ are formed in the reaction from the catalyzed reaction of trace water with HSiEt$_3$, or from other side-reactions. At the low catalyst loading used (0.2 mol%), even small amounts of H$_2$ can inhibit the reaction because of the very favorable equilibrium for the formation of the dihydride [CpW(CO)$_2$(IMes)(H)$_2$]$^+$ (Scheme 3.5). The initial rate of catalytic hydrosilylation of acetophenone was about three times faster in an open vial under an inert atmosphere than for an identical solution in a closed tube. In the open system, the H$_2$ generated can escape, whereas in the sealed tube, any H$_2$ produced can inhibit the rate of catalysis by forming more of the dihydride [CpW(CO)$_2$(IMes)(H)$_2$]$^+$ as the resting state, which has to re-enter the catalytic cycle through an unfavorable equilibrium with [CpW(CO)$_2$(IMes)(SiEt$_3$)H]$^+$.

As for hydrogenations, precious metal catalysts have dominated the study of catalytic hydrosilylations, with Rh and Pt catalysts being prevalent in catalytic hydrosilylations. Compared to hydrogenations, however, there are more examples of cheap metals that catalyze the hydrosilylation reaction. In addition to the tungsten catalysts discussed above, several examples of iron catalysts for hydrosilylation are discussed in Chapter 4. Examples of catalytic hydrosilylations with Mn, Ti and Cu catalysts are briefly mentioned below.

Cutler found that manganese carbonyl acyl complexes such as $(CO)_4(PPh_3)Mn(C=O)CH_3$ are excellent catalysts for the hydrosilylation of ketones [63] and esters [64]. Using 1 mol% $(CO)_4(PPh_3)Mn(C=O)CH_3$ as the catalyst precursor, a turnover frequency of 27 ± 5 min^{-1} was determined for hydrosilylation of acetone using $HSiMe_2Ph$. This Mn catalyst was shown to be much faster than Wilkinson's catalyst, $RhCl(PPh_3)_3$, for hydrosilylations using $HSiMe_2Ph$; the Mn and Rh catalysts exhibited similar activity using H_2SiPh_2. The Mn acetyl complex $(CO)_5Mn(C=O)CH_3$ was shown to react with silanes to provide the α-siloxyethyl complex $(CO)_5MnCH(OSiR_3)(CH_3)$ [65]. These reactions had an induction period, and mechanistic studies implicated the coordinatively unsaturated complex $(CO)_4MnSiR_3$ as the active catalyst.

Buchwald and coworkers developed a series of titanium complexes into highly efficient catalysts for hydrosilylation of ketones. Esters were catalytically converted to alcohols by catalysts formed through reaction of Cp_2TiCl_2 with n-BuLi, followed by reaction with $HSi(OEt)_3$ [66]. Subsequent studies with chiral Ti catalysts led to highly enantioselective hydrosilylation of ketones [67].

Lipshutz and coworkers have developed copper hydride complexes with diphosphine ligands that catalyze the asymmetric hydrosilylation of aryl ketones at low temperatures (-50 to $-78\,°C$) [68]. Nolan and coworkers discovered that copper complexes with NHC ligands are very efficient catalysts for the hydrosilylation of ketones, including hindered ketones such as di-cyclohexyl ketone and di-*tert*-butyl ketone [69].

3.12
Cp$_2$Mo Catalysts for Hydrolysis, Hydrogenations and Hydrations

Organometallic complexes derived from Cp_2Mo have emerged as catalysts for a broad range of reactions. In contrast to the majority of catalysts discussed in this book that proceed in organic solvents, many of these Cp_2Mo systems are conducted in water. In 1979, Köpf and Köpf-Maier reported that metallocene complexes such as Cp_2TiCl_2 and Cp_2MoCl_2 had anti-tumor activity [70]. This discovery had a great impact, as this class of organometallic complexes was recognized to have potential anticancer activity; prior studies had focused primarily on the use of platinum complexes. The studies of these complexes as therapeutic agents will not be discussed in detail here but a recent review gives an overview of the inorganic and medicinal chemistry of antitumor metallocene complexes [71].

Marks and coworkers found that neutral aqueous solutions of Cp_2MoCl_2 underwent hydrolysis of both of the chloride ligands (Scheme 3.6), while the Cp ligands were resistant to hydrolysis [72].

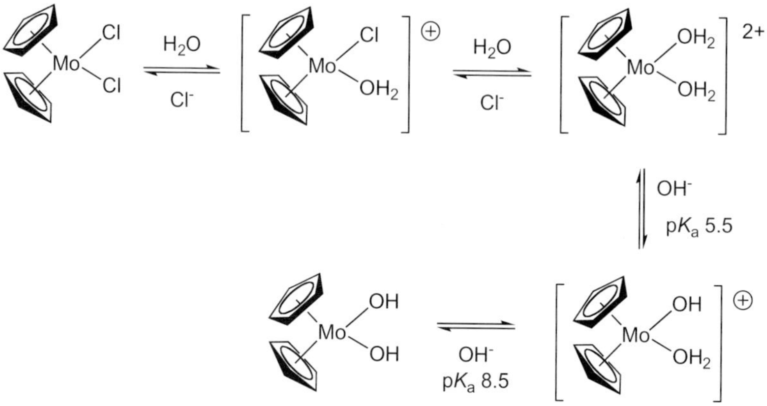

Scheme 3.6

The acidity of the two water ligands were determined to be $pK_a = 5.5$ and 8.5, indicating that at physiological pH the predominant species resulting from aquation of Cp_2MoCl_2 is $Cp_2Mo(OH)(OH_2)^+$. Tyler and coworkers found that $Cp_2Mo(OH)(OH_2)^+$ exists in aqueous solution in equilibrium with a dimeric form bridged by OH groups (shown in Equation 3.25 for the $CH_3C_5H_4$ analog) [73].

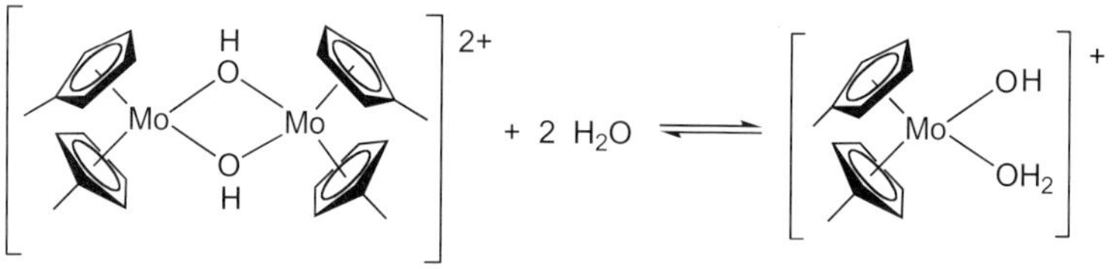

$$(3.25)$$

Kuo and Barnes found that Cp_2MoCl_2 catalyzes the phosphoester bond cleavage of dimethyl phosphate in acidic aqueous solutions (pD = 4.0) at 70 °C [74]. Acceleration of the P–O bond cleavage by the Mo catalyst was about 10^3–10^4 compared to the hydrolysis in water at pH 4 in the absence of catalysts. Hydrolysis of phosphate esters is important in biological systems, and the dimethyl phosphate compound mimics the diester functionality of DNA. The kinetics of the catalytic reaction exhibited a one half-order dependence on the molybdenum complex, which is consistent with the formation of a catalytically active monomeric species from dissociation of a dimeric complex (cf. Equation 3.25). These reactions are thought to proceed by the molybdenum functioning as a Lewis acid for binding the substrate, and activating it towards attack by hydroxide bound to the Mo.

There is a need to develop methods for P–S bond cleavage since some nerve agents used in chemical warfare are organophosphate toxins that contain P–S as well as P–O bonds.

3.12 Cp$_2$Mo Catalysts for Hydrolysis, Hydrogenations and Hydrations

$$\text{Ph}-\underset{\underset{\text{Ph}}{|}}{\overset{\overset{\text{O}}{\|}}{\text{P}}}-\text{S}-\langle\text{C}_6\text{H}_4\rangle-\text{OCH}_3 \xrightarrow[\text{H}_2\text{O}]{\text{Cp}_2\text{MoCl}_2 \text{ (cat.)}} \text{Ph}-\underset{\underset{\text{Ph}}{|}}{\overset{\overset{\text{O}}{\|}}{\text{P}}}-\text{OH} + \text{HS}-\langle\text{C}_6\text{H}_4\rangle-\text{OCH}_3 \quad (3.26)$$

Hydrolysis of the P–S bond of a thiophosphinate (3.26) is catalyzed by Cp$_2$MoCl$_2$ at temperatures as low as 20 °C [75]. A study of the hydrolysis of O,S-diethyl phenylphosphonothioate, a neurotoxin mimic, catalyzed by Cp$_2$MoCl$_2$, showed that the P–S cleavage occurred exclusively rather than P–O scission (3.27).

$$\text{Ph}-\underset{\underset{\text{OEt}}{|}}{\overset{\overset{\text{O}}{\|}}{\text{P}}}-\text{SEt} \xrightarrow[\text{H}_2\text{O}]{\text{Cp}_2\text{MoCl}_2 \text{ (cat.)}} \underset{\substack{\text{product of} \\ \text{P-S cleavage}}}{\text{Ph}-\underset{\underset{\text{OEt}}{|}}{\overset{\overset{\text{O}}{\|}}{\text{P}}}-\text{OH}} + \underset{\substack{\text{product of} \\ \text{P-O cleavage}}}{\text{Ph}-\underset{\underset{\text{OH}}{|}}{\overset{\overset{\text{O}}{\|}}{\text{P}}}-\text{SEt}} \quad (3.27)$$

Cleavage of the P–S bond is desired, since cleavage of the P–O bond in the chemical warfare nerve agent VX gives a product that is only slightly less toxic than VX itself [76]. The chemoselectivity found in this metal-catalyzed reaction contrasts favorably with the alkaline hydrolysis of the same compound, which generates about a 4:1 ratio of P–S cleavage to P–O cleavage products.

Tyler and coworkers found that molybdocene complexes catalyze incorporation of deuterium into the α-position of alcohols through a process that involves C–H bond activation in water [77, 78]. Catalytic intermolecular H/D exchange was studied for methanol, ethanol, benzyl alcohols and other alcohols at 80–102 °C. The proposed mechanism (Scheme 3.7) involves displacement of a D$_2$O ligand by an alcohol, with the alcoholic H (or D) being transferred to the OD ligand. Dissociation of the D$_2$O ligand and oxidative addition of the C–H bond of the alkoxide gives a molybdenum complex with ketone and hydride ligands. The ketone is thought to be η^2-bonded in this case, as shown in Scheme 3.7, with the metal interacting with both the carbon and oxygen, rather than the η^1-bonding established in [Cp(CO)$_2$(PR$_3$)M(O=CEt$_2$)]$^+$. The assignment of side-on, π-bonding of the ketone is based on neutral Cp$_2$Mo(η^2-ketone) complexes [79] and the crystal structure of Cp$_2$Mo(η^2-H$_2$CO) [80] but the possibility of η^1-bonding cannot be completely excluded; Templeton and coworkers found that the mode of bonding for ketones and aldehydes of Mo and W complexes can be influenced by several factors [81].

Tyler and coworkers reported catalytic transfer hydrogenation of 2-butanone using isopropyl alcohol [78]. Kuo and coworkers found that transfer hydrogenation of acetophenone is catalyzed by [Cp$_2$Mo(μ-OH)$_2$MoCp$_2$]$^{+2}$ (OTs$^-$)$_2$ at 75 °C using isopropyl alcohol as the solvent [82]. This catalytic reaction had a turnover rate of about 0.1 h^{-1} at 75 °C. The proposed mechanism is shown in Scheme 3.8, in which the formation of [Cp$_2$Mo(acetone)H]$^+$ proceeds by steps analogous to those shown in Scheme 3.7. Intramolecular hydride transfer from the Mo–H to the bound acetophenone is followed by reaction with water to release the alcohol product and regenerate the catalyst. The rate of catalysis was accelerated by the addition of

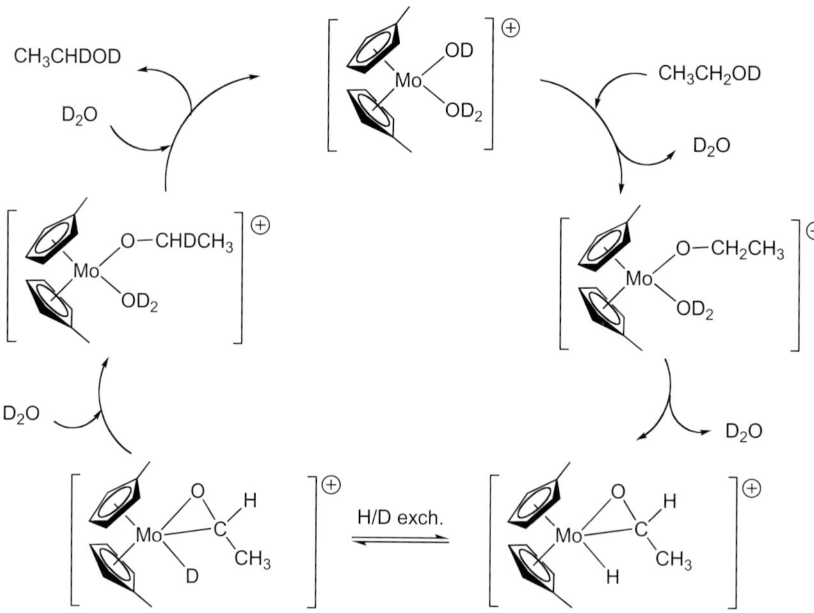

Scheme 3.7

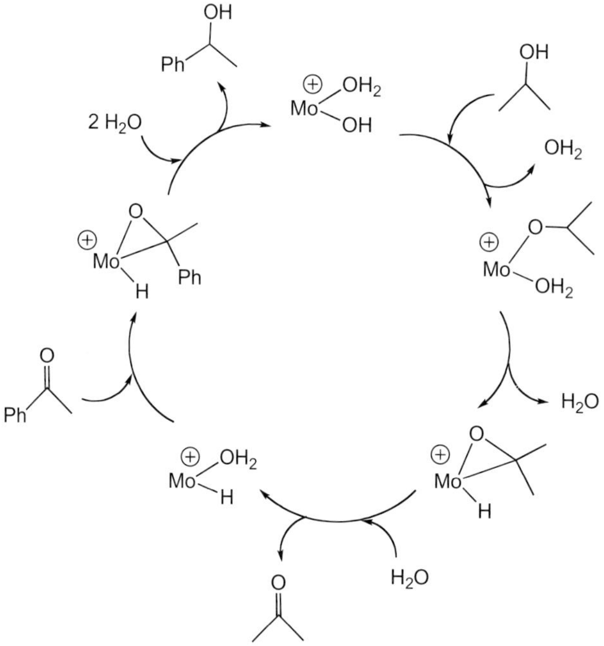

Scheme 3.8

3.12 Cp₂Mo Catalysts for Hydrolysis, Hydrogenations and Hydrations

KOH; the role of the base is not completely clear, but it may increase the rate of deprotonation of the bound alcohol in the production of the metal alkoxide.

Tyler and coworkers found that $[Cp'_2Mo(\mu\text{-}OH)_2MoCp'_2]^{+2}$ ($Cp' = \eta^5\text{-}CH_3C_5H_4$) is a catalyst precursor for the hydration of nitriles in aqueous solution [83]. Several nitriles were hydrated by their catalyst, including acrylonitrile, acetonitrile, isobutyronitrile, 4-cyanopyridine and methyl cycanoacetate. Equation 3.28 shows hydration of acrylonitrile to produce acrylamide, which is used on a large scale industrially due to many uses for polyacrylamide.

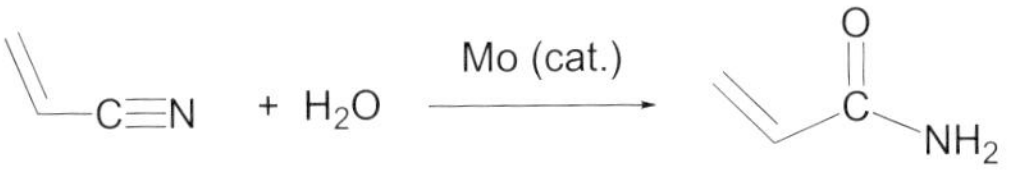

$$\text{CH}_2=\text{CH-C}\equiv\text{N} + \text{H}_2\text{O} \xrightarrow{\text{Mo (cat.)}} \text{CH}_2=\text{CH-C(O)NH}_2 \quad (3.28)$$

The hydration of acrylonitrile is particularly challenging since it requires chemoselective hydration of the C≡N group in the presence of the C=C bond. In addition, the catalytic conditions must avoid the polymerization of acrylamide. Use of $[Cp'_2Mo(\mu\text{-}OH)_2MoCp'_2]^{+2}$ as the catalyst precursor gave >99% chemoselectivity for hydration of the C≡N bond of acrylonitrile; an initial rate of $1.35\,h^{-1}$ was found at 75 °C. The active catalyst appears to be $Cp'_2Mo(OH)(OH_2)^+$ formed from the dimer (3.25). The proposed mechanism shown in Scheme 3.9 begins with

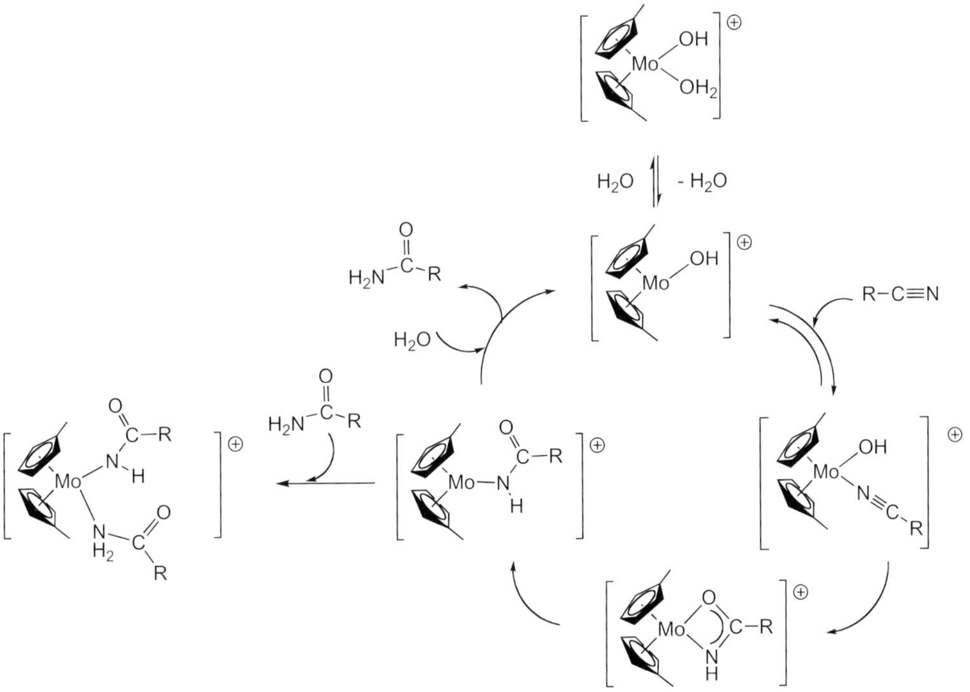

Scheme 3.9

dissociation of water and coordination of the nitrile. Intramolecular attack of the bound hydroxide on the bound nitrile was proposed as the next step, forming an η^2-amidate ligand; alternatives considered for this step involved intermolecular attack of water on the coordinated nitrile. Addition of water and reductive elimination to give the free amide completes the catalytic cycle. This catalytic reaction is reversibly inhibited by the nitrile, as shown by the equilibrim for nitrile binding in Scheme 3.9, and is irreversibly inhibited by the product amide, as shown at the lower left side of the proposed mechanism.

Tyler and coworkers also found that hydrolysis of esters is catalyzed by $Cp'_2Mo(OH)(OH_2)+$ [84]. The hydrolysis of ethyl acetate is shown in Equation 3.29, giving acetic acid and ethanol.

$$H_3C-C(=O)-OEt + H_2O \xrightarrow{\text{Mo (cat.)}} H_3C-C(=O)-OH + EtOH \qquad (3.29)$$

These catalytic reactions are thought to proceed similarly to those discussed above, by a pathway involving binding of the substrate to the metal and attack by hydroxide. The versatility of these systems based on Cp_2Mo is likely due to the ability of this water-soluble system to attain a cis-configuration of H_2O and OH^- ligands. Displacement of the water by the incoming substrate activates it towards attack by the hydroxide, leading to a remarkably versatile series of reactions catalyzed by these Mo complexes.

3.13
Conclusion

Homogeneous catalysts using Mo or W as catalyst precursors have been developed for hydrogenation of ketones, hydrosilylation of ketones, hydrolysis reactions, and hydration of nitriles. The hydrogenation catalysts were designed following reactivity studies in which the steps of the catalytic cycle were established under stoichiometric conditions. These studies have expanded our understanding of the fundamental reaction pathways that can be elicited from new reactions of cheap metals, and point the way for further development of new classes of catalysts.

Acknowledgements

Research at Pacific Northwest National Laboratory was funded by the Division of Chemical Sciences, Biosciences and Geosciences, Office of Basic Energy Sciences, U.S. Department of Energy. Pacific Northwest National Laboratory is operated by Battelle for the U.S. Department of Energy.

References

1 de Vries, J.G., and Elsevier, C.J. (2007) *Handbook of Homogeneous Hydrogenation*, Wiley-VCH Verlag GmbH, Weinheim, Germany.
2 Noyori, R. (2003) *Adv. Synth. Catal.*, **345**, 1.
3 Morris, R.H. (2007) Ruthenium and osmium, in *Handbook of Homogeneous Hydrogenation*, vol. **1** (eds J.G. de Vries and C.J. Elsevier), Wiley-VCH Verlag GmbH, Weinheim, Germany, Chapter 3, pp. 45–70.
4 Oro, L.A., and Carmona, D. (2007) Rhodium, in *Handbook of Homogeneous Hydrogenation* (eds J.G. de Vries and C.J. Elsevier), Wiley-VCH Verlag GmbH, Weinheim, Germany, Chapter 1, pp. 3–30.
5 Bullock, R.M. (1991) *Comments Inorg. Chem.*, **12**, 1–33.
6 Eisenberg, D.C., and Norton, J.R. (1991) *Isr. J. Chem.*, **31**, 55–66.
7 Kristjánsdóttir, S.S., and Norton, J.R. (1991) Acidity of hydrido transition metal complexes in solution, in *Transition Metal Hydrides* (ed. A. Dedieu), VCH, New York, (Chapter 9), pp. 309–359.
8 Landau, S.E., Morris, R.H., and Lough, A.J. (1999) *Inorg. Chem.*, **38**, 6060–6068; Fong, T.P., Forde, C.E., Lough, A.J., Morris, R.H., Rigo, P., Rocchini, E., and Stephan, T. (1999) *J. Chem. Soc., Dalton Trans.*, 4475–4486.
9 Moore, E.J., Sullivan, J.M., and Norton, J.R. (1986) *J. Am. Chem. Soc.*, **108**, 2257–2263.
10 Ciancanelli, R., Noll, B.C., DuBois, D.L., and DuBois, M.R. (2002) *J. Am. Chem. Soc.*, 2984–2992.
11 Berning, D.E., Noll, B.C., and DuBois, D.L. (1999) *J. Am. Chem. Soc.*, **121**, 11432–11447.
12 Price, A.J., Ciancanelli, R., Noll, B.C., Curtis, C.J., DuBois, D.L., and DuBois, M.R. (2002) *Organometallics*, **21**, 4833–4839.
13 Ellis, W.W., Ciancanelli, R., Miller, S.M., Raebiger, J.W., DuBois, M.R., and DuBois, D.L. (2003) *J. Am. Chem. Soc.*, **125**, 12230–12236.
14 Curtis, C.J., Miedaner, A., Ellis, W.W., and DuBois, D.L. (2002) *J. Am. Chem. Soc.*, **124**, 1918–1925.
15 Ellis, W.W., Raebiger, J.W., Curtis, C.J., Bruno, J.W., and DuBois, D.L. (2004) *J. Am. Chem. Soc.*, **126**, 2738–2743.
16 Miedaner, A., Raebiger, J.W., Curtis, C.J., Miller, S.M., and DuBois, D.L. (2004) *Organometallics*, **23**, 2670–2679.
17 Zhang, X.-M., Bruno, J.W., and Enyinnaya, E. (1998) *J. Org. Chem.*, **63**, 4671–4678.
18 DuBois, D.L., Blake, D.M., Miedaner, A., Curtis, C.J., DuBois, M.R., Franz, J.A., and Linehan, J.C. (2006) *Organometallics*, 4414–4419.
19 Raebiger, J.W., Miedaner, A., Curtis, C.J., Miller, S.M., Anderson, O.P., and DuBois, D.L. (2004) *J. Am. Chem. Soc.*, **126**, 5502–5514.
20 Curtis, C.J., Miedaner, A., Raebiger, J.W., and DuBois, D.L. (2004) *Organometallics*, **23**, 511–516.
21 Edidin, R.T., Sullivan, J.M., and Norton, J.R. (1987) *J. Am. Chem. Soc.*, **109**, 3945–3953.
22 Bullock, R.M. (1987) *J. Am. Chem. Soc.*, **109**, 8087–8089.
23 Cheng, T.-Y., Brunschwig, B.S., and Bullock, R.M. (1998) *J. Am. Chem. Soc.*, **120**, 13121–13137.
24 Cheng, T.-Y., and Bullock, R.M. (1999) *J. Am. Chem. Soc.*, **121**, 3150–3155. Cheng, T.-Y., and Bullock, R.M. (2002) *Organometallics*, 21, 2325–2331.
25 Darensbourg, M.Y., and Ash, C.E. (1987) *Adv. Organomet. Chem.*, **27**, 1–50.
26 Kao, S.C., and Darensbourg, M.Y. (1984) *Organometallics*, **3**, 646–647; Kao, S.C., Spillett, C.T., Ash, C., Lusk, R., Park, Y.K., and Darensbourg, M.Y. (1985) *Organometallics*, **4**, 83–91.
27 Ash, C.E., Hurd, P.W., Darensbourg, M.Y., and Newcomb, M. (1987) *J. Am. Chem. Soc.*, **109**, 3313–3317.
28 Kao, S.C., Gaus, P.L., Youngdahl, K., and Darensbourg, M.Y. (1984) *Organometallics*, **3**, 1601–1603.
29 Gaus, P.L., Kao, S.C., Darensbourg, M.Y., and Arndt, L.W. (1984) *J. Am. Chem. Soc.*, **106**, 4752–4755.

30 Gaus, P.L., Kao, S.C., Youngdahl, K., and Darensbourg, M.Y. (1985) *J. Am. Chem. Soc.*, **107**, 2428–2434.

31 Brunet, J.-J., Chauvin, R., and Leglaye, P. (1999) *Eur. J. Inorg. Chem.*, 713–716.

32 Darensbourg, M.Y. (1985) *Prog. Inorg. Chem.*, **33**, 221–274.

33 Tooley, P.A., Ovalles, C., Kao, S.C., Darensbourg, D.J., and Darensbourg, M.Y. (1986) *J. Am. Chem. Soc.*, **108**, 5465–5470.

34 Markó, L., and Nagy-Magos, Z. (1985) *J. Organomet. Chem.*, **285**, 193–203.

35 Darensbourg, D.J., and Rokicki, A. (1982) *Organometallics*, **1**, 1685–1693.

36 Kursanov, D.N., Parnes, Z.N., and Loim, N.M. (1974) *Synthesis*, 633–651.

37 Kursanov, D.N., Parnes, Z.N., Kalinkin, M.I., and Loim, N.M. (1985) *Ionic Hydrogenation and Related Reactions*, Harwood Academic Publishers, New York.

38 Kubas, G.J. (2001) *Metal Dihydrogen and σ-Bond Complexes: Structure, Theory, and Reactivity*, Kluwer Academic/Plenum Publishers, New York.

39 Ito, M., and Ikariya, T. (2007) *Chem. Comm.*, 5134–5142; Noyori, R., Yamakawa, M., and Hashiguchi, S. (2001) *J. Org. Chem.*, **66**, 7931–7944; Noyori, R., and Ohkuma, T. (2001) *Angew. Chem. Int. Ed.*, **40**, 40–73; Ikariya, T., Murata, K., and Noyori, R. (2006) *Org. Biomol. Chem.*, **4**, 393–406.

40 Bullock, R.M., and Song, J.-S. (1994) *J. Am. Chem. Soc.*, **116**, 8602–8612.

41 Bullock, R.M., Song, J.-S., and Szalda, D.J. (1996) *Organometallics*, **15**, 2504–2516.

42 Song, J.-S., Szalda, D.J., Bullock, R.M., Lawrie, C.J.C., Rodkin, M.A., and Norton, J.R. (1992) *Angew. Chem. Int. Ed. Engl.*, **31**, 1233–1235.

43 Song, J.-S., Szalda, D.J., and Bullock, R.M. (2001) *Organometallics*, **20**, 3337–3346.

44 Song, J.-S., Szalda, D.J., and Bullock, R.M. (1997) *Inorg. Chim. Acta*, **259**, 161–172.

45 Reynoud, J.-F., Leboeuf, J.-F., Leblanc, J.-C., and Moïse, C. (1986) *Organometallics*, **5**, 1863–1866.

46 Bullock, R.M., and Voges, M.H. (2000) *J. Am. Chem. Soc.*, **122**, 12594–12595.

47 Voges, M.H., and Bullock, R.M. (2002) *J. Chem. Soc., Dalton Trans.*, 759–770.

48 Ryan, O.B., Tilset, M., and Parker, V.D. (1990) *J. Am. Chem. Soc.*, **112**, 2618–2626; Smith, K.-T., and Tilset, M. (1992) *J. Organomet. Chem.*, **431**, 55–64.

49 Papish, E.T., Rix, F.C., Spetseris, N., Norton, J.R., and Williams, R.D. (2000) *J. Am. Chem. Soc.*, **122**, 12235–12242.

50 Kolthoff, I.M., and Chantooni, M.K., Jr. (1973) *J. Am. Chem. Soc.*, **95**, 8539–8546.

51 Smith, K.-T., Norton, J.R., and Tilset, M. (1996) *Organometallics*, **15**, 4515–4520.

52 Namorado, S., Antunes, M.A., Veiros, L.F., Ascenso, J.R., Duarte, M.T., and Martins, A.M. (2008) *Organometallics*, **27**, 4589–4599.

53 Namorado, S., Cui, J., Azevedo, C.G., Lemos, M.A., Duarte, M.T., Ascenso, J.R., Dias, A.R., and Martins, A.M. (2007) *Eur. J. Inorg. Chem.*, 1103–1113.

54 Kimmich, B.F.M., Fagan, P.J., Hauptman, E., and Bullock, R.M. (2004) *Chem. Commun.*, 1014–1015; Kimmich, B.F.M., Fagan, P.J., Hauptman, E., Marshall, W.J., and Bullock, R.M. (2005) *Organometallics*, **24**, 6220–6229.

55 Tanaka, K., and Toda, F. (2000) *Chem. Rev.*, **100**, 1025–1074; DeSimone, J.M. (2002) *Science*, **297**, 799–803; Cave, G.W.V., Raston, C.L., and Scott, J.L. (2001) *J. Chem. Soc., Chem. Commun.*, 2159–2169.

56 Bourissou, D., Guerret, O., Gabbaï, F.P., and Bertrand, G. (2000) *Chem. Rev.*, **100**, 39–91; Herrmann, W.A., and Köcher, C. (1997) *Angew. Chem. Int. Ed. Engl.*, **36**, 2162–2187; Herrmann, W.A. (2002) *Angew. Chem. Int. Ed.*, **41**, 1290–1309; Arduengo, A.J., III (1999) *Acc. Chem. Res.*, **32**, 913–921; Nolan, S.P. (ed.) (2006) *N-Heterocyclic Carbenes in Synthesis*, Wiley-VCH Verlag GmbH, Weinheim.

57 Huang, J., Schanz, H.-J., Stevens, E.D., and Nolan, S.P. (1999) *Organometallics*, **18**, 2370–2375; Huang, J., Jafarpour, L., Hillier, A.C., Stevens, E.D., and Nolan, S.P. (2001) *Organometallics*, **20**, 2878–2882; Scott, N.M., Clavier, H., Mahjoor, P., Stevens, E.D., and Nolan, S.P. (2008) *Organometallics*, **27**, 3181–3186.

58 Dioumaev, V.K., Szalda, D.J., Hanson, J., Franz, J.A., and Bullock, R.M. (2003) *Chem. Commun.*, 1670–1671; Wu, F.,

Dioumaev, V.K., Szalda, D.J., Hanson, J., and Bullock, R.M. (2007) *Organometallics*, **26**, 5079–5090.

59 van Rensburg, H., Tooze, R.P., Foster, D.F., and Slawin, A.M.Z. (2004) *Inorg. Chem.*, **43**, 2468–2470.

60 Gründemann, S., Albrecht, M., Kovacevic, A., Faller, J.W., and Crabtree, R.H. (2002) *J. Chem. Soc., Dalton Trans.*, 2163–2167; Muehlhofer, M., Strassner, T., and Herrmann, W.A. (2002) *Angew. Chem. Int. Ed.*, **41**, 1745–1747.

61 Ojima, I. (1989) The hydrosilylation reaction, in *The Chemistry of Organic Silicon Compounds* (eds S. Patai and Z. Rappoport), John Wiley & Sons, Inc., New York, pp. 1479–1526 and references cited therein; Ojima, I., Li, Z., and Zhu, J. (1998) Recent advances in the hydrosilylation and related reactions, in *The Chemistry of Organic Silicon Compounds*, vol. **2** (eds Z. Rappoport and Y. Apeloig), John Wiley & Sons, Inc., New York, pp. 1687–1792.

62 Dioumaev, V.K., and Bullock, R.M. (2003) *Nature*, **424**, 530–532.

63 Cavanaugh, M.D., Gregg, B.T., and Cutler, A.R. (1996) *Organometallics*, **15**, 2764–2769.

64 Mao, Z., Gregg, B.T., and Cutler, A.R. (1995) *J. Am. Chem. Soc.*, **117**, 10139–10140.

65 Gregg, B.T., and Cutler, A.R. (1996) *J. Am. Chem. Soc.*, **118**, 10069–10084.

66 Berk, S.C., Kreutzer, K.A., and Buchwald, S.L. (1991) *J. Am. Chem. Soc.*, **113**, 5093–5095.

67 Carter, M.B., Schiott, B., Gutierrez, A., and Buchwald, S.L. (1994) *J. Am. Chem. Soc.*, **116**, 11667–11670; Yun, J., and Buchwald, S.L. (1999) *J. Am. Chem. Soc.*, **121**, 5640–5644.

68 Lipshutz, B.H., Noson, K., and Chrisman, W. (2001) *J. Am. Chem. Soc.*, **123**, 12917–12918; Lipshutz, B.H., Noson, K., Chrisman, W., and Lower, A. (2003) *J. Am. Chem. Soc.*, **125**, 8779–8789; Lipshutz, B.H., and Frieman, B.A. (2005) *Angew. Chem. Int. Ed.*, **44**, 6345–6348.

69 Díez-González, S., Kaur, H., Zinn, F.K., Stevens, E.D., and Nolan, S.P. (2005) *J. Org. Chem.*, **70**, 4784–4796; Díez-González, S., and Nolan, S.P. (2008) *Acc. Chem. Res.*, **41**, 349–358.

70 Köpf, H., and Köpf-Maier, P. (1979) *Angew. Chem. Int. Ed. Engl.*, **18**, 477–478.

71 Abeysinghe, P.M., and Harding, M.M. (2007) *Dalton Trans.*, 3474–3482.

72 Kuo, L.Y., Kanatzidis, M.G., Sabat, M., Tipton, A.L., and Marks, T.J. (1991) *J. Am. Chem. Soc.*, **113**, 9027–9045.

73 Balzarek, C., Weakley, T.J.R., Kuo, L.Y., and Tyler, D.R. (2000) *Organometallics*, **19**, 2927–2931.

74 Kuo, L.Y., and Barnes, L.A. (1999) *Inorg. Chem.*, **38**, 814–817.

75 Kuo, L.Y., Blum, A.P., and Sabat, M. (2005) *Inorg. Chem.*, **44**, 5537–5541.

76 Yang, Y.-C. (1999) *Acc. Chem. Res.*, **32**, 109–115.

77 Balzarek, C., and Tyler, D.R. (1999) *Angew. Chem. Int. Ed.*, **38**, 2406–2408.

78 Balzarek, C., Weakley, T.J.R., and Tyler, D.R. (2000) *J. Am. Chem. Soc.*, **122**, 9427–9434.

79 Okuda, J., and Herberich, G.E. (1987) *Organometallics*, **6**, 2331–2336.

80 Herberich, G., and Okuda, J. (1985) *Angew. Chem. Int. Ed. Engl.*, **24**, 402.

81 Schuster, D.M., White, P.S., and Templeton, J.L. (2000) *Organometallics*, **19**, 1540–1548.

82 Kuo, L.Y., Finigan, D.M., and Tadros, N.N. (2003) *Organometallics*, **22**, 2422–2425.

83 Breno, K.L., Pluth, M.D., and Tyler, D.R. (2003) *Organometallics*, **22**, 1203–1211.

84 Breno, K.L., Pluth, M.D., Landorf, C.W., and Tyler, D.R. (2004) *Organometallics*, **23**, 1738–1746.

4
Modern Alchemy: Replacing Precious Metals with Iron in Catalytic Alkene and Carbonyl Hydrogenation Reactions

Paul J. Chirik

4.1
Introduction

Homogeneous catalysis has revolutionized the art of organic synthesis. It is now challenging to find a synthesis of any complex molecular target that does not rely on at least one, if not many, transition metal-mediated or catalyzed steps. Why have these transformations achieved such privileged status? In many cases, metal-catalyzed reactions open avenues to new methods unknown in traditional organic chemistry or impart levels of selectivity, especially enantioselectivity, that allow more atom economical syntheses of many valuable products.

One particularly powerful metal-catalyzed transformation is the hydrogenation of unsaturated organic molecules such as olefins, alkynes, aldehydes, ketones and imines. Wilkinson's ground-breaking discovery of the hydrogenation of simple alkenes with soluble $(Ph_3P)_3RhCl$ under readily accessed laboratory conditions spawned a new field of applied organometallic chemistry (Figure 4.1) [1]. Chemists, through rational ligand design and predictable synthetic chemistry, were able to optimize reactions and predictably improve levels of activity and selectivity. One early success of this approach was the introduction of configurationally stable, enantiopure, monodentate phosphines for the asymmetric hydrogenation of dehydroamino acids [2], which ultimately evolved into a commercial process for the production of L-DOPA [3]. Subsequent catalyst modifications included the synthesis of cationic diene complexes $[(R_3P)_2Rh(diene)]^+X^-$ (diene = 1,5-cyclooctadiene; norbornadiene; $X^- = PF_6^-$, ClO_4^-, BF_4^-, etc.) by Schrock and Osborn (Figure 4.1) [4, 5]. Introduction of C_2 symmetric, chiral bis(phosphine) ligands [6] produced metal complexes that exhibited high activities and produced near ideal enantioselectivities with several classes of substrates [7, 8]. It is important to realize that these discoveries relied on the availability of excellent synthetic precursors such as $[Rh(COD)Cl]_2$, $[Rh(COD)_2]^+$ (COD = 1,5-cyclooctadiene) and $[Rh(NBD)Cl]_2$ (NBD = norbornadiene). The compounds react reproducibly and usually quantitatively with the bis(phospine) ligands, often obviating isolation of the catalyst precursor.

Catalysis Without Precious Metals. Edited by R. Morris Bullock
© 2010 WILEY-VCH Verlag GmbH & Co. KGaA, Weinheim
ISBN: 978-3-527-32354-8

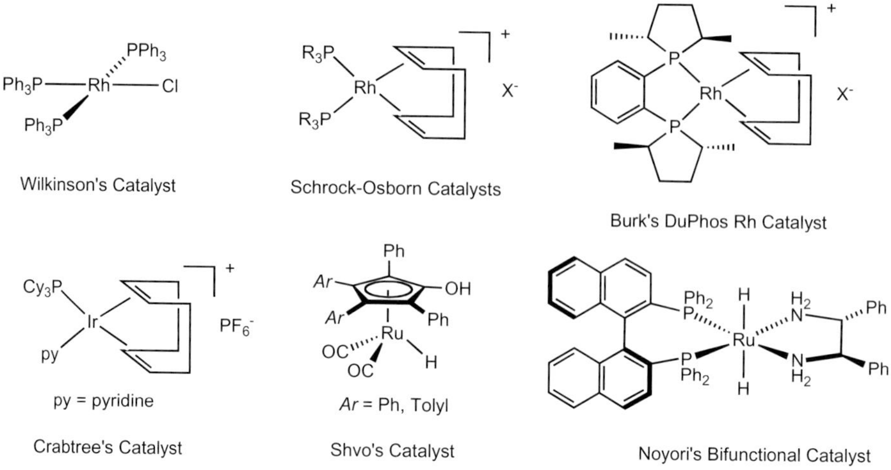

Figure 4.1 Commonly used precious metal catalysts for homogeneous hydrogenation.

Metals other than rhodium have also been explored for catalytic hydrogenation (Figure 4.1). One prominent example is "Crabtree's catalyst", [(COD)Ir(PCy$_3$)(py)]PF$_6$ [9], a compound that is effective for the hydrogenation of hindered alkenes as well as for substrate-directed reductions [10]. This basic pre-catalyst architecture has since been modified by Pfaltz [11] and Burgess [12] with phosphinooxazoline and related N-heterocyclic carbene ligands to generate iridium compounds that promote asymmetric hydrogenation of unfunctionalized tri- and tetrasubstituted olefins with high enantioselectivity–one of the most difficult and long-standing challenges in the field of homogeneous hydrogenation chemistry.

Ruthenium complexes have also played a prominent role in hydrogenation catalysis [13] and the discovery of bifunctional mechanisms (e.g., ionic hydrogenations/heterolytic H$_2$ splitting) has opened new avenues to outer sphere reduction chemistry [14]. As with rhodium hydrogenation chemistry, the origins of ruthenium catalysis can be traced to studies initially performed in the Wilkinson laboratory. In 1965, RuCl$_2$(PPh$_3$)$_3$ was reported to catalyze the hydrogenation of 1-hexene [15]. Additional experimentation established the necessity of ethanol as a co-solvent for efficient turnover and identified the ruthenium monohydride, (Ph$_3$P)$_3$Ru(H)Cl, as an important catalytic intermediate. These studies were also some of the first to suggest the potential utility of heterolytic H$_2$ cleavage in a catalytic hydrogenation cycle [16–18]. With this foundation, Noyori introduced (BINAP)Ru(OAc)$_2$ and expanded the substrate scope of asymmetric olefin hydrogenation to include synthetically important compounds such as (Z)-enamides, and allylic and homoallylic alcohols. Ultimately, these methods were applied to the synthesis of important targets such as morphine [19] and both enantiomers of citronellol [20]. In independent studies, Shvo, and later Casey, reported hydroxycyclopentadienyl ruthenium hydride compounds [21] that are active for aldehyde and imine reduction. Detailed mechanistic studies support concerted proton and hydride transfer to the substrate outside the coordination sphere of the metal [22, 23].

Additional catalyst development identified the positive effect of 1,2-diamines as additives in the (BINAP)Ru(OAc)$_2$-catalyzed enantioselective hydrogenations of ketones [24]. This discovery ultimately led to the synthesis of a class of (diphosphine)Ru(diamine)X$_2$ (X = H, halide) compounds [25] (Figure 4.1) which have emerged as some of the most active and selective hydrogenation catalysts ever reported [26]. Mechanistic studies by Noyori [14] and Morris [27] have established bifunctional hydrogen transfer to substrate from the *cis* Ru–H and N–H motifs and identified the importance of ruthenium hydridoamido complexes for the heterolytic splitting of H$_2$. This paradigm allows prediction of the absolute stereochemistry of the chiral alcohols produced from these reactions.

Homogeneous hydrogenation has emerged as one of the true success stories of organometallic chemistry and one that highlights how fundamental academic research may ultimately lead to a commercial process. There is, however, room for growth and improvement. Despite the potential for asymmetric hydrogenation, it remains an underutilized technology. The relatively high cost, environmental concerns, and uncertainty of the long-term supply of precious metals [28] inspire the search for equivalent or superior base metal alternatives [29].

Iron is an attractive precious metal surrogate due to its high natural abundance, low cost [29] and precedent in numerous biocatalytic reactions [30]. The challenge for synthetic chemists is one of "modern alchemy". How can the base metal be transmuted, through rational ligand design and careful control of electronic structure, to participate in the breadth of chemistry known for precious metals? The high standards set by Knowles, Noyori, Crabtree, Burke, Pfaltz, Burgess, and others mean that iron hydrogenation catalysis has a tough act to follow. As will be highlighted in this chapter, the early results in iron catalysis are promising and the field now has a strong foundation on which to grow and expand. Clearly the future is bright.

Using iron in catalysis, especially as a precious metal replacement, is not a new idea that has come from our current, so-called "age of sustainability". In 1909, Mittasch at BASF was charged with finding more economical catalysts as replacements for existing osmium and uranium compounds used in industrial ammonia synthesis [31]. An activated iron surface was discovered that became the foundation for modern Haber–Bosch technology [32]. Other early applications of iron catalysis include Gif chemistry [33], the Reppe cyclization [34], biomimetic chemistry [35], Lewis acid promoted reactions, and the foundations of modern cross-coupling methods [36]. Global sustainability efforts, coupled with the observation of unique and interesting reactivity, have resulted in a recent renaissance in iron catalysis [37–39]. Transformations including cycloadditions [40], cross couplings [41], oxidations [42], and olefin polymerization [43] are currently the subject of intense investigations from many different academic and industrial research groups. This chapter will focus on a specific and emerging class of iron-catalyzed reactions – namely the homogeneous hydrogenation of alkenes and organic carbonyl compounds. While the foundations of these efforts date back nearly fifty years, new accomplishments in ligand design and understanding of electronic structure have resulted in a resurgence of iron chemistry applied to homogeneous hydrogenation.

4.2
Alkene Hydrogenation

4.2.1
Iron Carbonyl Complexes

The organometallic chemistry of iron has long been dominated by coordinatively saturated carbonyl complexes [44]. Berthelot first synthesized $Fe(CO)_5$ in 1891 [45]. Therefore it is, not surprising that early efforts in iron-catalyzed olefin hydrogenation focused on $Fe(CO)_5$, $Fe_2(CO)_9$ and related clusters [46]. This interest has continued for the past four decades.

Inspired by the observation of $Fe(CO)_5$-catalyzed reduction of olefins during hydroformylation [47], Frankel and coworkers presented an early series of papers examining the hydrogenation of fatty esters such as methyl linoleate [48] and methyl linolenate [49]. At temperatures in excess of 180 °C and at a hydrogen pressure of 400 psi (~27.2 atm = 27.6 bar), mixtures of geometric and positional isomers of monoenoic fatty esters were observed as the primary products. Kinetic models were developed to account for the various organic products formed from these reactions [49]. One notable feature of this early work is the isolation and characterization of diene and triene complexes of iron tricarbonyl. In addition to being postulated as intermediates in the catalytic cycle, these species also catalyzed the hydrogenation of methyl linoleate under milder conditions than $Fe(CO)_5$, suggesting that carbonyl dissociation was the origin of the high kinetic barriers associated with catalytic turnover [50].

Cais and Maoz followed these seminal studies with investigations into the catalytic hydrogenation activity of $(diene)Fe(CO)_3$ and $(alkene)Fe(CO)_4$ complexes with substrates such as methyl sorbate [51]. These workers had a long-standing interest in the organometallic chemistry of diene complexes of iron tricarbonyl [52, 53]. Competing reactions such as alkene exchange and isomerization accompanied alkene hydrogenation [51]. As with the $Fe(CO)_5$-catalyzed hydrogenations previously reported, relatively high temperatures of 165 °C and high pressures of H_2 (~13.6 atm) were required for efficient turnover.

The proposed intermediacy of $[Fe(CO)_3]$, coupled with the high quantum efficiency observed with most mononuclear metal carbonyl complexes [54], inspired Schroeder and Wrighton to study photocatalytic olefin isomerization, hydrogenation and hydrosilylation reactions with $Fe(CO)_5$ [55]. Irradiation of $Fe(CO)_5$ with near ultraviolet (λ_{max} = 366 nm) light in the presence of excess 1-pentene furnished the thermodynamic mixture of *cis* and *trans*-2-pentene over the course of 2 h. From photolysis studies in neat olefin, it was estimated that each iron center produced at least 800 isomerization events. Conducting the irradiation in the presence of simple unactivated olefins and approximately 1 atm of H_2 at ambient temperature resulted in catalytic hydrogenation to the corresponding alkane (Scheme 4.1). Much like thermal catalytic hydrogenations with $(Ph_3P)_3RhCl$ or $HRuCl(PPh_3)_3$, the catalysis is sensitive to steric effects where less hindered alkenes such as propylene are selectively hydrogenated in the presence of cycloalkenes such as cyclopentene.

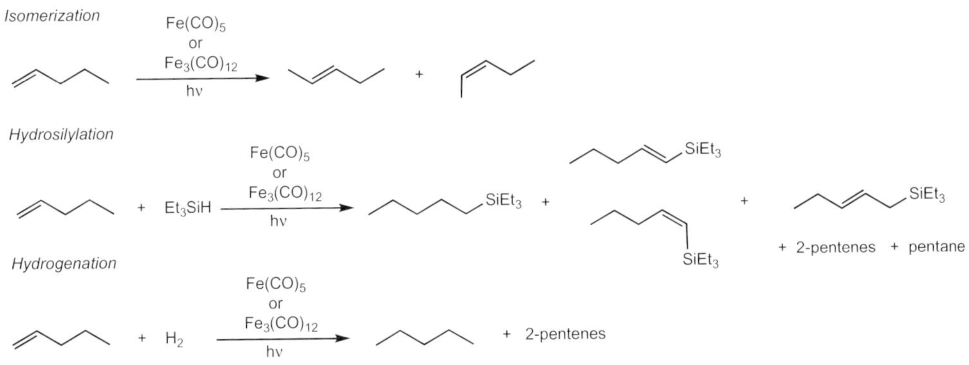

Scheme 4.1 $Fe(CO)_5$ photocatalytic isomerization, hydrosilylation and hydrogenation.

Following this discovery, attempts were made to gain more quantitative information about the activity of the photogenerated catalysts [56]. In the initial reports of 1-pentene hydrogenation and isomerization with irradiated $Fe(CO)_5$, quantum yields in excess of unity were observed. However, turnover did not persist in the dark following photoinitiation due to back reactions with the ejected CO and formation of catalytically inactive clusters, thereby complicating the determination of the turnover frequency. Using very high intensity excitation, Mitchener and Wrighton ultimately demonstrated that catalytic 1-pentene isomerization, hydrogenation, or hydrosilylation with either $Fe(CO)_5$ or $Fe_3(CO)_{12}$ proceeded with turnover rates greater than $10^3 h^{-1}$. For the hydrosilylation of 1-pentene with Et_3SiH, a mixture of hydrogenation, isomerization, hydrosilylation and dehydrogenative silane–alkene coupling was observed (Scheme 4.1).

Following the pioneering work of the Wrighton laboratory, Grant and coworkers [57] introduced a new photocatalytic initiation method where sustained near-ultraviolet pulsed radiation was used to produce catalytically active iron complexes on the nanosecond timescale. This technique was initially used to re-investigate $Fe(CO)_5$-catalyzed 1-pentene isomerization and hydrogenation and established quantum yields as high as 800 for isomerization and for hydrogenation. These large quantum yields established photogeneration of a highly active thermal catalyst.

The results of these studies also clearly ruled out the possibility of a binuclear mechanism [58] and identified two distinct phases for the reaction. The first part of the process involves an induction period where the rate of the reaction increases, followed by a steady state period where the reaction rate decreases monotonically as the system approaches equilibrium. Once equilibrium concentrations are established, addition of more 1-pentene with additional radiation proceeds without the induction period. These experiments also allowed estimation of the catalyst lifetime as 0.2 s.

In subsequent, more comprehensive studies, Grant and coworkers studied the gas-phase hydrogenation of ethylene using photoirradiated $Fe(CO)_5$ [59–61]. This substrate was chosen, in part, to separate catalytic hydrogenation from competing isomerization processes. Both $(CO)_4Fe(\eta^2\text{-}CH_2=CH_2)$ and $(CO)_3Fe(\eta^2\text{-}CH_2=CH_2)_2$ were identified [62, 63] in the reaction mixture, with the latter compound serving as a reservoir for the presumed active species, $(CO)_3Fe(\eta^2\text{-}CH_2=CH_2)$. Entry into the catalytic cycle occurs via thermal dissociation of one ethylene molecule (Scheme 4.2). Once formed, CO, H_2 and ethylene are in competition for the unsaturated iron center. Ethylene coordination reverts the active species back to a resting state, $(CO)_3Fe(\eta^2\text{-}CH_2=CH_2)_2$, while CO coordination and formation of $(CO)_4Fe(\eta^2\text{-}CH_2=CH_2)$ constitutes a deactivation pathway.

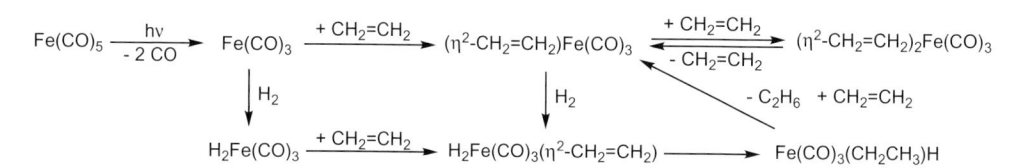

Scheme 4.2 Generally accepted mechanism for photocatalytic olefin hydrogenation with $Fe(CO)_5$.

Reaction with H_2 is productive and brings the iron compound into the catalytic cycle. The quantum yield rises with increasing ethylene and hydrogen pressures but reaches an asymptotic maximum with both reagents. The existence of a maximum for H_2 is due to a balance of two competing processes. First, hydrogen competes with CO for the presumed active species, $(CO)_3Fe(\eta^2\text{-}CH_2=CH_2)$, which causes the rate of turnover to increase to eventual saturation. Second, hydrogen initiates the catalytic cycle, where one intermediate forms an inactive ethylene complex. This process, which becomes more important with increasing hydrogen pressure, decreases catalytic efficiency. These studies clearly indicate that even for "simple" gas-phase reactions, such as the hydrogenation of ethylene, the number of competing elementary steps can be quite complex.

Subsequent studies by Hayes and Weitz [64] reported that addition of ethylene to $(CO)_3FeH_2$ or H_2 to $(CO)_3Fe(\eta^2\text{-}CH_2=CH_2)$ yielded singlet $(CO)_3Fe(\eta^2\text{-}CH_2=CH_2)H_2$ based on observation of a strong infrared band near 2051 cm^{-1}. Olefin insertion into an iron hydride followed by C–H reductive elimination yields ethane. The reductive elimination event could either be induced by addition of ethylene or occur directly from the iron alkyl hydride.

Because the identities of the catalytic intermediates were assigned by infrared spectroscopy, little is known about the actual structures of these important compounds. To gain more structural and geometric insights, Weitz and coworkers recently reported a comprehensive DFT study on the bonding and geometries of $(CO)_3FeH_2$, $(CO)_3Fe(\eta^2\text{-}CH_2=CH_2)H_2$ and $(CO)_3Fe(CH_2CH_3)H$ [65]. The results establish that $(CO)_3FeH_2$, formed from dihydrogen addition to triplet $Fe(CO)_3$, is an $S = 1$ dihydrogen complex, i.e., $(CO)_3Fe(\eta^2\text{-}H_2)$, with a computed H–H bond length between 0.72 and 0.83 Å. Interestingly, calculations on the related tetracarbonyl complex, $(CO)_4FeH_2$, establish a classical iron(II) dihydride, a result of a more electron-rich metal center upon coordination of an additional carbon monoxide ligand. This result may at first seem counterintuitive, as introduction of an additional π-accepting ligand could be expected to reduce the electron density at iron, disfavoring oxidative addition. However, natural bond order analysis reveals the opposite effect, demonstrating that σ-donation clearly outweighs the contributions from π-backbonding of the carbonyl ligand.

The structure of the catalytic resting state, $(CO)_3Fe(\eta^2\text{-}CH_2=CH_2)$, was also examined and found to have a triplet ground state approximately 9 kcal mol^{-1} lower in energy than the singlet. Prior calculations had favored a singlet ground state for this compound [66], despite experimental evidence for the triplet. The discrepancy between the two calculations was traced to inclusion of higher order spin states in the wavefunctions of open shell compounds. The energetics of ethylene exchange between $(CO)_3Fe(\eta^2\text{-}CH_2=CH_2)$ ($S = 1$) and free olefin were also examined, as was the coordination of ethylene to $(CO)_3Fe(\eta^2\text{-}H_2)$. The reductive elimination of ethane from $(CO)_3Fe(CH_2CH_3)H$ was also computed and found to be exothermic, even in the absence of added olefin.

Harris and coworkers re-examined iron carbonyl-catalyzed alkene isomerization [67]. Using nano- through microsecond time-resolved infrared spectroscopy and DFT calculations on the model system, $(CO)_4Fe(\eta^2\text{-}1\text{-hexene})$, the mechanism for catalytic olefin isomerization was revised to include a spin crossover process to singlet $(CO)_3Fe(\eta^2\text{-}1\text{-hexene})$ which undergoes C–H activation in 5–25 ns. While these comprehensive studies over the past few decades have thoroughly established the complex, competing pathways associated with $Fe(CO)_5$-photocatalytic isomerization and hydrogenation, using this finding to develop synthetically useful transformations used by practising synthetic chemists seems daunting.

4.2.2
Iron Phosphine Compounds

Aside from the thermal and photochemical activation of $Fe(CO)_5$ and related iron carbonyl compounds, the field of iron catalyzed hydrogenation lay mostly dormant for decades. As outlined in the previous section, comprehensive follow-up papers on $Fe(CO)_5$ chemistry had appeared but few new metal–ligand platforms for cata-

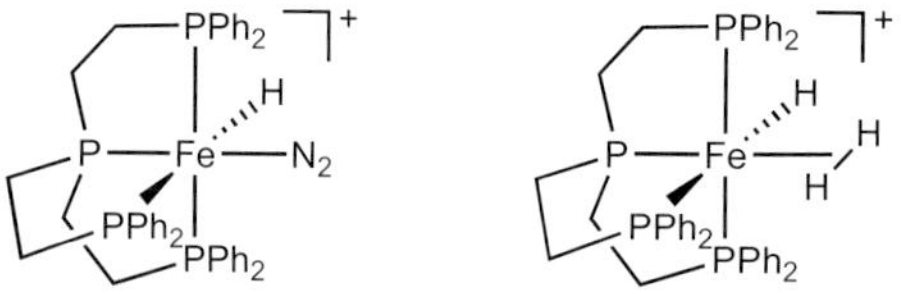

Figure 4.2 Bianchini's catalyst precursors, [PP$_3$Fe(H)(L)]BPh$_4$.

lytic hydrogenation had been introduced or explored. Transfer hydrogenation of dienes was reported in the presence of 10 mol% (Ph$_3$P)$_2$FeCl$_2$ at high temperatures, using polyhydroxybenzenes as the hydrogen source [68]. Tajima and Kunioka also reported that activation of [CpFe(CO)$_2$Cl] with AlEt$_3$ furnished a catalyst active for the partial hydrogenation of dienes [69]. In 1989, Bianchini renewed interest in the field with the communication of selective terminal alkyne hydrogenation to alkenes with phosphine-supported iron(II) dihydrogen and dinitrogen complexes (Figure 4.2) [70]. In the absence of H$_2$, the reductive dimerization of HC≡CSiMe$_3$ to 1,4-bis(trimethylsilyl)butadiene was observed. In a subsequent full paper [71] the scope of the alkyne hydrogenation reaction was examined more carefully with a family of terminal alkynes, RC≡CH (R = Ph, SiMe$_3$, ^{n}Pr, ^{n}Pent, CH=CH(OMe)). The rate of conversion slowed with the increased steric bulk of the alkyne substituent, and in each case selective reduction to the alkene occurred. The selectivity for alkene over alkane formation was rationalized by the reluctance of the iron hydride to insert olefins. Small alkenes such as ethylene were shown to coordinate to the metal center, but no insertion was observed even under forcing conditions.

More detailed studies to explore the mechanism of [PP$_3$Fe(H)(L)]BPh$_4$ (PP$_3$ = P(CH$_2$CH$_2$PPh$_2$)$_3$; L = H$_2$, N$_2$)-catalyzed alkyne hydrogenation were performed [71]. Kinetic studies on the hydrogenation of PhC≡CH to styrene with both the iron dihydrogen and dinitrogen precursors were carried out in dichloroethane and monitored using a gas uptake apparatus. A rate law first order in iron complex, first order in substrate and zero order in dihydrogen was established. The temperature dependence of the rate constants was also measured and yielded activation parameters of $\Delta H^\ddagger = 11.2(8)$ kcal mol^{-1} and $\Delta S^\ddagger = -27(3)$ e.u. for the N$_2$ precursor, and $\Delta H^\ddagger = 12(2)$ kcal mol^{-1} and $\Delta S^\ddagger = -24(6)$ e.u. for the dihydrogen complex. The similarity of the values suggests a common mechanism shared between the two different catalyst precursors (Scheme 4.3), a notion corroborated by the observation of rapid conversion of the N$_2$ complex to the dihydrogen compound upon addition of H$_2$. Attempts to observe paramagnetic intermediates by EPR spectroscopy were unsuccessful and the dihydrogen complex was the only NMR observable compound during turnover. The activation parameters are consistent with substrate coordination as the rate-determining step followed by insertion of the alkyne into the iron hydride.

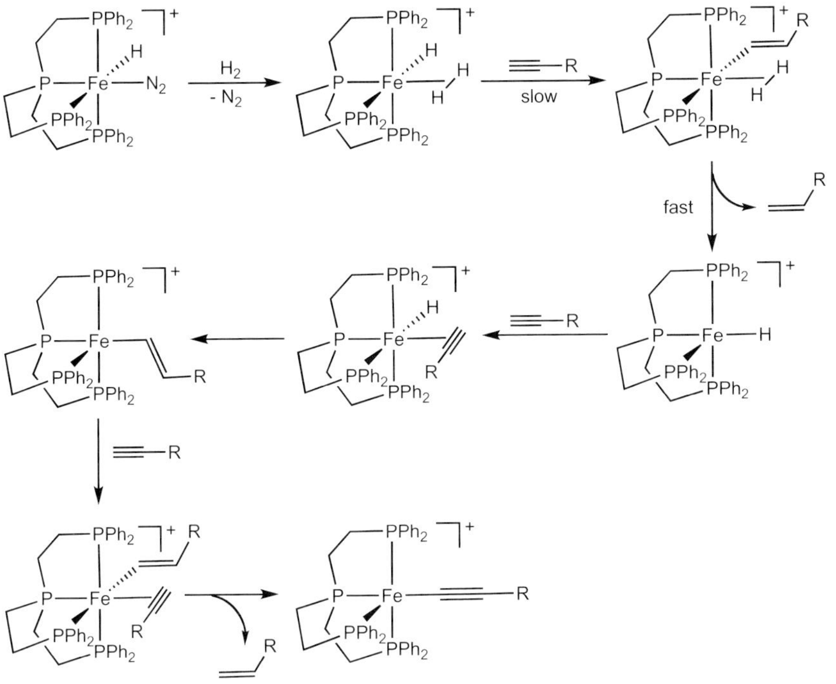

Scheme 4.3 Proposed mechanism for [PP$_3$Fe(H)(L)]BPh$_4$ alkyne hydrogenation.

Despite the ubiquity of iron phosphine complexes, few aside from the Bianchini compounds have been shown to be active for hydrogenation catalysis. In 2004, Daida and Peters [72] reported a series of pseudo-tetrahedral iron(II) alkyl complexes, [PhBP$_3^{iPr}$]Fe-R (PhBP$_3^{iPr}$ = [PhB(CH$_2$P^{i}Pr$_2$)$_3$]$^-$, R = CH$_3$, CH$_2$Ph, CH$_2$CMe$_3$), that are readily hydrogenated in the presence of various phosphines to yield the corresponding Fe(IV) trihydride phosphine complexes (Scheme 4.4).

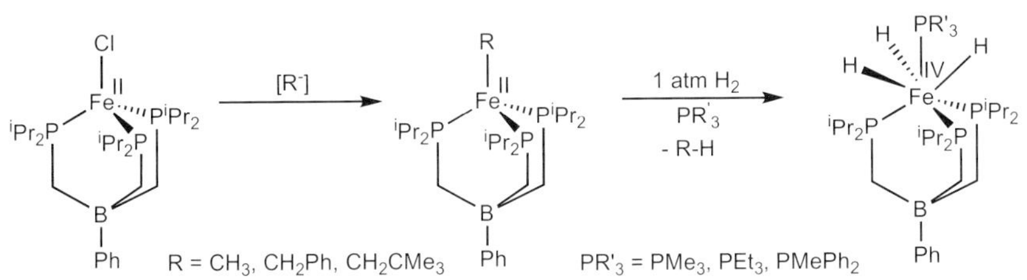

Scheme 4.4 Hydrogenation of [PhBP$_3^{iPr}$]Fe-R in the presence of phosphines.

Treatment of the PMe$_3$ derivative with 1 atm of ethylene yielded the iron(II) ethyl complex, [PhBP$_3^{iPr}$]FeCH$_2$CH$_3$, along with ethane and free phosphine. The observation of free alkane suggested that iron(II) alkyls or the iron(IV) trihydride complexes could serve as olefin hydrogenation catalysts. The iron(II) methyl, as well as three different phosphine trihydrides, [PhBP$_3^{iPr}$]FeH$_3$(PR$_3'$) (PR$_3'$ = PMe$_3$, PEt$_3$, PMePh$_2$), were evaluated for the catalytic hydrogenation of styrene. Under 1 atm of H$_2$ at ambient temperature and 10 mol% of iron complex, turnover frequencies (determined by ^{1}H NMR spectroscopy) between 0.5 and 7.7 h^{-1} were observed. The iron alkyls proved to be more active than the phosphine trihydride complexes, a result of competitive phosphine coordination. In addition to styrene, the catalytic hydrogenation of ethylene, 1-hexene and cyclooctene was also reported. For 1-hexene, olefin isomerization was competitive with reduction to the alkane. Studies using higher H$_2$ pressures in an attempt to suppress the isomerization reaction were not described.

Alkyne hydrogenation was also studied. With [PhBP$_3^{iPr}$]FeH$_3$(PMe$_3$), 2-pentyne was successfully hydrogenated to pentane with a small amount of cis-2-pentene observed by ^{1}H NMR spectroscopy as an intermediate during turnover. Attempts to hydrogenate terminal alkynes such as phenylacetylene proved problematic as competing reductive dimerization, trimerization and cyclotrimerization processes were identified.

Daida and Peters also conducted a series of experiments aimed at elucidating the mechanism of iron-catalyzed hydrogenation and provided additional support for the competency of iron(IV) hydride species. Deuteration of norbornene was carried out with [PhBP$_3^{iPr}$]FeH$_3$(PMe$_3$) and *exo*, *exo*-2,3-d_2-norbornane was identified as the sole product, as expected for a *cis*, *syn* addition of D$_2$ [73]. Effort was also devoted to determining the catalyst resting state and turnover-limiting step. Using styrene hydrogenation as a representative example, preliminary kinetic studies established a rate law zero order in olefin and first order with respect to both dihydrogen and the iron precursor. These data, in combination with the independent synthesis of [PhBP$_3^{iPr}$]FeCH$_2$CH$_2$Ph from the corresponding trihydride, support the iron(II) alkyls as the catalytic resting state and the hydrogenation of this species as the turnover-limiting step. When phosphine is present, as in the case of [PhBP$_3^{iPr}$]FeH$_3$(PMePh$_2$), ^{31}P NMR spectroscopic studies support displacement of phosphine by the incoming olefin as the turnover-limiting step.

One challenge in this chemistry was establishing the identity of the reactive iron hydride species. While the classical iron(IV) trihydride, [PhBP$_3^{iPr}$]FeH$_3$, was detected in solution, conversion to a reactive iron(II) dihydrogen hydride complex, [PhBP$_3^{iPr}$]Fe(H$_2$)H, similar to Bianchini's results, is also plausible. Based on numerous studies with the [[PhBP$_3^{iPr}$]Fe] platform [74], the authors favor an iron(II)/iron(IV) redox couple for the catalytic cycle involving a traditional oxidative addition, olefin insertion and reductive elimination sequence. This work should encourage future workers to explore the utility of the Fe(II)–Fe(IV) couple, in addition to the more traditional Fe(0)–Fe(II) approach, in catalytic hydrogenation chemistry.

4.2.3
Bis(imino)pyridine Iron Complexes

Inspired by the thermal and photochemical alkene isomerization, hydrosilylation and hydrogenation methods reported with $Fe(CO)_5$, our laboratory initiated a program aimed at developing easily synthesized, readily tuned iron hydrogenation catalysts that function under mild conditions in solution. Several practical criteria guided ligand design. Because mechanistic hypotheses and computational data implicated $[Fe(CO)_3]$ as a key catalytic intermediate [65, 75], π-acidic tridentate chelates were targeted. Aryl-substituted bis(imino)pyridine ligands (PDI), 2,6-$(ArN=CMe)_2C_5H_3N$ (Ar = aryl group), are ideal candidates as libraries of ligands that are readily accessible by straightforward aniline condensation with 2,6-diacetylpyridine [76]. In principle, the three-coordinate bis(imino)pyridine iron complex, [(PDI)Fe] is isolobal with $[Fe(CO)_3]$ (Scheme 4.5). Furthermore, Brookhart, Gibson and others had demonstrated the catalytic ethylene polymerization activity of bis(imino)pyridine iron dihalide complexes when treated with methylaluminoxane (MAO), suggesting that catalytic C–C bond-forming reactions in macromolecular synthesis were at least possible [77, 78] (see Chapter 5).

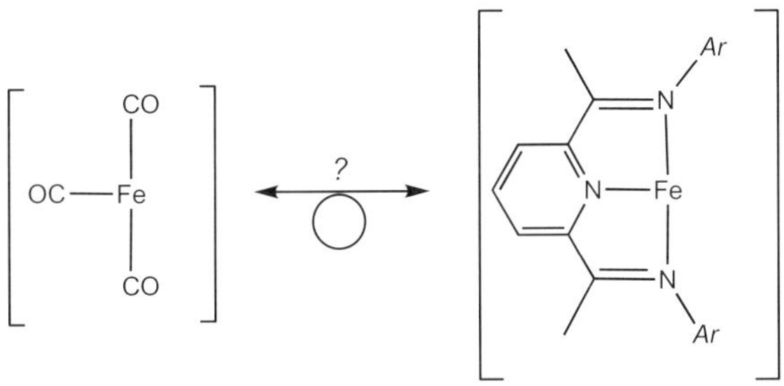

Scheme 4.5 Isolobal relationship between $[Fe(CO)_3]$ and [(PDI)Fe].

One additional challenge we anticipated in developing iron-catalyzed olefin hydrogenation and other classes of reduction reactions was coaxing the first row transition metal to participate in two-electron redox changes (e.g., oxidative addition/reductive elimination) in lieu of one-electron and potentially deleterious radical processes. Previous work by Toma [79], Wieghardt [80] and later Gambarotta and Budzelaar [81], firmly established the ability of this class of chelate to accept anywhere between one and three electrons. Such complexes with low formal oxidation states can be deceiving. For example, Gambarotta has prepared $(^{iPr}PDI)AlMe_2$, a complex that may first appear to be Al^{2+} but is best described as Al^{3+} with a bis(imino)pyridine radical anion [82].

Based on the potential isolobal relationship between $[Fe(CO)_3]$ and $[(^{iPr}PDI)Fe]$, we sought to prepare a formally iron(0) bis(imino)pyridine complex with a thermally labile ligand. Dinitrogen seemed an ideal target ligand that could readily be replaced by alkene or dihydrogen during a potential catalytic cycle. Sodium amalgam reduction of (^{iPr}PDI)FeBr$_2$ under a dinitrogen atmosphere furnished the desired bis(dinitrogen) complex, (^{iPr}PDI)Fe(N$_2$)$_2$ (Scheme 4.6) [83]. More comprehensive investigations into the electronic structure using a combination of NMR, Mössbauer and infrared spectroscopy augmented by DFT calculations, established an intermediate spin iron(II) compound complexed by a two-electron reduced, triplet diradical bis(imino)pyridine chelate [84]. Antiferromagnetic coupling of the chelate diradical with the intermediate spin, $S = 1$ iron center produces the experimentally observed $S = 0$ ground state. The computational studies also indicated a nearly isoenergetic $S = 1$ excited state whereby the intermediate spin iron center is complexed by a singlet chelate diradical. Mixing of this state with the ground state via spin–orbit coupling gives rise to temperature independent paramagnetism, a phenomenon detected by NMR spectroscopy for this and many other bis(imino)pyridine iron compounds with neutral ligands [84, 85]. Importantly, during the synthesis of (^{iPr}PDI)Fe(N$_2$)$_2$ from the corresponding iron dihalide complex, the formal two-electron reduction occurs at the bis(imino)pyridine ligand rather than the iron center (Scheme 4.6).

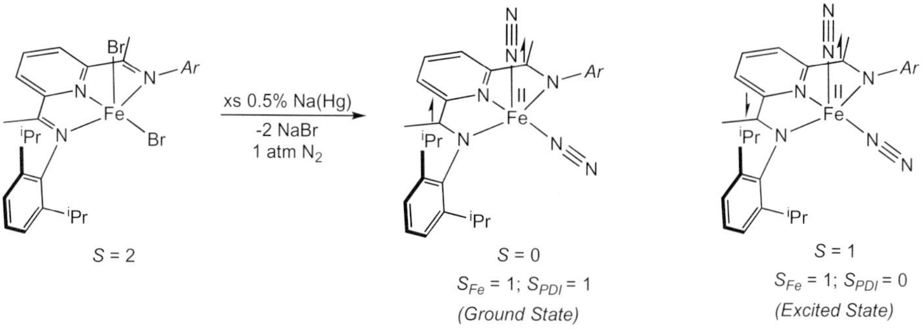

Scheme 4.6 Synthesis and electronic structure of (^{iPr}PDI)Fe(N$_2$)$_2$.

Studies into the solution behavior of (^{iPr}PDI)Fe(N$_2$)$_2$ demonstrated that one of the dinitrogen ligands is labile [83], offering promise for catalytic hydrogenation activity under mild thermal conditions. Indeed, (^{iPr}PDI)Fe(N$_2$)$_2$ is active for the hydrogenation of a variety of olefins under conditions readily accessed by most synthetic laboratories. Our initial report focused on the hydrogenation of simple unactivated alkenes. In these studies, each catalytic reaction was conducted using a 1.25 M solution of olefin in toluene using 0.3 mol% (^{iPr}PDI)Fe(N$_2$)$_2$ under 4 atm of H$_2$ at 22 °C (Scheme 4.7). These procedures are also effective for the hydrogenation of internal alkynes to alkanes. Cis-alkenes were observed as intermediates but could not be isolated in pure form. Importantly, the catalytic hydrogenation and hydrosilylation reactions with (^{iPr}PDI)Fe(N$_2$)$_2$ exhibit some of the highest turnover frequencies reported to date with an iron catalyst.

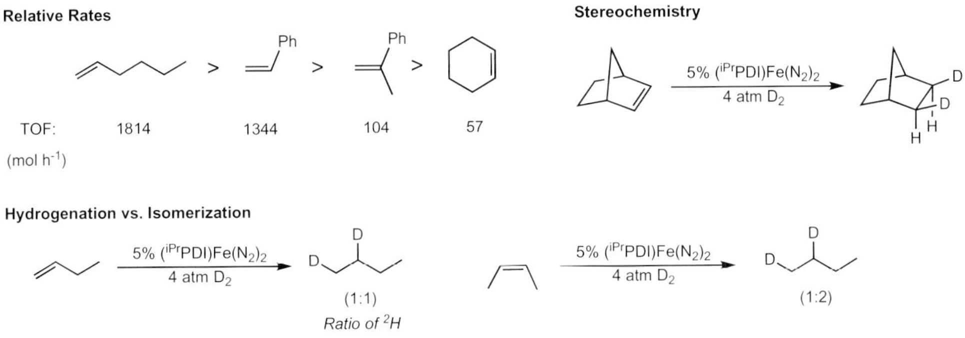

Relative Rates

TOF: 1814 > 1344 (Ph) > 104 (Ph) > 57 (cyclohexene)
(mol h⁻¹)

Stereochemistry

norbornene →[5% (^iPr^PDI)Fe(N₂)₂, 4 atm D₂]→ exo,exo-2,3-d₂-norbornane

Hydrogenation vs. Isomerization

1-butene →[5% (^iPr^PDI)Fe(N₂)₂, 4 atm D₂]→ D-CH₂-CHD-CH₂-CH₃ (1:1) Ratio of ²H

cis-2-butene →[5% (^iPr^PDI)Fe(N₂)₂, 4 atm D₂]→ D-CH₂-CHD-CH₂-CH₂D (1:2)

Scheme 4.7 Catalytic hydrogenation of unactivated olefins with (^iPr^PDI)Fe(N₂)₂.

As is well precedented with various precious metal catalysts, the relative rate of hydrogenation is largely dictated by steric effects, where less hindered olefins are reduced more rapidly than their more crowded counterparts. Attempts to hydrogenate more hindered, unfunctionalized alkenes such as 1-methylcyclohexene and tetramethylethylene have been unsuccessful. Preparative-scale hydrogenations of alkenes containing less substituted double bonds were carried out in neat olefin using *only* 40 ppm of the iron compound. The catalytic reactions were repeated several times and the catalyst recycled as long as anhydrous conditions were maintained. The high activity of (^iPr^PDI)Fe(N₂)₂ in non-polar media contrasts the behavior of most traditional precious metal catalysts where polar solvents are often required for optimal activity [86]. Solvent-free reactions are advantageous as they obviate the need for solvent–product separation and reduce waste.

A series of isotopic labeling experiments was also conducted in order to gain insight into the stereochemistry of the H_2 addition as well as to gauge potential competing isomerization reactions. Catalytic addition of D_2 to norbornene using 5 mol% of (^iPr^PDI)Fe(N₂)₂ yielded *exo, exo*-2,3-d_2-norbornane exclusively, consistent with *syn* addition of D_2 (H_2). Competing isomerization reactions were probed by deuteration of both 1- and *cis*-2-butene. No competitive isomerization was observed for D_2 addition to the terminal olefin. In contrast, deuterium was also detected in the methyl group of the butane produced from the deuteration of *cis*-2-butene, demonstrating that alkyl migration of the intermediate iron secondary alkyl was competitive with hydrogenation under 4 atm of D_2 (Scheme 4.7).

Attempts were also made to observe or isolate catalytically relevant intermediates. Addition of one equivalent of PhC≡CPh to (^iPr^PDI)Fe(N₂)₂ yielded a red paramagnetic solid identified as the iron η^2-alkyne complex, which was also characterized by X-ray diffraction. Treatment of the isolated product with 1 atm of H_2 initially yielded stilbene, followed by 1,2-diphenylethane, establishing its catalytic competency (Scheme 4.8). Our laboratory has recently conducted experiments to elucidate the stability of related bis(imino)pyridine olefin complexes [87]. Treatment of (^iPr^PDI)Fe(N₂)₂ with 1-hexene under a dinitrogen atmosphere resulted in slow conversion to n-hexane and the paramagnetic, NMR-silent intramolecular

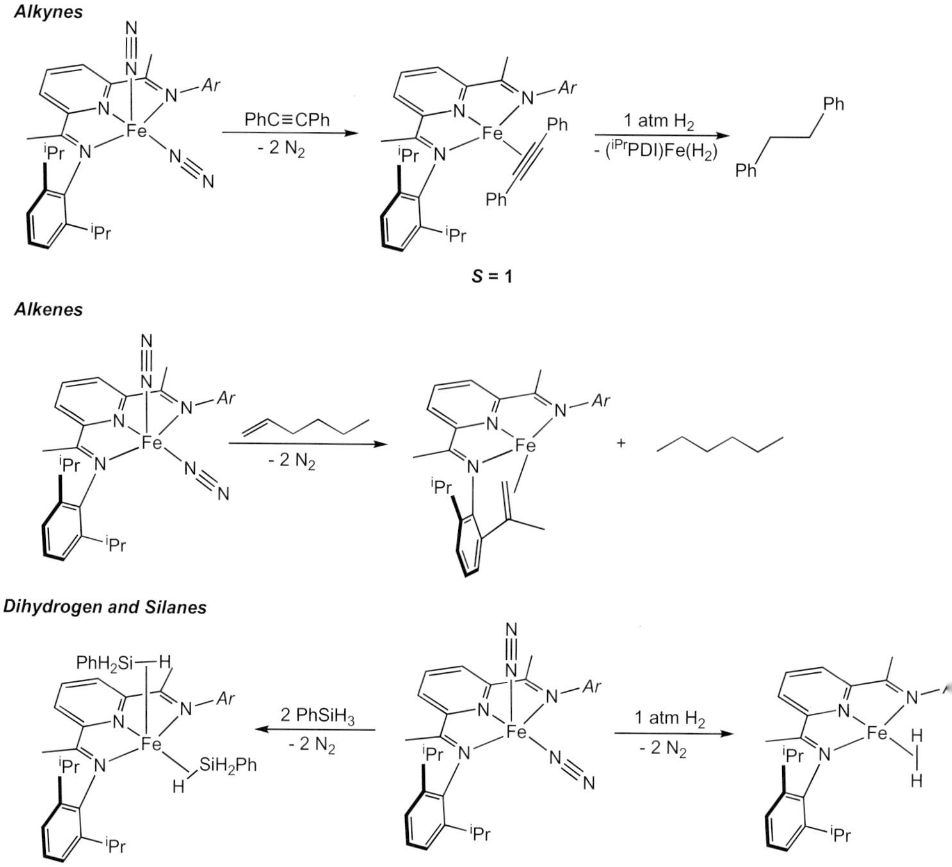

Scheme 4.8 Synthesis and characterization of catalytically competent intermediates in bis(imino)pyridine iron-catalyzed hydrogenation and hydrosilylation reactions.

olefin compound arising from dehydrogenation of one of the isopropyl aryl substituents.

Exposure of a benzene-d_6 solution of (^{iPr}PDI)Fe(N$_2$)$_2$ to 1 atm of H$_2$ produced a color change from green to brown, signaling formation of the tentatively assigned bis(imino)pyridine iron dihydrogen complex, (^{iPr}PDI)Fe(η^2-H$_2$). Definitive characterization of this compound has proven extremely challenging as even trace amounts of N$_2$ induced reversion to (^{iPr}PDI)Fe(N$_2$)$_2$. However, Toepler pump experiments on the isolated compound clearly establish 1 equiv of coordinated H$_2$. In contrast, addition of 2 equiv of PhSiH$_3$ yielded the bis(silane) complex, (^{iPr}PDI)Fe(H-SiH$_2$Ph)$_2$, which was isolated as a green crystalline solid that was fully characterized as a silane σ-complex by X-ray diffraction and multinuclear NMR spectroscopy. The strong affinity for alkyne binding and olefin transfer hydrogenation supports a mechanism involving an unsaturated iron complex, whereby substrate coordination precedes H$_2$ addition. However, additional experiments, currently ongoing in our laboratory, are needed to further substantiate this hypothesis.

More recent efforts in our laboratory have focused on exploring the substrate scope of bis(imino)pyridine iron catalyzed alkene hydrogenation [88]. Allyl amines, $CH_2=CHCH_2NR_2$, were readily hydrogenated with 5 mol% of $(^{iPr}PDI)Fe(N_2)_2$ and the observed turnover frequencies increase with the steric protection on the nitrogen atom (Scheme 4.9). Alkoxy-substituted alkenes such as ethyl vinyl ether, ethyl allyl ether and diallyl ether are hydrogenated with turnover frequencies indistinguishable from α-olefins. For carbonyl-substituted alkenes, the hydrogenation activity was exquisitely sensitive to the position and type of functional group. Esters such as dimethyl itaconate and *trans*-methyl cinnamate were effectively hydrogenated while ketones such as 5-hexen-2-one required heating to 65 °C. Others, such as allyl and vinyl acetate, underwent such rapid C–O bond cleavage that no turnover was observed with 4 atm of H_2 [89].

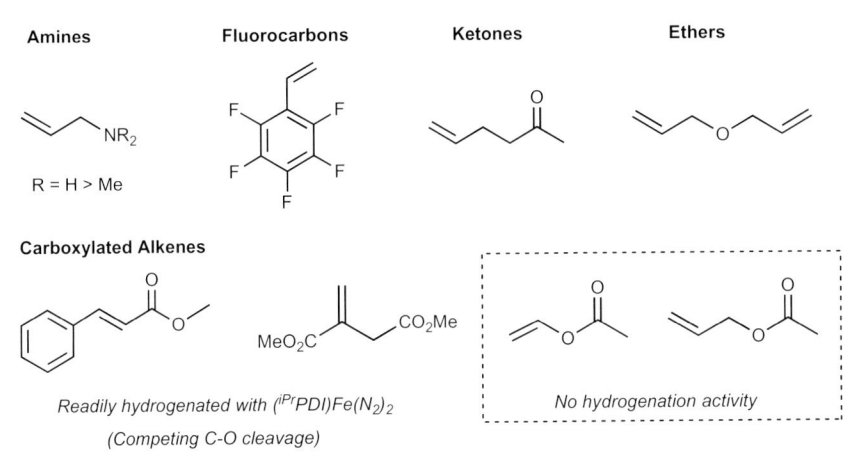

Scheme 4.9 Functional group tolerance in $(^{iPr}PDI)Fe(N_2)_2$-catalyzed olefin hydrogenation.

Stoichiometric experiments were conducted with the functionalized alkenes to further explore iron–substrate interactions and gauge relative coordination affinities. Several bis(imino)pyridine iron amine and ketone compounds were isolated and studied using a combination of X-ray diffraction, NMR and Mössbauer spectroscopy and established electronic structures similar to that for $(^{iPr}PDI)Fe(N_2)_2$. In the absence of H_2, diallyl ether and allyl ethyl ether underwent facile C–O bond cleavage and yielded a near equimolar mixture of the corresponding iron allyl and alkoxide complexes [89]. Our group has recently published a comprehensive study on these types of reactions and discovered rare examples of C–O bond cleavage in saturated esters [89].

Given the wealth of interesting reactivity observed with $(^{iPr}PDI)Fe(N_2)_2$, attention has been devoted to improving the performance of bis(imino)pyridine iron compounds in catalytic hydrogenation. Challenges inspiring catalyst design are the hydrogenation of hindered, unactivated olefins, improvement in overall activity, increased functional group tolerance and development of enantioselective reactions. In an attempt to improve activity, bis(imino)pyridine iron compounds bearing smaller aryl substituents are under investigation. Sodium amalgam reduction of a toluene or pentane slurry of the ethyl rather than isopropyl variant, (^{Et}PDI)FeBr$_2$, resulted in isolation of the bis(chelate) iron complex, (^{Et}PDI)$_2$Fe (Scheme 4.10) [90]. The solid state structure, established by X-ray diffraction, is a four-coordinate complex containing two κ^2-bis(imino)pyridine chelates. This reduction protocol appears general as several examples of bis(chelate)iron compounds have now been synthesized and crystallographically characterized [90]. For those compounds with smaller alkyl or aryl imino substituents, six-coordinate complexes were formed. Elucidation of the electronic structure of both the four- and six-coordinate examples established high spin iron(II) centers anti-ferromagnetically coupled to two one-electron reduced bis(imino)pyridine anions.

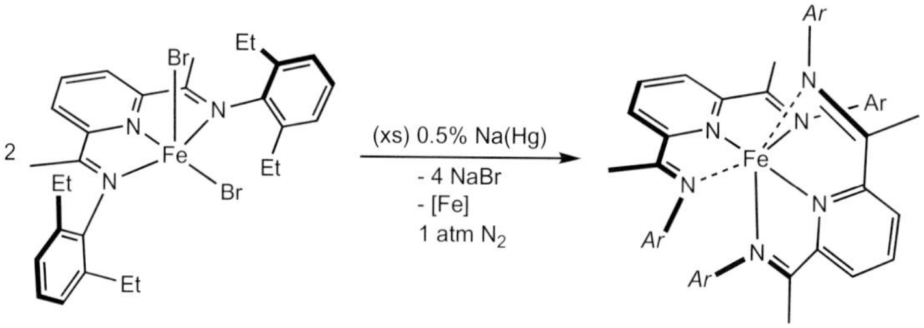

Scheme 4.10 Synthesis of bis(imino)pyridine iron bis(chelate) complexes.

Altering the bis(imino)pyridine chelate backbone while maintaining the 2,6-diisopropyl aryl substituents has yielded another catalytically active iron dinitrogen complex. In-plane ligand modifications are of interest to explore the influence of redox potential (e.g., redox-activity) on catalytic performance. Synthesis of (^{iPr}BPDI)Fe(N$_2$)$_2$ (^{iPr}BPDI = 2,6-(2,6-^{i}Pr$_2$-C$_6$H$_3$N=CPh)$_2$C$_5$H$_3$N) was accomplished using the standard sodium amalgam reduction method at 4 atm of N$_2$ (Scheme 4.11) [91]. Comparing the productivity of (^{iPr}BPDI)Fe(N$_2$)$_2$ to (^{iPr}PDI)Fe(N$_2$)$_2$ in the hydrogenation of 1-hexene established higher turnover frequencies for the phenylated catalyst. However, (^{iPr}BPDI)Fe(N$_2$)$_2$ produced slower rates for more hindered olefins such as cyclohexene and (+)-(R)-limonene. This discrepancy arises from competitive catalyst deactivation by irreversible formation of η^6-aryl and η^6-phenyl compounds with the [(^{iPr}BPDI)Fe] compound (Scheme 4.11).

The competition for coordination of the different arenes was examined as a function of solvent. For relatively non-coordinating solvents such as pentane, hexane, diethyl ether and mesitylene, exclusive formation of the η^6-aryl compound was observed. In more coordinating media, such as benzene, toluene, THF or cyclohexene, increased amounts of the η^6-phenyl complex were obtained. These

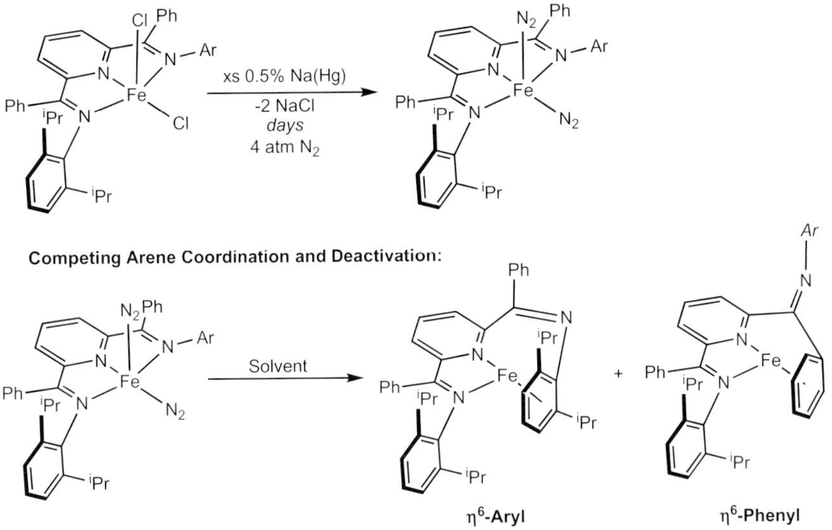

Scheme 4.11 Synthesis, reactivity and catalytic activity of $(^{iPr}BPDI)Fe(N_2)_2$.

coordination preferences were rationalized on the basis of donor solvents to stabilize the κ^2-bis(imino)pyridine intermediate involved in η^6-phenyl product formation. The observation of η^6-phenyl product in the presence of cyclohexene raises the question as to whether such κ^2-bis(imino)pyridine iron complexes play a significant role in the catalytic hydrogenation cycle.

4.2.4
α-Diimine Iron Complexes

Inspired by the observation of efficient hydrogenation catalysis with iron compounds bearing redox-active tridentate ligands, our laboratory also explored related chemistry with reduced iron compounds with bidentate chelates [92]. Aryl-substituted α-diimines were attractive supporting ligands due to their ease of synthesis, established one- and two-electron redox-activity [92–95] and precedent in iron-catalyzed C–C bond forming reactions [96–98].

Mimicking the conditions used to prepare $(^{iPr}PDI)Fe(N_2)_2$, reduction of a toluene or benzene slurry of $(^{iPr}DI)FeCl_2$ ($^{iPr}DI = ArN=C(Me)C(Me)=NAr$; $Ar = 2,6\text{-}^{i}Pr_2\text{-}C_6H_3$) with excess 0.5% sodium amalgam under 1 atm of N_2 did not yield a dinitrogen complex but rather the robust η^6-arene complexes, $(^{iPr}DI)Fe(\eta^6\text{-}C_6H_5R)$ (R = Me, H), were isolated as bright red–orange, diamagnetic crystalline solids in high yield (Scheme 4.12) [92]. Performing the reduction in diethyl ether or saturated hydrocarbons furnished the tetrahedral bis(α-diimine) compound, $(^{iPr}DI)_2Fe$. While these complexes exhibit interesting spectroscopic features that further establish the redox-activity of this ligand class [92, 93], they are inactive for olefin hydrogenation.

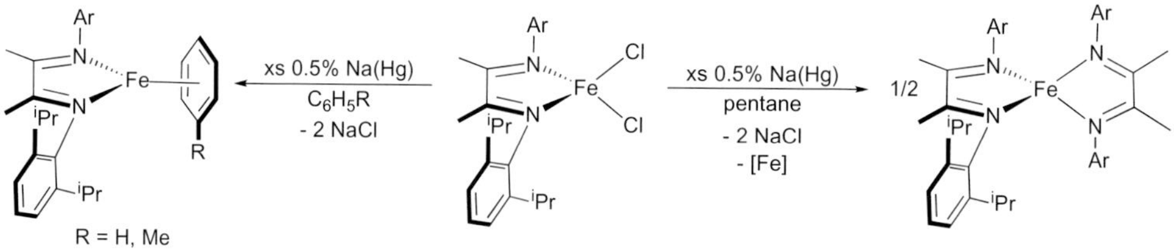

Scheme 4.12 Reduction chemistry of α-diimine iron complexes.

The inactivity of the arene and bis(chelate) complexes prompted the synthesis of (α-diimine)iron complexes with neutral ligands that could be readily removed by hydrogenation. Inspired by well-established 1,5-cyclooctadiene and norbornadiene complexes of rhodium and iridium and their utility as catalyst precursors for olefin hydrogenation [5], these unsaturates were explored as potential supporting ligands. Sodium amalgam reduction of a pentane slurry of (^{iPr}DI)FeCl$_2$ in the presence of PhC≡CPh furnished the $S = 1$ (α-diimine)iron η^2-alkyne complex, (^{iPr}DI)Fe(η^2-PhCCPh). Interestingly, gently warming a benzene-d_6 solution of (^{iPr}DI)Fe(η^2-PhCCPh) to 60 °C for 2 days resulted in isomerization to the catalytically inactive, diamagnetic η^6-diphenylacetylene complex, (^{iPr}DI)Fe(η^6-PhCCPh) (Scheme 4.13). Because no (^{iPr}DI)Fe(η^6-C$_6$D$_6$) complex was observed, the rearrangement appears to be intramolecular. To avoid this complication, (^{iPr}DI)Fe(η^2-RCCR) where R = Me, SiMe$_3$ were synthesized and isolated as paramagnetic red–brown solids.

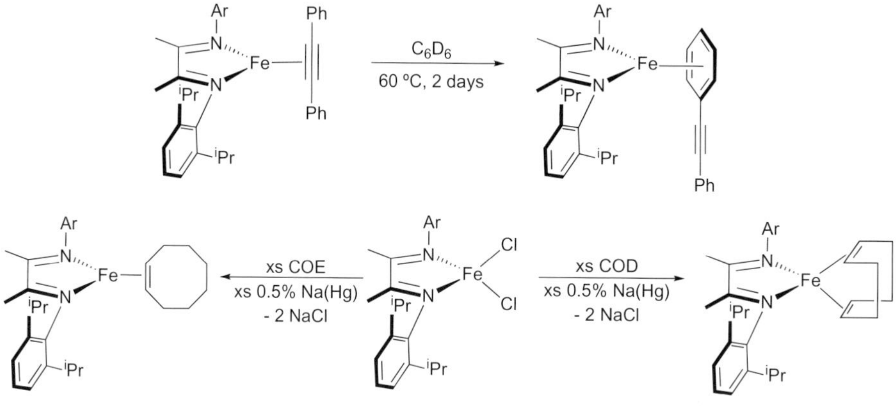

Scheme 4.13 α-Diimine iron acetylene and olefin complexes.

α-Diimine iron complexes supported by olefins such as 1,5-cyclooctadiene (COD) and cyclooctene (COE) were prepared by sodium amalgam reduction of the ferrous dichloride precursor. Excess olefin, typically five equivalents, was necessary to prevent formation of bis(chelate) iron complexes. Both (^{iPr}DI)Fe(COD) and (^{iPr}DI)Fe(COE) are paramagnetic with magnetic moments consistent with $S = 1$ ground states. The mono(olefin) complex, (^{iPr}DI)Fe(COE), can be synthesized by partial hydrogenation of (^{iPr}DI)Fe(COD) in benzene-d_6. Continued hydrogenation resulted in isolation of the η^6-arene complex, (^{iPr}DI)Fe(η^6-C_6D_6), with liberation of cyclooctane, suggesting that these complexes may be useful precursors for catalytic hydrogenation reactions.

The activity of the α-diimine iron acetylene and olefin complexes was assayed using the catalytic hydrogenation of 1-hexene as a model substrate. Each catalytic reaction was conducted with a 1.25 M solution of 1-hexene in pentane containing 0.3 mol% of the desired iron complex with 4 atm of dihydrogen. The alkyne complexes were poor catalytic precursors; only partial conversion to hexane was observed after heating at 60 °C for 24 h. A significant (~12%) amount of isomerization to internal hexenes accompanies the hydrogenation reaction. Both (^{iPr}DI)Fe(COD) and (^{iPr}DI)Fe(COE) are more efficient catalyst precursors for 1-hexene hydrogenation with turnover frequencies of 90 mol h^{-1}. The turnover frequencies between the two compounds are indistinguishable because the hydrogenation of (^{iPr}DI)Fe(COD) to (^{iPr}DI)Fe(COE) was fast relative to turnover. Importantly, every α-diimine iron complex examined to date exhibits inferior catalytic activity compared to the catalysts containing the tridentate bis(imino)pyridine ligand. These results, along with those from the (^{iPr}BPDI)Fe(N$_2$)$_2$, demonstrate that the activity is not the only criterion to be considered in catalyst design. While faster turnover may initially be observed, deactivation by rapid formation of bis(chelate) or η^6-arene species results in an overall less productive catalyst.

4.3
Carbonyl Hydrogenation

4.3.1
Hydrosilylation

The metal-catalyzed reduction of carbonyl compounds by hydrogenation, hydrosilylation or transfer hydrogenation has enjoyed widespread application in organic synthesis and has been used industrially in the synthesis of fine chemicals, perfumes and important drug targets [13, 99]. As with most reactions of this type, precious metal catalysts, particularly those of Rh(I) and Ru(II), have traditionally dominated the catalytic landscape. However, compared to olefin hydrogenation, several base metal methods for catalytic carbonyl reduction have been developed. Highly active and selective titanium [100] and copper [101] catalysts for hydrosilylations, and, most recently, enantioselective organocatalytic methods [102] for hydrogenation of the C=C bonds of α,β-unsaturated ketones, have been reported.

Much like olefin hydrogenation, early work in iron-catalyzed carbonyl reduction focused on Fe(CO)$_5$ and related complexes. When mixed with various ligands such as terpyridines, porphyrins, aminophosphines and phosphines, [Fe$_3$(CO)$_{12}$] is active for the reduction of ketones to alcohols upon heating to 80–100 °C [103]. Over the past few years there has been a resurgence in iron-catalyzed reductions and several research groups from around the globe have initiated intense programs in this area. In 2006, Beller and coworkers described the transfer hydrogenation of ketones using iron salts and tridentate amine ligands using 2-propanol as the hydrogen source [104]. With iron porphyrin complexes, successful transfer hydrogenation of more challenging α-alkoxy- and α-aryloxy-substituted acetophenones was achieved [105]. Shortly after Beller's 2006 paper, Furata and Nishiyama reported that simple iron salts such as Fe(OAc)$_2$, in combination with various N-donor ligands and (EtO)$_3$SiH, are effective for ketone hydrosilylation at 65 °C in THF (Scheme 4.14) [106]. A subsequent study from the same group explored thiophene derivatives as supporting ligands [107]. Improved yields compared to the original TMEDA (N,N,N',N'-tetramethylethylenediamine) catalysts were observed for a wide spectrum of ketones. Beller and coworkers later modified this protocol with the introduction of phosphine ligands, most effectively PCy$_3$, resulting in catalysts for the hydrosilylation of aldehydes using PMHS (polymethylhydrosiloxane) [108]. Introduction of privileged C_2-symmetric chiral phosphines such as Me$_2$DuPhos produced high enantioselectivities for a select group of hindered ketones [109]. While impressive advances in Fe(OAc)$_2$-catalyzed reactions have been made in a very short amount of time, a general procedure for enantioselective reductions awaits discovery. In addition, little is known about the active species or the mechanism of catalysis for these transformations.

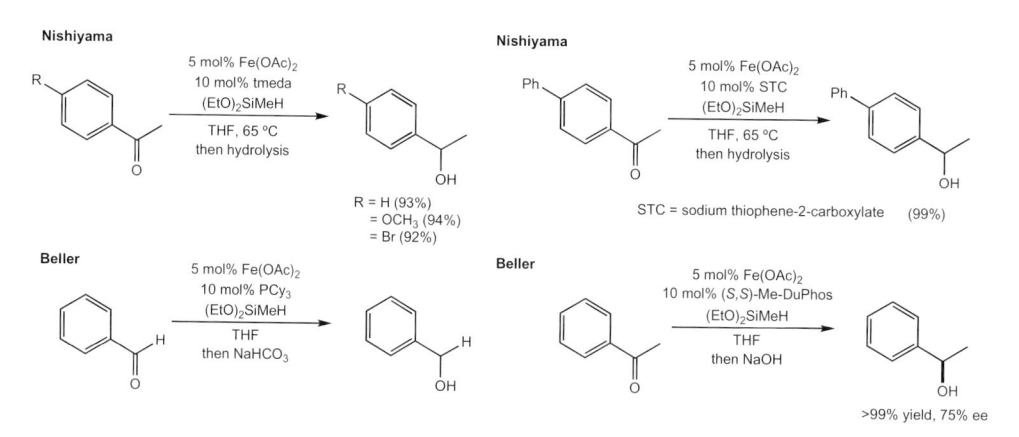

Scheme 4.14 Fe(OAc)$_2$-catalyzed methods for the hydrosilylation of aldehydes and ketones.

More well-defined iron compounds have also been used as catalyst precursors for aldehyde and ketone hydrosilylation. Nikonov [110] reported that [(η^5-C$_5$H$_5$)Fe(PR$_3$)(CH$_3$CN)$_2$]$^{2+}$ is an active hydrosilyation catalyst, although the substrate scope has yet to be explored. Using a ligand displacement reaction originally pioneered by Cámpora and coworkers [111], our laboratory reported that aryl- and alkyl-substituted bis(imino)pyridine iron dialkyl complexes are active for the hydrosilylation of both aldehydes and ketones using $PhSiH_3$ and Ph_2SiH_2 [112]. In contrast to the $Fe(OAc)_2$ chemistry, PMHS and $(EtO)_3SiH$ are ineffective. Broad functional group tolerance was observed including Me_2N-, F_3C- and CH_3O-substituted acetophenones. Unlike olefin hydrogenation with the bis(imino)pyridine iron dinitrogen complex where decomposition of the iron compound was observed, selective hydrosilylation of the carbonyl group in α, β-unsaturated ketones was also accomplished. In recent work, Gade and coworkers reported a class of chiral iron bis(pyridylimino) complexes for the enantioselective hydrosilylation of ketones with $(EtO)_2MeSiH$ [113]. Enantiomeric excesses up to 83% were observed for substituted acetophenone derivatives, while dialkyl ketones exhibited lower selectivity. Moreover, our laboratory has recently observed enantioselective carbonyl reduction with pyridine bis(oxazoline) iron dialkyl complexes [114].

4.3.2
Bifunctional Complexes

The tremendous success enjoyed by metal–ligand bifunctional catalysts in homogeneous carbonyl hydrogenation [7] has inspired the search for more environmentally benign and inexpensive iron equivalents [115]. Casey and Guan have recently reported a Shvo-type cyclopentadienyl iron hydride catalyst that efficiently hydrogenates ketones at ambient temperature and 3 atm of H_2 [116]. Selective reduction of ketones was achieved in the presence of alkenes (Scheme 4.15). The reactions are quantitative by NMR spectroscopy and isolated yields are slightly lower. While only preliminary mechanistic studies were reported, the reaction likely proceeds via an ionic hydrogenation where proton transfer from the cyclopentadienyl O–H group is coupled to delivery of the iron hydride. *In situ* infrared experiments

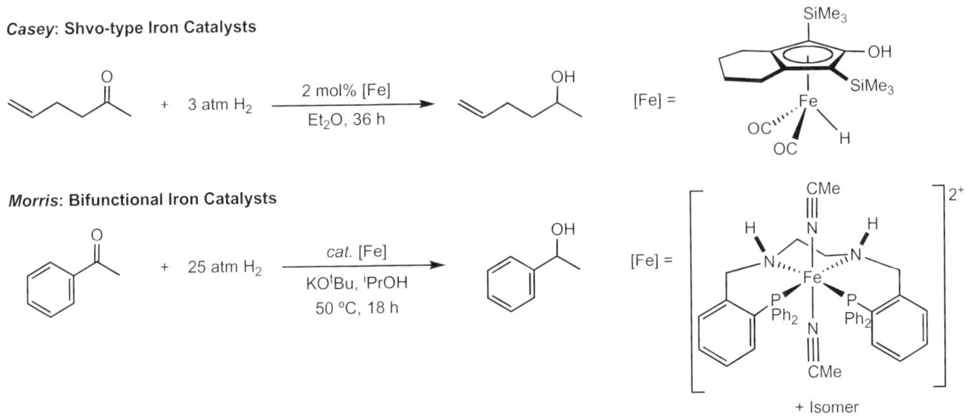

Scheme 4.15 Bifunctional iron catalysts for aldehyde and ketone hydrogenation.

suggest that hydrogen transfer is the turnover-limiting step in the catalytic cycle. While still in their infancy, complexes of this type offer tremendous promise for the future of iron hydrogenation catalysis.

Other efforts in bifunctional iron hydrogenation catalysis have been led by Bob Morris and his coworkers at the University of Toronto. This group played a pioneering and indispensable role in elucidating the ionic hydrogenation mechanism in the Noyori catalysts [14] and it is, therefore, not surprising they are instrumental in developing the growing field of iron catalysis. Dicationic iron(II) complexes bearing tetradentate diiminophosphine ligands were synthesized and isolated as the *trans*-bis(acetonitrile) adducts [117]. Under 25 atm of H_2 at 50 °C in the presence of KO^{t}Bu, certain examples were active for the hydrogenation of acetophenone with a TOF of ~5 h^{-1} and 27% enantiomeric excess. Variations in the electronic properties of the acetophenone were examined and it was found that electron-donating groups such as *p*-Me and *p*-OMe slowed turnover relative to the parent substrate while electron-withdrawing *p*-Cl accelerated it. Catalytic transfer hydrogenation of aldehydes, ketones and certain imines was also observed with another variant using ^{i}PrOH as the hydrogen source. Enantiomeric excesses as high as 71% were initially observed although higher values have recently been reported with a carbonyl-substituted catalyst [118]. It is interesting to note that the precatalysts used in this study lack N–H bonds on the ligand framework, suggesting that if a bifunctional mechanism were operative, hydrogenation of the coordinated imines to generate the active iron compound must occur.

In a recent, subsequent full paper [119], the synthesis of tetradentate ligands with N–H bonds and the preparation, and in some cases crystallographic characterization, of the corresponding iron dications were described. Iron complexes with N–H bonds in the diamine backbone were included in this study and found to be among the most active for acetophenone hydrogenation. In general, catalysts with hydrogens in the axial positions of the tetradentate ligand backbone were more active than those with isopropyl or phenyl substituents, likely due to steric crowding around the iron center. Standard conditions for each catalyst evaluation were 25 atm of H_2 at 50 °C with substrate to catalyst ratios of 225 : 1 and base to catalyst ratios of 25 : 1. In some cases, lower temperatures and hydrogen pressures could be used, albeit with reduced activity.

The mechanism of the catalytic hydrogenation was studied computationally using DFT and a truncated model complex. Key assumptions in the computed mechanism are that all of the iron compounds are low-spin iron(II) and hence diamagnetic and that the catalytic cycle is the same as previously computed for the corresponding ruthenium system [120]. The proposed mechanism is presented in Scheme 4.16. The computed activation barriers for the heterolytic splitting of H_2 between the iron and ruthenium complexes were indistinguishable with values of 20.8 and 20.9 kcal mol^{-1}, respectively. For both metals, the transition structure for H_2 splitting reveals that H–H elongation occurs simultaneously with a change in nitrogen–ligand geometry from trigonal planar to tetrahedral. If the barriers for H_2 splitting are nearly the same for both metals, why then is ruthenium much more active? The authors postulate that the hydrogen transfer step is the answer,

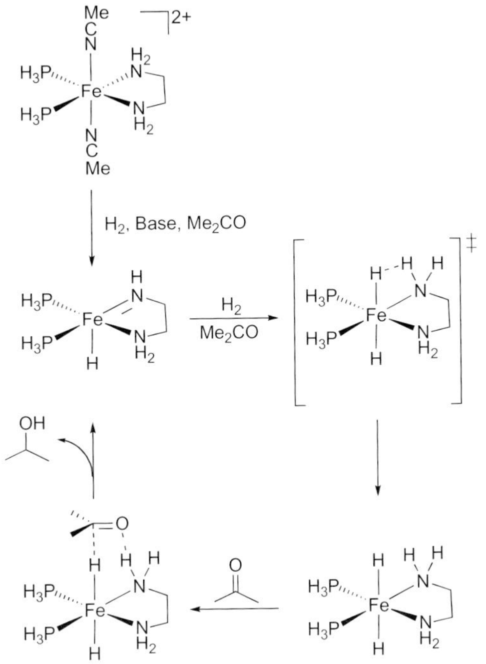

Scheme 4.16 Proposed mechanism for ketone hydrogenation with bifunctional iron complexes based on DFT calculations on a model complex.

with the ruthenium hydride being more "hydridic" (e.g., $H^{\delta-}$) in the transition state where the C–H and O–H bonds form. These bifunctional iron catalysts are still in their infancy but early work appears promising. Future exploration of various ligand architectures is likely to improve both the activity and selectivity and will, perhaps, make iron catalysts competitive with their venerable ruthenium counterparts.

4.4
Outlook

Catalysis with inexpensive, relatively environmentally benign first row transition metals is an area undergoing rapid growth in both inorganic and organometallic chemistry. Global efforts in sustainability, coupled with increasing prices and concerns over long-term supplies of precious metals, have inspired a flurry of activity in iron catalysis applied to a diverse array of catalytic bond forming methodologies. The past five years alone have witnessed an intense effort and many diverse catalytic reactions, ranging from hydrogenation to cross coupling to oxidations, have been studied.

This is only the beginning. While many exciting breakthroughs have been reported, only recently have new ligand architectures been explored. The stability, activity and selectivity of most iron olefin and carbonyl hydrogenation catalysts lags behind more traditional rhodium and ruthenium compounds. However, these latter metals have enjoyed over four decades of optimization made possible by the availability of excellent synthetic precursors, reliable reaction chemistry and predictable two-electron redox changes. As chemists move away from strong field carbonyl, cyclopentadienyl and phosphine ligands, the control of one-electron chemistry and manipulation of the electronic structure of the first row metal becomes more challenging and interesting. However, when the appropriate ligand environment is implemented, new and unexpected chemistry, perhaps unique to iron, can be uncovered. Thus the future of the field should not only focus on replacing existing precious metal technologies but also seek to develop new transformations specific to the first row metal and its variable electronic structures. Once these goals are realized iron may indeed undergo transmutation from base to precious metal status – at least in the minds of synthetic chemists! Only time will tell.

References

1 Osborn, J.A., Jardine, F.H., Young, J.F., and Wilkinson, G. (1966) *J. Chem. Soc. A*, 1711.
2 Knowles, W.S., Sabacky, M.J., and Vineyard, B.D. (1972) *Chem. Commun.*, 10.
3 Knowles, W.S. (2002) *Angew. Chem. Int. Ed.*, **41**, 1998.
4 Osborn, J.A., and Schrock, R.R. (1971) *J. Am. Chem. Soc.*, **93**, 2397.
5 Schrock, R.R., and Osborn, J.A. (1976) *J. Am. Chem. Soc.*, **98**, 2134.
6 Dang, T.P., and Kagan, H.B. (1971) *Chem. Commun.*, 481.
7 Noyori, R. (2002) *Angew. Chem. Int. Ed.*, **41**, 2008.
8 Burk, M.J. (2000) *Acc. Chem. Res.*, **33**, 363.
9 Crabtree, R.H. (1979) *Acc. Chem. Res.*, **12**, 331.
10 (a) Crabtree, R.H., and Davis, M.W. (1986) *J. Org. Chem.*, **51**, 2655; (b) Paquette, L., Peng, X., and Nondar, D. (2002) *Org. Lett.*, **4**, 937; (c) Barriault, L., and Deon, D.H. (2001) *Org. Lett.*, **3**, 1925; (d) Stork, G., and Kahne, D.E. (1983) *J. Am. Chem. Soc.*, **105**, 1072; (e) Hoveyda, E.F. (1993) *Chem. Rev.*, **93**, 1307.
11 Roseblade, S.J., and Pfaltz, A. (2007) *Acc. Chem. Res.*, **40**, 1402.
12 (a) Perry, M.C., Cui, X., Powell, M.T., Hou, D.-R., Reibenspies, J.H., and Burgess, K. (2003) *J. Am. Chem. Soc.*, **125**, 113; (b) Cui, X., and Burgess, K. (2003) *J. Am. Chem. Soc.*, **123**, 14212; (c) Cui, X., Ogle, J.W., and Burgess, K. (2005) *Chem. Commun.*, 672; (d) Zhou, J., Zhu, Y., and Burgess, K. (2007) *Org. Lett.*, **9**, 1391.
13 Ikariya, T., and Blacker, A.J. (2007) *Acc. Chem. Res.*, **40**, 1300.
14 (a) Noyori, R., and Ohkuma, T. (2001) *Angew. Chem. Int. Ed.*, **40**, 40; (b) Noyori, R., Kitamura, M., and Ohkuma, T. (2004) *Proc. Natl. Acad. Sci. U.S.A.*, **101**, 5356; (c) Clapham, S.E., Hadzovic, A., and Morris, R.H. (2004) *Coord. Chem. Rev.*, **248**, 2201; (d) Ikariya, T., Murata, K., and Noyori, R. (2006) *Org. Biomol. Chem.*, **4**, 393.
15 Evans, D., Osborn, J.A., Jardine, F.H., and Wilkinson, G. (1965) *Nature*, **208**, 1203.
16 Hallman, P.S., McGarvery, B.R., and Wilkinson, G. (1968) *J. Chem. Soc. A*, 3143.

17 Skapski, A.C., and Troughto, P.G. (1968) *Chem. Commun.*, 1230.
18 Nishimur, S., Ichino, T., Akimoto, A., and Tsuneda, K. (1973) *Bull. Chem. Soc. Jpn.*, **46**, 279.
19 Noyori, R., Ohta, M., Hsiao, Y., Kitamura, M., Ohta, T., and Takaya, H. (1986) *J. Am. Chem. Soc.*, **108**, 7117.
20 Takaya, H., Ohta, T., Sayo, N., Kumobayashi, H., Akutagawa, S., Inoue, S., Kasahara, I., and Noyori, R. (1987) *J. Am. Chem. Soc.*, **109**, 1596.
21 (a) Blum, Y., Czarkie, D., Rahamim, Y., and Shvo, Y. (1985) *Organometallics*, **4**, 1459; (b) Casey, C.P., Vos, T.E., and Bikzhanova, G.A. (2003) *Organometallics*, **22**, 901.
22 Casey, C.P., Singer, S.W., Powell, D.R., Hayashi, R.K., and Kavana, M. (2001) *J. Am. Chem. Soc.*, **123**, 1090.
23 For an alternative mechanistic possibility see: Csjernyik, G.; Éll, A.H.; Fadini, L.; Pugin, B.; Bäckvall, J.-E. (2002) *J. Org. Chem.*, **67**, 1657.
24 Ohkuma, T., Ooka, H., Hashiguchi, S., Ikariya, T., and Noyori, R. (1995) *J. Am. Chem. Soc.*, **117**, 2675.
25 Ohkuma, T., Koizumi, M., Doucet, H., Pham, T., Kozawa, M., Murata, K., Katayama, E., Yokozawa, T., Ikariya, T., and Noyori, R. (1998) *J. Am. Chem. Soc.*, **120**, 13529.
26 Ducet, H., Okhuma, T., Murata, K., Yokozama, T., Kozawa, M., Katayama, E., England, A.F., Ikariya, T., and Noyori, R. (1998) *Angew. Chem. Int. Ed. Engl.*, **37**, 1703.
27 Kamaluddin, A.-R., Faatz, M., Lough, A.J., and Morris, R. (2001) *J. Am. Chem. Soc.*, **123**, 7473.
28 Kettler, P.B. (2003) *Org. Proc. Res. Devel.*, **7**, 342.
29 Enthaler, S., Junge, K., and Beller, M. (2008) *Angew. Chem. Int. Ed.*, **47**, 3317.
30 Lippard, S.J., and Berg, J.M. (1994) *Principles of Bioinorganic Chemistry*, University Science Books, Mill Valley, CA.
31 (a) Smil, V. (2001) *Enriching the Earth: Fritz Haber, Carl Bosch and the Transformation of World Food Production*, MIT Press, Cambridge, MA; (b) Mittasch, A., and Kuss, E. (1928) *Z. Elektrochem. Angew. Physik. Chem.*, **34**, 159.
32 Ertl, G. (2008) *Angew. Chem. Int. Ed.*, **47**, 3524.
33 Barton, D.H.R. (1992) *Acc. Chem. Res.*, **25**, 504.
34 Reppe, W., and Vetter, H. (1953) *Justus Liebigs Ann. Chem.*, **582**, 133.
35 (a) Costas, M., Chen, K., and Que, L., Jr. (2000) *Coord. Chem. Rev*, **200–202**, 517; (b) Costas, M., Mehn, M.P., Jensen, M.P., and Que, L., Jr. (2004) *Chem. Rev.*, **104**, 939; (c) Tshuva, E.Y., and Lippard, S.J. (2004) *Chem. Rev.*, **104**, 987.
36 (a) Kharasch, M.S., Morrison, R., and Urry, W.H. (1944) *J. Am. Chem. Soc.*, **66**, 368; (b) Kharasch, M.S., and Fuchs, C.F. (1945) *J. Org. Chem.*, **10**, 292; (c) Tamura, M., and Kochi, J. (1971) *J. Am. Chem. Soc.*, **93**, 1487; (d) Tamura, M., and Kochi, J. (1971) *Bull. Chem. Soc. Jpn.*, **44**, 3063; (e) Kochi, J. K. (1974) *Acc. Chem. Res.*, **7**, 351.
37 Ritter, S.K. (2008) Iron's rising star. *Chem. Eng. News*, July 28, **86**, pp. 53–57.
38 Bolm, C., Legros, J., Paih, J., and Zani, L. (2004) *Chem. Rev.*, **104**, 6217.
39 Enthaler, S., Junge, K., and Beller, M. (2008) *Iron Catalysis in Organic Chemistry*, Wiley-VCH Verlag GmbH, Weinheim, Chapter 4.
40 Bouwkamp, M.W., Bowman, A.C., Lobkovsky, E., and Chirik, P.J. (2006) *J. Am. Chem. Soc.*, **128**, 13340.
41 For a recent review see: Sherry, B.D.; Fürstner, A. (2008) *Acc. Chem. Res.*, **41**, 1500.
42 (a) Chen, M.S., and White, M.C. (2007) *Science*, **318**, 783; (b) Legros, J., and Bolm, C. (2003) *Angew. Chem.*, **115**, 5645; (c) Legros, J., and Bolm, C. (2004) *Angew. Chem.*, **43**, 4225; (d) Korte, A., Legros, J., and Bolm, C. (2004) *Synlett*, 2397; (e) Bryliakov, K.P., and Talsi, E.P. *Angew. Chem. Int. Ed.*, 2004, **39**, 5228; (f) Rose, E., Ren, Q.Z., and Andrioletti, B. (2004) *Chem. Eur. J.*, **10**, 224; (g) Gelalcha, F.G., Bitterlich, B., Anilkumar, G., Tse, M.K., and Beller, M. (2007) *Angew. Chem. Int. Ed.*, **46**, 7293; (h) Anilkumar, G., Bitterlich, B., Gadissa Gelalcha, F., Tse, M.K., and

Beller, M. (2007) *Chem. Commun.*, 289; (i) Oldenburg, P.D., Shteinman, A.A., and Que, L., Jr. (2005) *J. Am. Chem. Soc.*, **127**, 15672; (j) Bukowski, M.R., Comba, P., Lienke, A., Limberg, C., de Laorden, C.L., Mas-Ballesté, R., Merz, M., and Que, L., Jr. (2006) *Angew. Chem. Int. Ed.*, **45**, 3446.

43 Bianchini, C., Giambastiani, G., Rios, I.G., Mantovani, G., Meli, A., and Segarra, A.M. (2006) *Coord. Chem. Rev.*, **250**, 1391.

44 Kerber, R.C. (1995) Mononuclear Iron Compounds with η^1-η^6-Hydrocarbon Ligands, in *Comprehensive Organometallic Chemistry II*, vol. **7** (eds E.W. Abel, F.G.A. Stone, and G. Wilkinson), Pergamon, New York, Chapter 2, pp. 101–229.

45 Partington, J.R. (1964) *A History of Chemistry*, vol. **4**, MacMillan & Co, Ltd, New York, p. 308.

46 For early studies using iron salts activated by aluminum compounds see: Takegami, Y.; Ueno, T.; Fujii, T. (1969) *Bull. Chem. Soc. Jpn.*, **42**, 1663.

47 Orchin, M. (1953) *Adv. Catal.*, **5**, 385.

48 Frankel, E.N., Emken, E.A., Peters, H.M., Davison, V.K., and Butterfield, R.O. (1964) *J. Org. Chem.*, **29**, 3292.

49 Frankel, E.N., Emken, E.A., and Davison, V.K. (1965) *J. Org. Chem.*, **30**, 2739.

50 Harmon, R.E., Gupta, S.K., and Brown, D.J. (1973) *Chem. Rev.*, **73**, 21.

51 Cais, M., and Maoz, N. (1971) *J. Chem. Soc. A*, 1811.

52 Dell, D., Maoz, N., and Cais, M. (1969) *Israel J. Chem.*, **7**, 783.

53 Cais, M., and Maoz, N. (1966) *J. Organomet. Chem.*, **5**, 370.

54 Wrighton, M. (1974) *Chem. Rev.*, **74**, 401.

55 Schroeder, M.A., and Wrighton, M.S. (1976) *J. Am. Chem. Soc.*, **98**, 551.

56 Mitchener, J.C., and Wrighton, M.S. (1981) *J. Am. Chem. Soc.*, **103**, 975.

57 Whetten, R.L., Fu, K.-J., and Grant, E.R. (1982) *J. Am. Chem. Soc.*, **104**, 4270.

58 Swartz, G.L., and Clark, R.J. (1980) *Inorg. Chem.*, **19**, 3191.

59 Weiller, B.H., Miller, M.E., and Grant, E.R. (1987) *J. Am. Chem. Soc.*, **109**, 352.

60 Weiller, B.H., and Grant, E.R. (1987) *J. Am. Chem. Soc.*, **109**, 1051.

61 Miller, M.E., and Grant, E.R. (1987) *J. Am. Chem. Soc.*, **109**, 7951.

62 Fleckner, H., Grevels, F.-W., and Hess, D. (1984) *J. Am. Chem. Soc.*, **106**, 2027.

63 Wuu, Y.-M., Bentsen, J.G., Brinkley, C.G., and Wrighton, M.S. (1987) *Inorg. Chem.*, **26**, 530.

64 Hayes, D., and Weitz, E. (1991) *J. Phys. Chem.*, **95**, 2723.

65 Kismartoni, L.C., Weitz, E., and Cedeño, D.L. (2005) *Organometallics*, **24**, 4714.

66 Cedeño, D.L., and Weitz, E. (2003) *Organometallics*, **22**, 2652.

67 Sawyer, K.R., Glascoe, E.A., Cahoon, J.F., Schlegel, J.P., and Harris, C.B. (2008) *Organometallics*, **27**, 4370.

68 Nishiguchi, T., and Fukuzumi, K. (1972) *Bull. Chem. Soc. Jpn.*, **45**, 1656.

69 Tajima, Y., and Kunioka, E. (1968) *J. Org. Chem.*, **33**, 1689.

70 Bianchini, C., Mell, A., Peruzzini, M., Vizza, F., and Zanobini, F. (1989) *Organometallics*, **8**, 2080.

71 Bianchini, C., Mell, A., Peruzzini, M., Frediani, P., Bohanna, C., Esteruelas, M.A., and Oro, L.A. (1992) *Organometallics*, **11**, 138.

72 Daida, E.J., and Peters, J.C. (2008) *Inorg. Chem.*, **43**, 7474.

73 Marchand, A.P., and Marchand, N.W. (1971) *Tetrahedron Lett.*, **18**, 621.

74 Betley, T.A., and Peters, J.C. (2004) *J. Am. Chem. Soc.*, **108**, 6252.

75 Snee, P.T., Payne, C.K., Mebane, S.D., Kotz, K.T., and Harris, C.B. (2001) *J. Am. Chem. Soc.*, **123**, 6909.

76 Gibson, V.C., Redshaw, C., and Solan, G.A. (2007) *Chem. Rev.*, **107**, 1745.

77 (a) Small, B.L., Brookhart, M., and Bennett, A.M.A. (1998) *J. Am. Chem. Soc.*, **120**, 4049; (b) Britovsek, G.J.P., Gibson, V.C., Kimberley, B.S., Maddox, P.J., McTavish, S.J., Solan, G.A., White, A.J.P., and Williams, D.J. (1998) *Chem. Commun.*, 849.

78 Small, B.L., and Brookhart, M. (1999) *Macromolecules*, **32**, 2120.

79 (a) Kuwabara, I.H., Comninos, F.C.M., Pardini, V.L., Viertler, H., and Toma, H.E. (1994) *Electrochim. Acta*, **39**, 2401; (b) Toma, H.E., and Chavez-Gil, T.E. (1997) *Inorg. Chim. Acta*, **257**, 197.

80 de Bruin, B., Bill, E., Bothe, E., Weyhermüller, T., and Wieghardt, K. (2000) *Inorg. Chem.*, **39**, 2936.
81 Knijnenburg, Q., Gambarotta, S., and Budzelaar, P.H.M. (2006) *Dalton Trans.*, 5442.
82 Scott, J., Gambarotta, S., Korobkov, I., Knijnenburg, Q., de Bruin, B., and Budzelaar, P.H.M. (2005) *J. Am. Chem. Soc.*, **127**, 17205.
83 Bart, S.C., Lobkovsky, E., and Chirik, P.J. (2004) *J. Am. Chem. Soc.*, **126**, 13794.
84 Bart, S.C., Chlopek, K., Bill, E., Bouwkamp, M.W., Lobkovsky, E., Neese, F., Wieghardt, K., and Chirik, P.J. (2006) *J. Am. Chem. Soc.*, **128**, 13901.
85 Bart, S.C., Lobkovsky, E., Bill, E., Wieghardt, K., and Chirik, P.J. (2007) *Inorg. Chem.*, **46**, 7055.
86 For an example of a rhodium catalyst that operates effectively in non-polar media see: Betley, T.A., and Peters, J.C. (2003) *Angew. Chem. Int. Ed.*, **42**, 2385.
87 Trovitch, R.J., Lobkovsky, E., and Chirik, P.J. (2008) *J. Am. Chem. Soc.*, **130**, 11631.
88 Trovitch, R.J., Lobkovsky, E., Bill, E., and Chirik, P.J. (2008) *Organometallics*, **27**, 1470.
89 Trovitch, R.J., Lobkovsky, E., Bouwkamp, M.W., and Chirik, P.J. (2008) *Organometallics*, **27**, 6264.
90 Wile, B.M., Trovitch, R.J., Bart, S.C., Tondreau, A.M., Lobkovsky, E., Milsmann, C., Bill, E., Wieghardt, K., and Chirik, P.J. (2009) *Inorg. Chem.*, **48**, 4190–4200.
91 Archer, A.M., Bouwkamp, M.W., Cortez, M.-P., Lobkovsky, E., and Chirik, P.J. (2006) *Organometallics*, **25**, 4269.
92 Bart, S.C., Hawrelak, E.J., Lobkovsky, E., and Chirik, P.J. (2005) *Organometallics*, **24**, 5518.
93 Muresan, N., Lu, C.C., Ghosh, M., Peters, J.C., Abe, M., Henling, L.M., Weyhermüller, T., Bill, E., and Wieghardt, K. (2008) *Inorg. Chem.*, **47**, 4579.
94 Muresan, N., Chlopek, K., Weyhermüller, T., Neese, F., and Wieghardt, K. (2007) *Inorg. Chem.*, **46**, 5327.
95 Le Floch, P., Knoch, F., Kremer, F., Mathey, F., Scholz, J., Scholz, W., Thiele, K.H., and Zenneck, U. (1998) *Eur. J. Inorg. Chem.*, **1**, 119.
96 tom Dieck, H., Stamp, L., Diercks, R., and Müller, C. (1985) *New. J. Chem.*, **9**, 289.
97 Gibson, V.C., O'Reilly, R.K., Reed, W., Wass, D.F., White, A.J.P., and Williams, D.J. (2002) *Chem. Commun.*, 1850.
98 Gibson, V.C., O'Reilly, R.K., Wass, D.F., White, A.J.P., and Williams, D.J. (2003) *Macromolecules*, **36**, 2591.
99 Nishiyama, H. (2004) *Transition Metals for Organic Synthesis* (eds M. Beller and C. Bolm), Wiley-VCH Verlag GmbH, Weinheim, pp. 182–191.
100 Yun, J., and Buchwald, S.L. (1999) *J. Am. Chem. Soc.*, **121**, 5640.
101 (a) Lipshutz, B.H., Noson, K., Chrisman, W., and Lower, A. (2003) *J. Am. Chem. Soc.*, **125**, 8779; (b) Lipshutz, B.H., and Frieman, B.A. (2005) *Angew. Chem. Int. Ed.*, **44**, 6345.
102 Ouellet, S.G., Walji, A.M., and MacMillan, D.W.C. (2007) *Acc. Chem. Res.*, **40**, 1327.
103 (a) Jothimony, K., Vancheesan, S., and Kuriacose, J.C. (1985) *J. Mol. Catal.*, **32**, 11; (b) Jothimony, K., and Vancheesan, S. (1989) *J. Mol. Catal.*, **52**, 301.
104 Enthaler, S., Hagemann, B., Erre, G., Junge, K., and Beller, M. (2006) *Chem. Asian J.*, **1**, 598.
105 Enthaler, S., Spilker, B., Erre, G., Junge, K., Tse, M.K., and Beller, M. (2008) *Tetrahedron*, **64**, 3867.
106 Nishiyama, H., and Furuta, A. (2007) *Chem. Comm.*, 760.
107 Furuta, A., and Nishiyama, H. (2008) *Tetrahedron Lett.*, **49**, 110.
108 Shaikh, N.S., Junge, K., and Beller, M. (2007) *Org. Lett.*, **9**, 5429.
109 Shaikh, N.S., Enthaler, S., Junge, K., and Beller, M. (2008) *Angew. Chem. Int. Ed.*, **47**, 2497.
110 Gutsulyak, D.V., Kuzmina, L.G., Howard, J.A.K., Vyboishchikov, S.F., and Nikonov, G.I. (2008) *J. Am. Chem. Soc.*, **130**, 3732.
111 Cámpora, J., Naz, A.M., Palma, P., Álvarez, E., and Reyes, M.L. (2005) *Organometallics*, **24**, 4878.

112 Tondreau, A.M., Lobkovsky, E., and Chirik, P.J. (2008) *Org. Lett.*, **10**, 2789.
113 Langlotz, B.K., Wadepohl, H., and Gade, L.H. (2008) *Angew. Chem. Int. Ed.*, **47**, 4670.
114 Tondreau, A.M., Darmon, J.M., Wile, B.M., Floyd, S.K., Lobkovsky, E., and Chirik, P.J. (2009) *Organometallics*, **28**, 3928–3940.
115 Bullock, R.M. (2007) *Angew. Chem. Int. Ed.*, **46**, 7360.
116 Casey, C.P., and Guan, H. (2007) *J. Am. Chem. Soc.*, **129**, 5816.
117 Sui-Seng, C., Freutel, F., Lough, A.J., and Morris, R.H. (2008) *Angew. Chem. Int. Ed.*, **47**, 940.
118 Mikhailine, A., Lough, A.J., and Morris, R.H. (2009) *J. Am. Chem. Soc.*, **131**, 1394.
119 Sui-Seng, C., Haque, F.N., Pütz, A.-M., Reuss, V., Meyer, N., Lough, A.J., Zimmer-De Iuliis, M., and Morris, R.H. (2009) *Inorg. Chem.*, **48**, 735.
120 Li, T., Bergner, I., Haque, F.N., Zimmer-De Iuliis, M., Song, D., and Morris, R.H. (2007) *Organometallics*, **27**, 5940.

5
Olefin Oligomerizations and Polymerizations Catalyzed by Iron and Cobalt Complexes Bearing Bis(imino)pyridine Ligands

Vernon C. Gibson and Gregory A. Solan

5.1
Introduction

The capacity of 2,6-bis(arylimino)pyridine-iron(II) and -cobalt(II) halide complexes, on activation with methylalumoxane (MAO), to act as exceptionally active catalysts (comparable with Group 4 metallocenes) for the conversion of ethylene to high density polyethylene (HDPE) was first disclosed independently by Brookhart, Bennett and ourselves in 1998 (Figure 5.1) [1–7]. The significance of this discovery was further highlighted by the complete absence at the time of examples of iron and cobalt complexes in olefin polymerization applications. In comparison with early transition metal-based catalysts such as metallocenes, these late transition metal systems offer some distinct advantages which range from their relative ease of handling and manipulation, their reduced oxophilicity, to the cost-effectiveness and low toxicity of the metal centers employed. Moreover, by simple modification of the 2,6-bis(arylimino)pyridine ligand frame, exceedingly efficient iron and cobalt oligomerization catalysts can also be generated, facilitating the formation of highly linear α-olefins with near ideal Schulz–Flory distributions. Since these initial findings the past decade has witnessed a dramatic growth in reports associated with this class of late transition metal catalysts, and clear structure–activity relationships have emerged.

This chapter is concerned with highlighting some of the more notable advances that have come to light as a result of identifying key factors that influence catalyst performance, particularly those related to precatalyst structure. The importance of the initiator and the role played in chain transfer is probed. Current mechanistic understanding is examined from both a spectroscopic and a computational viewpoint while efforts to prepare well-defined iron or cobalt alkyl catalysts are discussed. Efforts to heterogenize these homogeneous catalysts are briefly reviewed, as is their use in multi-component catalysis.

The reader is referred to a series of reviews for more general overviews of post-metallocene α-olefin polymerization catalysts [8–12], while recent reviews relating to the versatility of 2,6-bis(imino)pyridines and to iron- and cobalt-based catalysis itself have also been documented [13–15].

Catalysis Without Precious Metals. Edited by R. Morris Bullock
© 2010 WILEY-VCH Verlag GmbH & Co. KGaA, Weinheim
ISBN: 978-3-527-32354-8

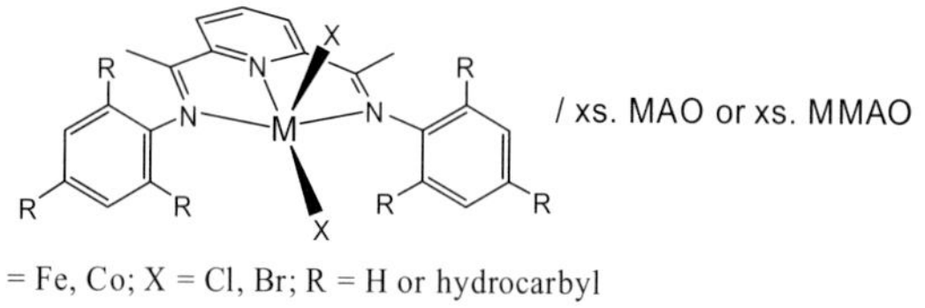

M = Fe, Co; X = Cl, Br; R = H or hydrocarbyl

Figure 5.1 Ethylene polymerization and oligomerization catalysts first developed [1–7].

5.2
Precatalyst Synthesis

5.2.1
Ligand Preparation

The acid-catalyzed condensation reaction of a 2,6-pyridinedicarbonyl compound with the corresponding aniline in a 1:2 molar ratio using an alcohol (e.g., ethanol, propanol, methanol) as solvent is generally employed to synthesize symmetrical bis(arylimino)pyridine ligands, [2,6-(ArN=CR)$_2$C$_5$H$_3$N] (L1 Ar = aryl group) (Scheme 5.1). The majority of examples of L1 are based on the imino-carbon substituents (R) being an H or Me group and indeed more than a hundred derivatives of L1, differing in the steric and electronic properties of the aryl groups, have now been prepared [1–70]. Other variations of R that have been disclosed include Ph [38, 49], OR, SR [52, 53] and other alkyl variants [51]. Examples of L1 containing a non-hydrogen substituent at the 4-position of the central pyridyl unit, [2,6-(ArN=CMe)$_2$-4-X-C$_5$H$_2$N] (X = t-butyl [55]), allyl [69, 70], O(ω-alkenyl) [57, 58], benzyl [69, 70], Cl [56], 2-methyl-2-phenylpropyl [69, 70], can also be accessed, as can 4,4′-linked bis(arylimino)pyridines [54].

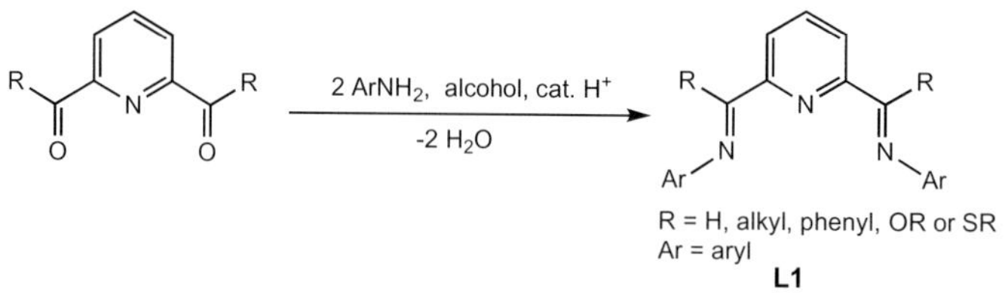

R = H, alkyl, phenyl, OR or SR
Ar = aryl
L1

Scheme 5.1 General route to symmetrical 2,6-bis(arylimino)pyridines (L1).

By the sequential condensation reactions of 2,6-diacetylpyridine with two different anilines unsymmetrical 2,6-bis(arylimino)pyridines, [2-(ArN=CMe)-6-(Ar′N=CMe)C$_5$H$_3$N] (L2), can prepared [26, 38, 43, 59, 60, 71]. The substitution pattern at the imino-nitrogen atoms is not restricted to aryl groups with symmetrical 2,6-bis(alkylimino)pyridines, [2,6-{(alkyl)N=CR}$_2$C$_5$H$_3$N] (L3) [38, 46, 72, 73] and 2,6-bis(hydrazone)pyridines, [2,6-{(R$_1$R$_2$N)N=CR}$_2$C$_5$H$_3$N] (L4) [74–76] also

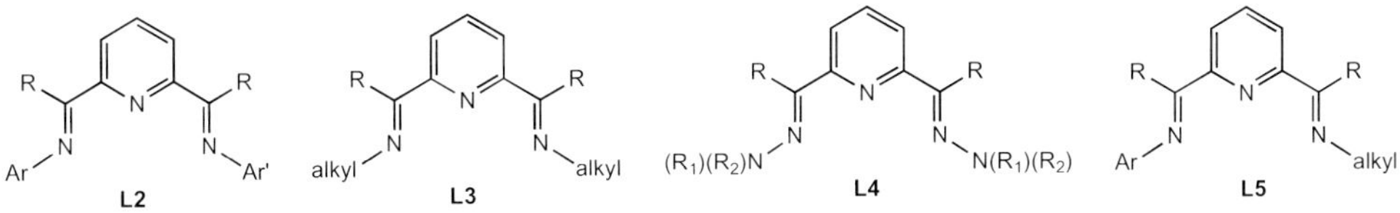

Figure 5.2 Structural diversity at the imino-nitrogen position in 2,6-bis(imino)pyridines.

accessible using the general route employed for L1 (Figure 5.2). Using the successive condensation approach unsymmetrical 2-(arylimino)-6-(alkylimino)pyridines [2-(ArN=CMe)-6-{(alkyl)N=CMe}C$_5$H$_3$N] (L5) have also recently been reported [77, 78].

While the structures of L1–L5 depicted in Figures 5.1 and 5.2 are representative of the configuration adopted on coordination to a metal center, single crystal X-ray diffraction studies reveal a mutually trans-configuration for the nitrogen donors to be the preferred arrangement of the free ligand in the solid state [26, 42, 45, 79–83]. A further significant feature of bis(imino)pyridines that is not discussed in detail here is the intriguing variety of new chemistry that has been simultaneously developed, including deprotonation [84–87], alkylation [86–92] and reduction [93, 94]; several recent reviews have covered these topics [13, 15, 95].

5.2.2
Complexation with MX$_2$ (M = Fe, Co)

The neutral five-coordinate precatalysts [2,6-(ArN=CR)$_2$C$_5$H$_3$N]MX$_2$ (**1** M = Fe; **2** M = Co; X = Cl or Br) can be obtained in high yield by reaction of L1 (and also L2–L5) with an anhydrous or hydrated divalent iron or cobalt halide in tetrahydrofuran or n-butanol (Scheme 5.2) [1–7]. Using this facile synthetic approach an impressive library of examples of **1** and **2** have been developed (for a comprehensive list the reader is referred to a series of recent reviews [13–15]), displaying a range of different symmetries (e.g., C_{2v}, C_2, C_s or C_1). It should be noted that if L1 lacks sufficient steric bulk (e.g., Ar = Ph), bis(ligand) salts of the type [{2,6-(ArN=CR)$_2$C$_5$H$_3$N}$_2$M][MX$_4$] (**3**) can be formed [33, 37].

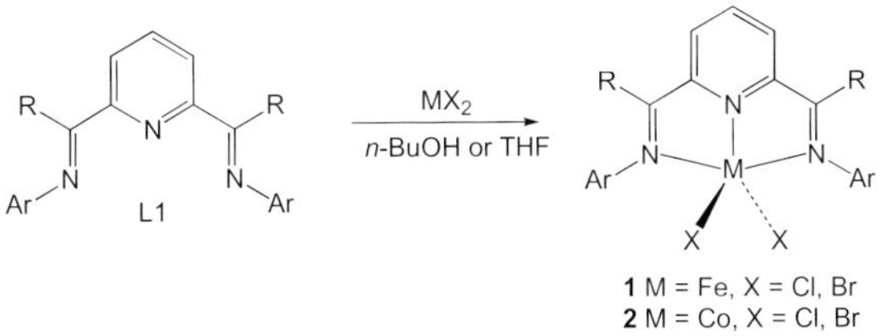

1 M = Fe, X = Cl, Br
2 M = Co, X = Cl, Br

Scheme 5.2 Synthesis of bis(arylimino)pyridine iron(II) and cobalt(II) halides.

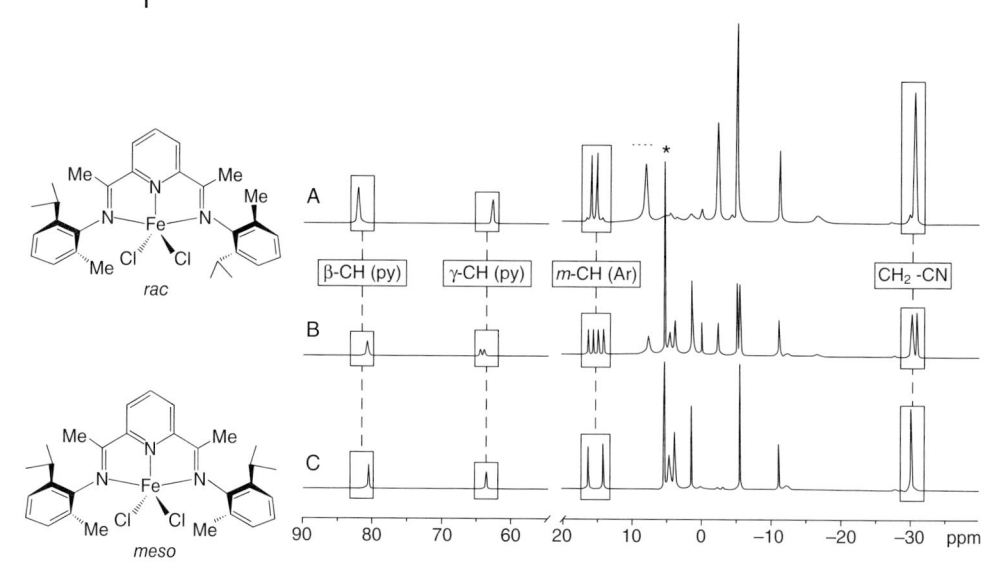

Figure 5.3 Paramagnetically shifted ^{1}H NMR spectra of: (A) complex rac-**1**, (B) rac-**1** + meso-**1** and (C) meso-**1** (data from Ref. [96], with permission of the copyright holders, was used to generate the figure).

Both **1** and **2** are paramagnetic and display high spin configurations in accordance with four [$S = 2$ (**1**)] and three unpaired electrons [$S = 3/2$ (**2**)], respectively. ^{1}H NMR spectroscopy further supports the paramagnetic nature of the complexes, with highly shifted resonances apparent. In a number of cases full signal assignment of the resulting ^{1}H NMR spectra, on the basis of integration and proximity to the paramagnetic center, has proved possible [6]. Interestingly, ^{1}H NMR spectroscopy has recently been used to distinguish the isolable rac- and meso-diastereomers of [2,6-(ArN=CR)$_2$C$_5$H$_3$N]FeCl$_2$ (**1** R = Me, Ar = 2-Me-6-^{i}PrC$_6$H$_3$) (Figure 5.3) [96]. Use of IR spectroscopy and fast atom bombardment (FAB) mass spectrometry further confirms the identity of **1** and **2** with characteristic (C=N)$_{imine}$ stretches found in their IR spectrum, while fragmentation peaks corresponding to the loss of halide ligands are seen in their mass spectra.

Single crystal X-ray diffraction studies have been performed on over 50 examples of **1** and **2** and indicate that the geometry at the five-coordinate metal centers can be anywhere between square-pyramidal and trigonal bipyramidal, depending in large degree on the aryl group substitution pattern [97]. Interestingly, in the case of [2,6-(ArN=CR)$_2$C$_5$H$_3$N]FeCl$_2$ (**1** R = ^{i}Pr, Ar = 2,4,6-Me$_3$C$_6$H$_2$), both distorted trigonal bipyramidal (molecule A) and square pyramidal (molecule B) geometries were found in one structure determination, suggesting a weak energetic bias for one geometry over the other in these systems [51] (Figure 5.4). In all cases the bis(arylimino)pyridine ligand binds as a tridentate ligand with the imino and pyridine moieties essentially planar; the aryl groups are inclined perpendicularly

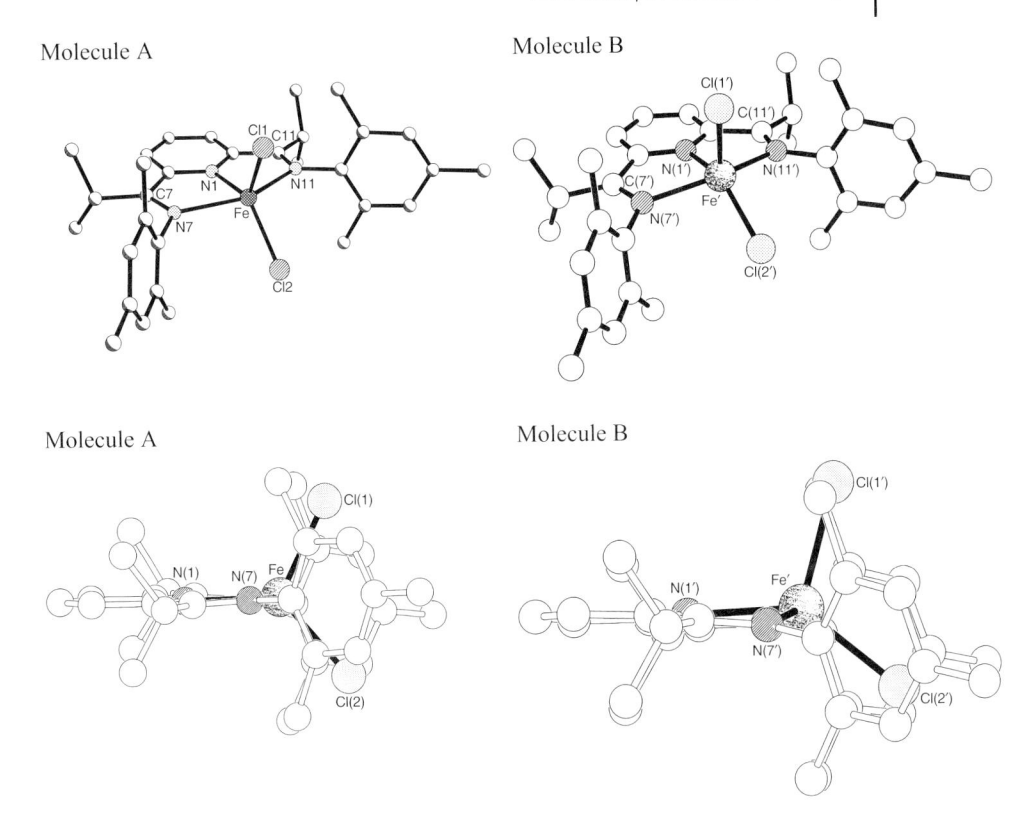

Figure 5.4 Molecular structure of **1** (R = ^{i}Pr, Ar = 2,4,6-Me$_3$C$_6$H$_2$, X = Cl) showing views of the independent molecules A (trigonal bipyramidal) and B (square pyramidal) (data from Ref. [51], with permission of the copyright holders, was used to generate the figure).

to the bis(imino)pyridine plane which is likely enforced by the close proximity of the imino-carbon substituent to the neighboring ortho-substituent on the aryl group.

5.3
Precatalyst Activation and Catalysis

5.3.1
Olefin Polymerization

The polymerization of ethylene or α-olefins to give polyolefins represents a major industrial process with a worldwide production of more than 103 million tons reported in 2005 [98]. While early transition metals have led the way for metal-mediated polymerization processes, the great potential of late transition metal catalysts was first recognized in the mid-1990s when Brookhart and coworkers employed sterically demanding α-diimine Group 10 metal catalysts to mediate the transformation [10, 11]. The introduction of strategically placed steric bulk proved key to supressing detrimental chain transfer processes that had, until that time, impeded the development of late transition metal polymerization catalysts. Extension of these concepts of steric control while matching the electronic requirements of the relatively electron poor Group 8 and 9 metal centers inspired the synthesis of bis(imino)pyridine iron and cobalt precatalysts **1** and **2**.

5.3.1.1 Catalytic Evaluation

The first demonstration that bis(arylimino)pyridine iron(II) (**1**) and cobalt(II) halides (**2**) could act as precatalysts for the polymerization of ethylene was realized when toluene solutions of specific examples of **1** or **2** were treated with excess (100–2000 equiv.) MAO or MMAO [MMAO contains 20–25% Al^{i}Bu$_3$] and the resultant catalysts exposed to an ethylene atmosphere (1–10 bar) at ambient temperature (Scheme 5.3) [1–7]. A considerable exotherm was observed, with solid polyethylene precipitating from the reaction. Analysis of the polymer revealed the formation of HDPE with melting points in the range 134–139 °C. The molecular weights of the HDPEs can be high and the molecular weight distribution (M_w/M_n) can vary from narrow, broad to bimodal, depending on the precatalyst and the reaction conditions (*vida infra*).

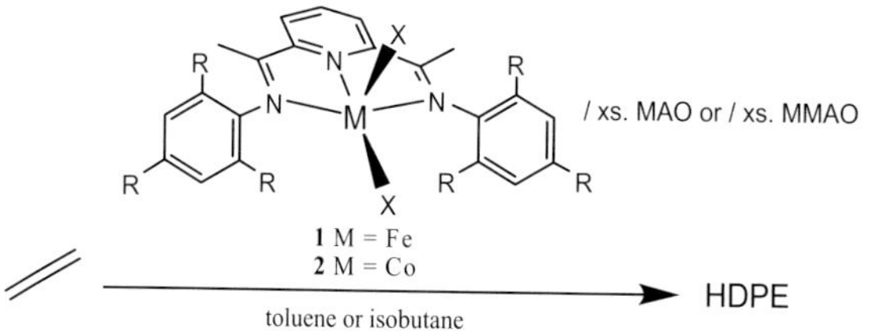

Scheme 5.3 Ethylene polymerization using **1** and **2** in the presence of MAO or MMAO.

Under related reaction conditions bis(arylimino)pyridine ferric halides can also serve as active precatalysts for ethylene polymerization [6]. While MAO or MMAO continues to be the preferred co-catalyst/activator, a range of alternative activators have also been screened (Table 5.1).

5.3.1.2 Steric Versus Electronic Effects

The substitution pattern displayed by the precatalyst is crucial to the catalyst behaving as a polymerization catalyst. With two bulky substituents at the 2,6-positions of the N-aryl groups [e.g., ^{i}Pr (**a**), Me (**b**)] access to high-molecular weight HDPE was achievable for both iron and cobalt systems (Figure 5.5) [1, 4–6], with the less bulky 2,6-Me$_2$ derivative giving relatively lower molecular weight polymer for iron (**1b**). In general, the iron catalysts are an order of magnitude more active than their

Table 5.1 Alternative co-catalysts that have been employed to activate bis(arylimino)pyridine iron dihalides (**1**).

Co-catalyst	Reference	Co-catalyst	Reference
AlMe$_3$	[99–102]	AlMe$_3$/B(C$_6$F$_5$)$_3$	[99]
AlEt$_3$	[103, 104]	AlMe$_3$/CPh$_3$B(C$_6$F$_5$)$_4$	[99]
Al^{i}Bu$_3$	[100–103, 105]	AlEt$_3$/CPh$_3$B(C$_6$F$_5$)$_4$	[106]
Al(octyl)$_3$	[100, 103]	Al^{i}Bu$_3$/CPh$_3$B(C$_6$F$_5$)$_4$	[100]
AlEt$_2$Cl	[101]	Et$_2$AlOAlEt$_2$	[104, 106]
Al(hexyl)$_3$	[103]	^{i}Bu$_2$AlOAl^{i}Bu$_2$	[101, 104]
AlMe$_3$/Al^{i}Bu$_3$	[103]	AlEt$_3$/CPh$_3$Al(O^{t}BuF)$_4$	[107]

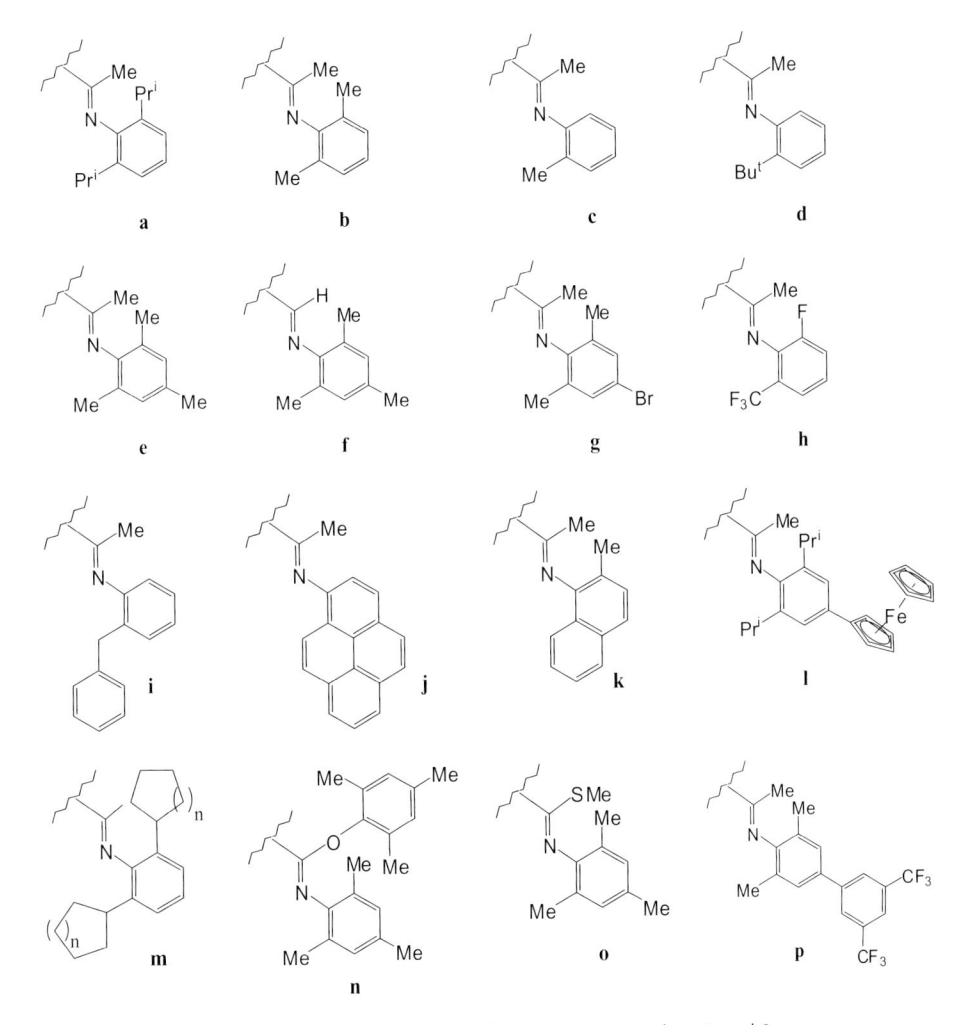

Figure 5.5 Selected iminoaryl units in ethylene polymerization precatalysts **1** and **2**.

cobalt counterparts. On the other hand when a single ortho-substituent is present [Me (**c**)], the iron and cobalt systems perform as selective oligomerization catalysts (*vide infra*) [2, 16]. However, if the steric bulk of the ortho-position is too large [*t*-Bu (**d**)], the catalysts generate only polymer [1–7].

The importance of electronic properties for catalytic performance was initially demonstrated when a methyl group was appended to the para-position of iron-containing **1b** to give **1e**; a significant increase in the activity of the polymerization catalyst was observed [2, 5, 6]. On changing from the ketimine (R = Me) to the aldimine (R = H) derivative (for a given aryl group) of either **1** or **2**, a reduction in polymer molecular weight was observed in conjunction with a lowering in the activity of the catalyst (e.g., **1e** versus **1f**) [1–7]. Introduction of a *para*-bromo group (**1g**) also raises the activity, with a simultaneous reduction in polymer molecular weight [41]. The influence of the *para*-bromo group has been likened to that of a *para*-isopropyl, and it has been suggested that electronic effects are beneficial for increased activity for iron systems whereas steric effects appear more important for cobalt [24].

Significantly, the use of an *ortho*-aryl CF_3 group in the cobalt systems can lead to substantial increases in activity (and catalyst lifetimes) upon activation with MAO [39]. Notably, the cobalt-based catalyst bearing an *ortho*-CF_3 group in combination with an *ortho*-F substituent (**h**) displays an activity > 100 000 g mmol^{-1} h^{-1} bar^{-1}, which is comparable in performance to the most active iron catalysts (e.g., **1e**/MAO).

Inclusion of a benzyl group (**i**) at the ortho-positions has been shown to furnish a very high activity iron catalyst (40 800 g mmol^{-1} h^{-1} bar^{-1} *cf.* 1655 g mmol^{-1} h^{-1} bar^{-1} for **1a** under similar catalytic conditions), producing linear polyethylene. By contrast, related ligands possessing either an *ortho*-phenyl or *ortho*-myrtanyl group are far less active [45]. Intriguingly, use of a pyrenyl group (**j**) gave an iron catalyst exhibiting an activity of 13 480 g mmol^{-1} h^{-1} bar^{-1} that is claimed to yield (for the first time) branched polyethylene [43]. For iron systems based on napthyl groups, introduction of an *ortho*-methyl group (**k**) greatly enhances the activity of the system [45].

With the intent of using an electrochemically active group to modify the electronic properties of **1** and **2**, a ferrocenyl unit was attached to the para-position of the *N*-aryl groups. The resulting system (**1l**) displayed an activity of 6300 g mmol^{-1} h^{-1} bar^{-1} affording a polymer with a molecular weight of 900 000; the corresponding cobalt system (**2l**) was an order of magnitude less active [22]. Such iron and cobalt systems can be readily chemically oxidized with ferrocenium hexafluorophosphate to afford [{2,6-(2,6-^{i}Pr$_2$-4-FcArNCMe)$_2$C$_5$H$_3$N}MCl$_2$](PF$_6$)$_2$ (**4l**, M = Fe; **5l**, M = Co). However, no difference in polymerization performance has been observed between **4l** and **5l** when compared with the corresponding reduced species; this has been attributed to the *in situ* reduction of the dicationic ferrocenium species by the excess alkylaluminum co-catalyst.

The presence of cycloalkyl substituents in the 2,6-positions of the *N*-aryl groups (**m**) has been found to increase the temperature stability of the iron catalysts over their alkyl analogues (e.g., **1b** or **1c**) [29]. On the basis of a quantum mechanical

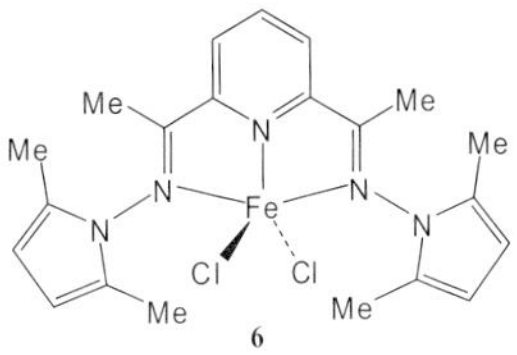

Figure 5.6 Bis(hydrazone)pyridine iron complexes, **6** and **7**.

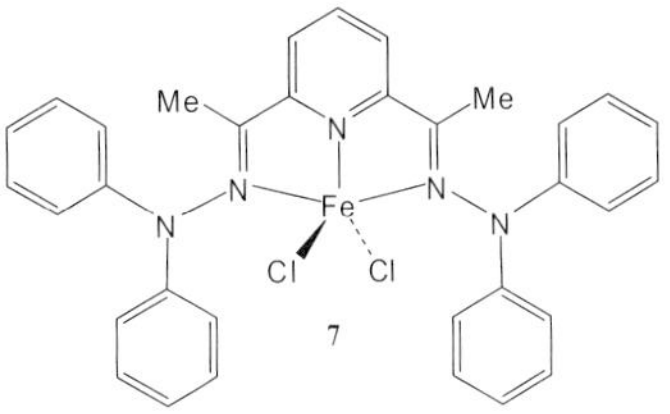

study the extra stability is not associated with the MAO-induced activation reaction but rather the greater stability of the cycloaliphatic-containing precatalyst.

Precatalysts containing ether and thioether groups as the imine-carbon substituents such as [2,6-{(2,4,6-Me$_3$C$_6$H$_2$)N=CR}$_2$C$_5$H$_3$N]FeCl$_2$ [**1n** R = 2,4,6-Me$_3$C$_6$H$_2$O; **1o** R = MeS], afford highly active catalysts for ethylene polymerization on treatment with MAO (Figure 5.5). Unexpectedly, substitution of the mesityl aryloxide in **1n** with an O–Me group leads to an inactive system [52].

Polymerization catalysts can also be obtained from precatalysts that are devoid of *N*-aryl groups. For example, iron catalysts composed of the bis-hydazones [2,6-(R'N=CMe)$_2$C$_5$H$_3$N]FeCl$_2$ [R' = 2,5-dimethylpyrrolyl (**6**), NPh$_2$ (**7**)] and MAO afford low molecular weight solid polyethylene with pyrrolyl-based **6** being more active than **7** but less active than the most closely related bis-aryl based systems (e.g., **1a**/MAO or **1b**/MAO) [74–76]. Considerable manipulation of the bis(imino) pyridine skeleton has been reported over the last ten years, with the introduction of various heteroatoms (e.g., P, N, O) within the planar conjugated ligand backbone a key feature of the more productive derivatives [47, 50, 108–112]; the molecular weights and activities still, however, have not in most cases reached the performance characteristics of the parent iron and cobalt polymerization catalysts (Figure 5.6).

5.3.1.3 Effect of MAO Concentration

Early in the development of the iron and cobalt polymerization catalysis the important role played by the co-catalyst in catalyst performance was recognized. In general, an increase in MAO concentration results in an increase in the activity of the catalysts, which can be attributed to an increase in the number of active sites [1–7]. For iron catalysts alone, larger MAO concentrations can lead to broadening of the molecular weight distribution and, in some cases, a bimodal distribution is observed. For example, systematic variation in the [MAO]/[Fe] was carried out using **1a** at 25 °C and 1 bar ethylene pressure from 250 : 1 to 2000 : 1, and the GPC traces collected (Figure 5.7) [6]. At 250 equiv of MAO a broad unimodal molecular weight distribution is observed. As the ratio of [MAO]/[Fe] is increased to 500 : 1 a broader distribution is apparent. At 1000 equiv the distribution has clearly become bimodal with a low molecular weight fraction centered at 2800 and a high molecular weight fraction at 46 000. On reaching 2000 equiv of MAO, the lower molecular weight peak is most prominent. Analysis of the low molecular weight

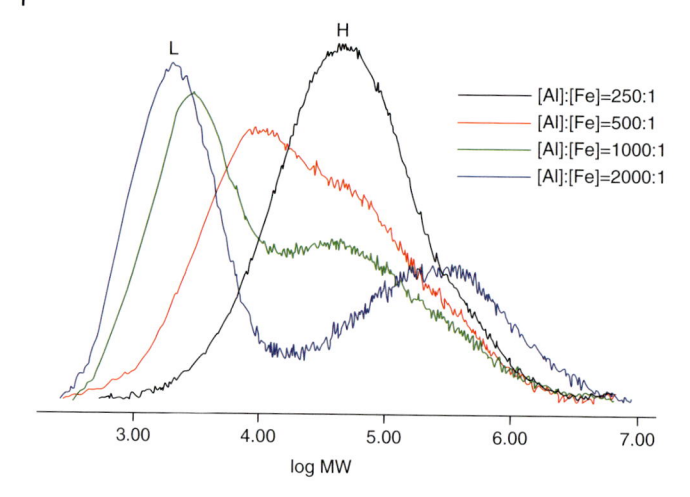

Figure 5.7 Effect of MAO concentration on modality of polymer formed using **1a** (figure was reproduced from Ref. [6], with permission of the copyright holders).

peak revealed it to be a saturated paraffin while the higher molecular weight species contains vinyl end groups. The formation of a low molecular weight saturated polymer at high MAO concentrations has been ascribed to the increased content of trimethylaluminum (present in commercial MAO solutions) which can participate preferentially in chain transfer to aluminum over β-H elimination (see Section 5.4.2). Indeed, a similar bimodal distribution containing large fractions of low molar mass products can be obtained using **1a** with trimethylaluminum as the sole activator [99–102]. Notably, there is no evidence that cobalt catalysts terminate by a similar chain transfer to the aluminum mechanism.

5.3.1.4 Effects of Pressure and Temperature

For both iron and cobalt, ethylene pressure can have a profound influence on both the yield and the molecular weight distribution of the polymer. In general, the yield of polyethylene increases linearly with ethylene pressure (Figure 5.8), while the activity of the catalyst and molecular weight remain essentially unchanged. Such a dependence of yield on ethylene concentration is consistent with a first-order rate dependence on monomer, while the invariance of molecular weight with ethylene pressure implies the chain transfer is first order in monomer concentration [2, 3, 6]. It should also be noted that with iron catalysts that can promote bimodal distributions (e.g. **1a**, Figure 5.7), the effect of increased pressure can result in narrowing of the distribution as a result of a decrease in the lower molecular weight component [6]. By contrast, raising the temperature (>80°C) of the polymerization results in a reduction in productivity and also a lowering in the molecular weight for both iron and cobalt, which can be attributed to catalyst deactivation and also to the lower solubility of ethylene at higher temperatures.

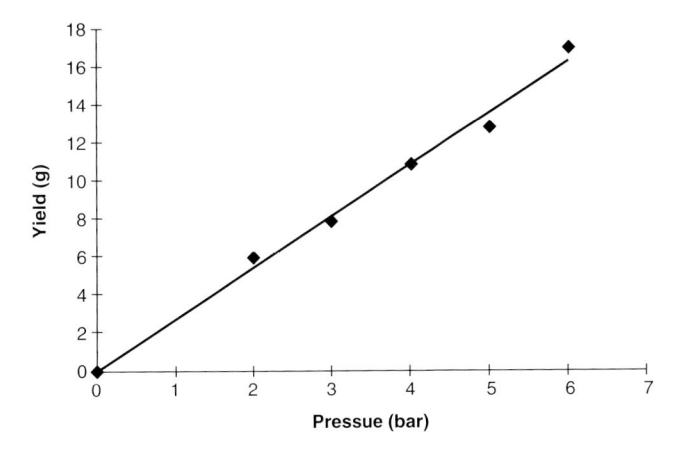

Figure 5.8 Effect of ethylene pressure on the yield of polymer using iron catalyst **1a** (figure was reproduced from Ref. [6], with permission of the copyright holders).

5.3.1.5 α-Olefin Monomers

Iron and cobalt catalysts that have proved highly active for ethylene polymerization have also been screened for the polymerization of α-olefins. However, significantly lower productivities are observed for the polymerization of propylene [26, 113, 114] and this reflects the inability of the ethylene polymerization catalysts to incorporate higher α-olefins into the growing polymer chains. Fink et al. have shown that the low activities can be considerably increased by using $CPh_3B(C_6F_5)_4/AlR_3$ (R = ^{i}Bu, Et) as the co-catalyst; the combination of [2,6-{(2-^{i}Pr-6-MeC$_6$H$_3$)N=CMe}$_2$C$_5$H$_3$N]FeCl$_2$ (**1q**)/$CPh_3B(C_6F_5)_4$/Al^{i}Bu$_3$ proving the most active system [106]. Inspection of the polypropylene formed indicates prevailingly isotactic propylene formed by a chain-end control mechanism. Recently, the screening of a C_2-symmetric diastereomerically pure rac-[2,6-{(2-^{i}Pr-6-MeC$_6$H$_3$)N=CMe}$_2$C$_5$H$_3$N]FeCl$_2$ using MMAO as activator has shown some evidence that enantiomorphic site control can be a competing mechanism (Scheme 5.4) [96].

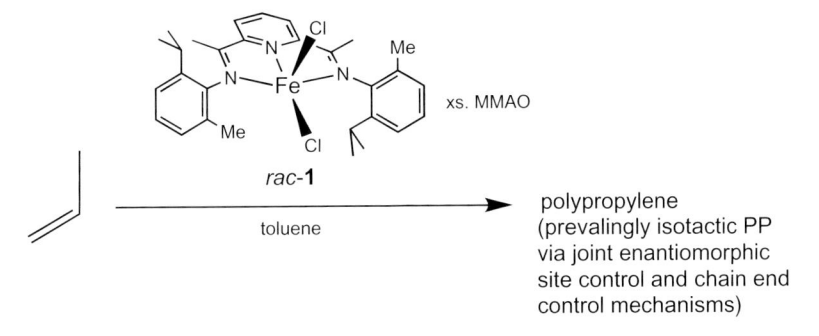

Scheme 5.4 Propylene polymerization using rac-**1** in the presence of MMAO.

5.3.2
Olefin Oligomerization

As with ethylene polymerization, the metal-mediated conversion of ethylene to short chain linear α-olefins (range: C_4–C_{20}) represents an important industrial process. Such oligomers find considerable use in the manufacture of detergents, and plasticizers and in the production of linear low density polyethylene (LLDPE). An additional benefit of bis(imino)pyridine iron and cobalt catalysts is the ability to tune the ligand environment to allow the formation of exclusively linear α-olefins with activities and selectivities comparable with other well-known late transition metal catalysts (e.g., the SHOP catalyst) [115].

5.3.2.1 Catalytic Evaluation
Generation of the active catalyst proceeds in a similar manner to that employed to form the polymerization catalyst, by treating the corresponding precatalyst with MAO or MMAO in toluene (Scheme 5.5). The resultant oligomers are toluene soluble and can be isolated, depending on the molecular weight, as waxes with high selectivity for linear α-olefins (>95%) with no branching apparent.

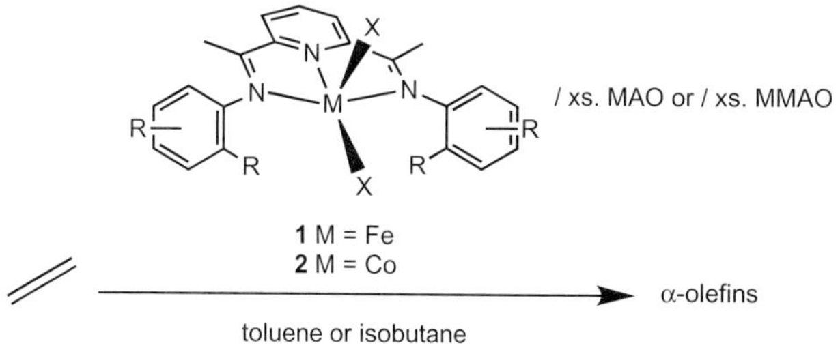

Scheme 5.5 Ethylene oligomerization using **1** and **2** in the presence of MAO or MMAO.

5.3.2.2 Substituent Effects
Among the first reported iron and cobalt catalysts that perform as selective oligomerization catalysts, are those containing a single ortho-substituent (Figure 5.9) [2, 16]. As with the polymerization catalysts, both the nature of the imino-carbon substitutent and aryl group substitution pattern were found to influence the catalytic performance. In general, the iron catalysts showed higher activities than their cobalt counterparts, a trend that mirrors the general behavior of iron versus cobalt ethylene polymerization catalysts. In the same way, iron ketimine catalyst **1c** was found to be significantly more active than its aldimine analogue (**1r**). Methyl substituents in meta- (**s**) and para- (**t**) positions on the aryl group also significantly affect the activities whereby the *meta*-methyl-substituted precatalyst **1s** shows the highest activity of this series.

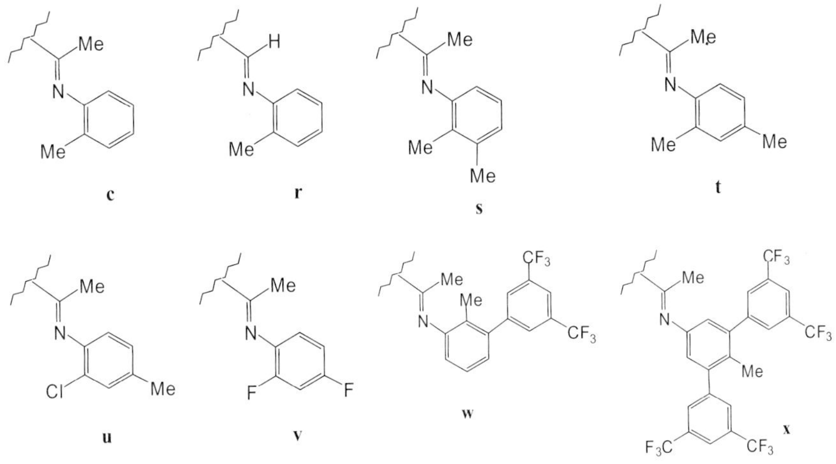

Figure 5.9 Selected iminoaryl units in oligomerization precatalysts **1** and **2**.

A wide variety of suitably substituted aryl groups has since been examined with the intent of obtaining optimal activities for ethylene oligomerization (Figure 5.9). In particular, combinations of halide, fluorocarbyl and hydrocarbyl have been thoroughly investigated. For example, the use of a chloro substituent in an ortho-position and a methyl either meta or para (**u**) has led to some of the most active oligomerization catalysts [36]. Evaluation of difluoro-substituted ligands (**v**, one F always ortho) has also been performed (with MMAO) for both iron and cobalt, and high oligomerization activities were found for the iron systems (**1v**), whereas the cobalt systems (**2v**) were inactive [36, 37]. The fluoro or fluorocarbyl substituent need not be bound directly to the N-aryl group, with a number of highly active iron oligomerization systems based on more remote substitutions; see for example the 2,3-pattern of *ortho*-methyl and *meta*-aryl(CF_3) in **1w** or 3,5-pattern of *meta*-aryl(CF_3)s in **1x** [42].

Attempts to increase the temperature stability of an iron oligomerization catalyst have shown that addition of boryl groups to one of the N-aryl groups has beneficial effects. For example, [2-{(2,4,6-Me$_3$C$_6$H$_2$)N=CMe}-6-{(4-B-(OCMe$_2$CMe$_2$O)C$_6$H$_4$)N=CMe}C$_5$H$_3$N]FeCl$_2$/MMAO catalyst is more thermally stable, but less productive, than **1c**/MMAO [116]. In addition the boryl-based precatalyst makes less solids than **1c**/MMAO.

Steric variations can also be used to force bis(hydrazone)pyridyl iron complexes to behave as oligomerization catalysts. For example, substituting the NPh$_2$ groups in **7** for the less sterically bulky NPhMe and NMe$_2$ (**8**) groups generates catalysts that afford toluene soluble α-olefin polymers [94–96] (Figure 5.10). Interestingly, the C_1-symmetric complex [2-{2,6-^{i}Pr$_2$C$_6$H$_3$)N=CMe}-6-{(PhCH(Me))N=CMe}C$_5$H$_3$N)]FeCl$_2$ (**9**), upon activation with MAO, has been found to promote

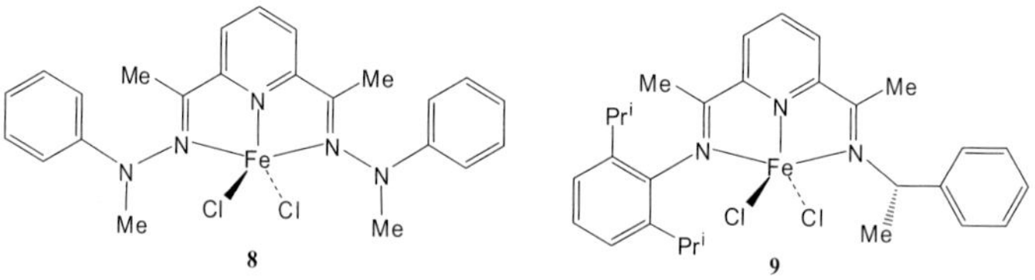

Figure 5.10 Oligomerization precatalysts **8** and **9**; with one or no N-aryl groups.

simultaneous oligomerization and polymerization of ethylene to give α-olefins and HDPE, respectively [60]; a similar dual catalytic performance has been recently reported for [2-{2,6-^{i}Pr$_2$C$_6$H$_3$)N=CMe}-6-{(H$_2$C=CH$_2$(CH$_2$)$_n$CH$_2$)N=CMe}C$_5$H$_3$N)] FeCl$_2$/MAO (n = 1–4) [78].

5.3.2.3 Schulz–Flory Distributions

In all the above ethylene oligomerizations, Schulz–Flory distributions of oligomers are observed which can be quantified by the α value, which represents the probability of chain propagation [117, 118]. The magnitude of α can be experimentally determined by the ratio of two oligomer fractions Eq. (5.1). The α value is usually preferred to be in the range of about 0.6 to about 0.8 to make α-olefins of the most commercial interest.

$$\alpha = \frac{\text{rate of propagation}}{\text{rate of propagation} + \text{rate of chain transfer}} = \frac{\text{moles of } C_{n+2}}{\text{moles of } C_n} \tag{5.1}$$

Typically, for the bis(arylimino)pyridine catalysts (**1c, 1r–1x**) the value of α ranges from 0.50 to 0.85, with lower values corresponding to reduced levels of higher olefins in the resultant distribution. Unlike the differences in activity for **1c, 1s** and **1t**, additional methyl groups in the meta- and para-positions of the aryl rings have very little effect on the α value [α = 0.79 (**1c**), 0.79 (**1s**), 0.78 (**1t**)]. On the other hand, steric effects play a more important role with aldimine **1r** displaying an α value of 0.50 while the more sterically encumbered N-(1-napthyl) analogue gives a value of 0.63 [16]. In addition to structural variations, increasing temperature can have the effect of reducing α.

5.3.2.4 Poisson Distributions

To complement the Schulz–Flory distributions of α-olefins achievable with iron catalysts, it is also possible to obtain a Poisson distribution of α-olefins using a modification of the catalytic system. In particular, addition of excess diethylzinc to a catalyst composed of **1a**/MAO results in polyethylene chain growth at zinc (Scheme 5.6) [119–121] leading to a Poisson distribution of zinc alkyl products. Significantly, the grown alkyl chains can then be displaced from the zinc centers by an olefin exchange reaction catalyzed by, for example, [Ni(acac)$_2$] to give linear α-olefins with a Poisson distribution of chain lengths.

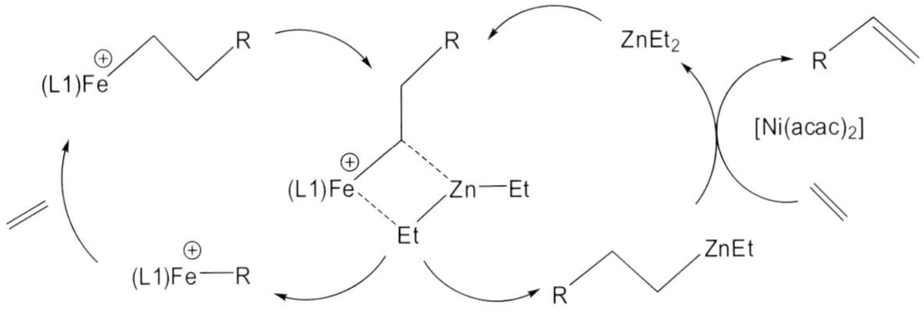

Scheme 5.6 Chain growth on zinc by **1a**/MAO and proposed mechanism for the process.

5.3.2.5 α-Olefin Monomers

Linear (head-to-head) dimerization of α-olefins such as 1-butene, 1-hexene, 1-decene and Chevron Phillips' C20–24 α-olefin mixture can be promoted when less sterically hindered examples of **1** (e.g., **1c**) are employed in combination with MAO (Scheme 5.7) [48]. The mechanism for dimerization is thought to involve an initial 1,2-insertion into an iron-hydride bond followed by a 2,1-insertion of the second alkene, and then chain transfer to give the dimers. Structurally related cobalt systems have also been shown to promote dimerization, albeit with lower activities [62]. Oligomerization of the α-olefins propene, 1-butene and 1-hexene has additionally been achieved with the CF_3-containing iron and cobalt systems **1h** and **2h** yielding highly linear dimers [39].

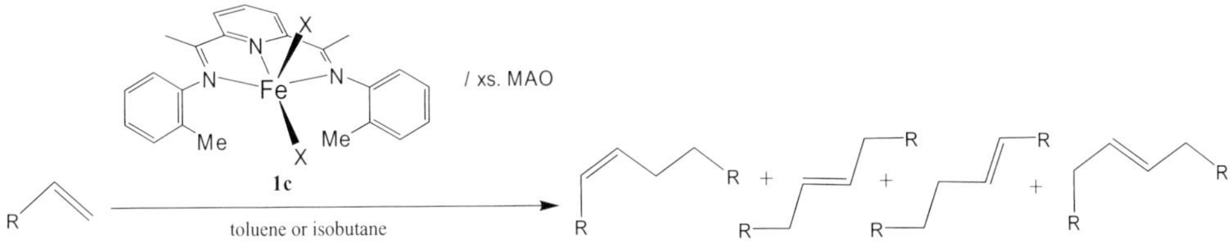

Scheme 5.7 Use of **1c**/MAO to mediate the head-to-head dimerization of α-olefins.

5.4
The Active Catalyst and Mechanism

5.4.1
Active Species

Originally the active species for both iron and cobalt catalysts (generated using MAO), by analogy with metallocenes, was considered to be a highly reactive monomethylated cobalt(II) or iron(II) cation of the type [L1]MMe$^+$ bearing a weakly coordinating counter-anion such as [X-MAO]$^-$ (X = halide, Me). Unequivocal identification of the active species still remains, particularly for iron, elusive due in large measure to the excess of activator employed that can cloud spectroscopic assignment, the paramagnetic nature of the iron center, along with the likely redox and chemical participation of the active species. Nevertheless, numerous spectroscopic investigations have been directed towards identifying the active species for each metal center and, in a number of cases, conflicting conclusions offered (*vide infra*).

5.4.1.1 Iron Catalyst

Both ^{1}H NMR and ^{2}H NMR spectroscopies have been employed to study intermediates formed via activation of ferrous-based polymerization precatalysts with MAO (also studied were AlMe$_3$, AlMe$_3$/B(C$_6$F$_5$)$_3$ and AlMe$_3$/CPh$_3$B(C$_6$F$_5$)$_4$) [99, 100, 122, 123]. With **1b**/MAO, different types of ion-pair complexes are accessible, depending on the number of equivalents of MAO employed. For an Al/Fe ratio of less than 200:1 [2,6-{(2,6-Me$_2$C$_6$H$_3$)N=CMe}$_2$C$_5$H$_3$N]Fe^{II}(µ-Cl)(µ-Me)AlMe$_2$]$^+$[Me-MAO]$^-$ (**10**) has been identified, while at ratios of greater than 500:1 [{(2,6-Me$_2$C$_6$H$_3$)N=CMe}$_2$C$_5$H$_3$N]Fe^{II}(µ-Me)$_2$AlMe$_2$]$^+$[Me-MAO]$^-$ (**11**) has been shown (Scheme 5.8) [103–106]. An EPR spectroscopic study by the same group suggested that the iron center remains in the ferrous state throughout the activation, in agreement with their NMR study [122]. However, this assignment of the formal oxidation state for the active catalyst still remains the subject of some debate. For example, we have shown, using Mössbauer and EPR spectroscopy, that 100% conversion to a species exhibiting a +3 oxidation state occurs on treating **1a** with MAO [93]. On the other hand, Scott *et al.* have postulated a complex that is formally zero valent [124]. In a recent density functional theory study by de Bruin and coworkers they have again concluded that a trivalent oxidation state is most likely, in this case based on an oligomerization catalyst composed of **1c** and MAO [125].

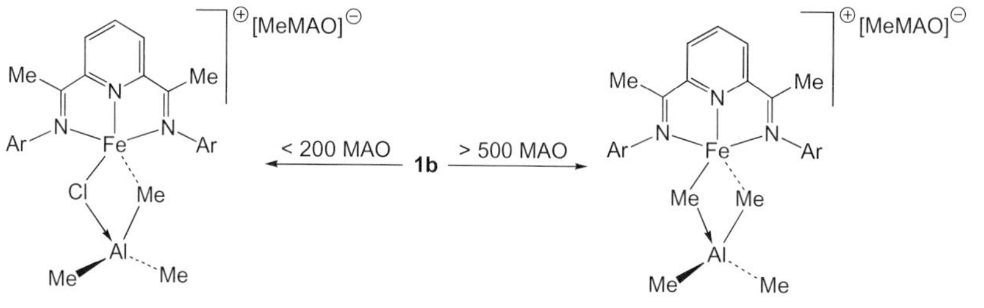

Scheme 5.8 Effects of Al/Fe ratio on activation of **1b** with MAO (Ar = 2,6-Me$_2$C$_6$H$_3$).

A variety of other techniques have also been used to probe the active iron catalyst including ESI mass spectrometry and UV–visible spectroscopy. For example, ESI mass spectrometry has shown the presence of both a four coordinate [2,6-{(2,6-^{i}Pr$_2$C$_6$H$_3$)N=CMe}$_2$C$_5$H$_3$N]FeMe$^+$ cation and an iron hydride species [2,6-{(2,6-^{i}Pr$_2$C$_6$H$_3$)N=CMe}$_2$C$_5$H$_3$N]FeH$^+$ [126] on activation of **1a** with MAO in THF.

Furthermore, using low Fe/MAO ratios, it has been shown that the activation reaction of **1a** by MAO does not reach completion, with a cationic monochloride complex [2,6-{(2,6-^{i}Pr$_2$C$_6$H$_3$)N=CMe}$_2$C$_5$H$_3$N]FeCl$^+$ being the only species observed. On the other hand UV–visible spectroscopic studies have revealed that the alkyl and hydride cations exist as THF adducts. The importance of elapsed time following MAO addition on the composition of the active species has also been investigated by UV–visible spectroscopy [127].

5.4.1.2 Cobalt Catalyst

The mode of activation of MAO-initiated bis(imino)pyridine cobalt halide catalysts has also been the subject of independent investigations which have shown some compelling findings. By taking examples of L1 labeled with deuterium at their ketimine positions, ^{2}H NMR spectroscopy has been used to monitor the reactions of the resulting cobalt complexes with MAO and subsequent interaction with ethylene. With regard to **2a-d$_6$**/MAO, initial reduction of the cobalt(II) precatalyst to cobalt(I) halide followed by conversion to a cobalt(I) methyl, and ultimately to a cobalt(I) cationic species **12**, has been demonstrated [128, 129]. Ethylene addition to the cationic species results in an η^2-bound adduct [{2,6-{(2,6-^{i}Pr$_2$C$_6$H$_3$) N=C(CD$_3$)}$_2$C$_5$H$_3$N}Co(η^2-C$_2$H$_4$)][MeMAO] (**13**), which represents the immediate precursor to the active species **14** (Scheme 5.9). By performing the polymerization with **2a** and perdeuterated MAO (synthesized by the controlled hydrolysis of Al$_2$Me$_6$-d$_{18}$) polymers could be obtained with CD$_3$ end groups, indicating that an abstracted methide group is reincorporated into the polymer. Based on this evidence it is proposed that a methide group from [MeMAO]$^-$ can attack the cobalt center, with concomitant attack by the Lewis acid on the bound ethylene to afford the zwitterionic cobalt dialkyl active species **14** [129].

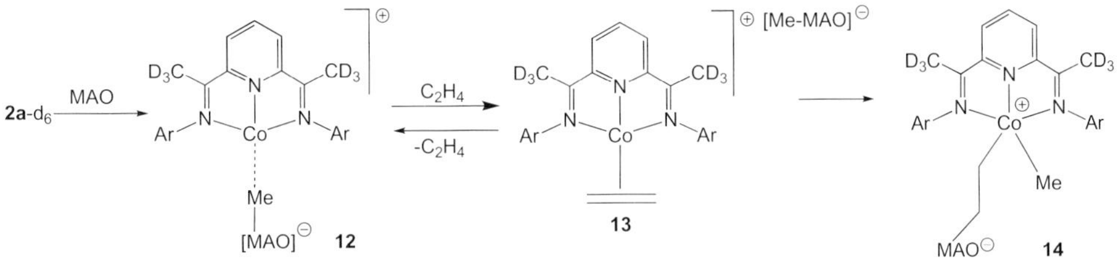

Scheme 5.9 Proposed mechanism for activation **2a** with MAO [129].

5.4.2
Propagation and Chain Transfer Pathways/Theoretical Studies

The propagation mechanism for both iron and cobalt catalysts is considered to follow a Cossee-type mechanism [130] (pathway **A** in Scheme 5.10), in which sequential ethylene coordination and migratory insertion of the bound ethylene into a metal alkyl bond occur before chain transfer terminates the reaction. The

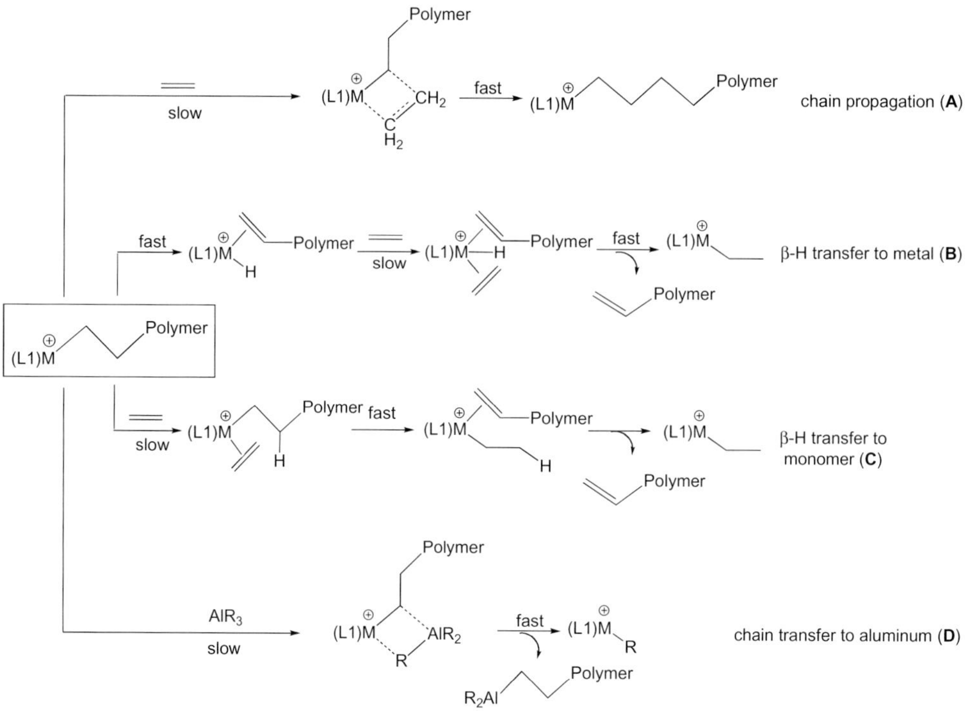

Scheme 5.10 Proposed chain propagation and transfer process in bis(imino)pyridine iron and cobalt catalysts.

first-order rate dependence on ethylene pressure/concentration observed lends some support for the Cossee-type (see Section 5.3.1.3) [6]. However, the molecular weight was found to remain constant with variation in ethylene pressure, in accordance with the overall rate of chain transfer being first order in ethylene. On the basis of these observations, it was concluded that β-H transfer to metal (**B**) or to monomer (**C**) (kinetically indistinguishable) represents the dominant chain-transfer process [6]. Uniquely for iron, however, chain transfer to aluminum (**D**) can also be an operative termination pathway (see Section 5.3.1.3). The preference for cobalt systems to undergo uniquely β-H transfer has been the basis of a combined experimental and theoretical study [131].

In an attempt to rationalize the experimental observations, a number of theoretical studies have been performed to probe the mode of propagation and chain transfer. As with oxidation state considerations for the active catalyst, uncertainty also exists about the precise electronic structure of the iron species. In an initial full *ab initio* study on the diisopropylphenyl Fe(II) catalysts derived from **1a**, Gould and coworkers [132] determined the key structures operating for the first monomer insertion and showed that intermediates along the reaction coordinate have low spin ($S = 0$) configurations. Ziegler *et al.* have carried out density functional theory

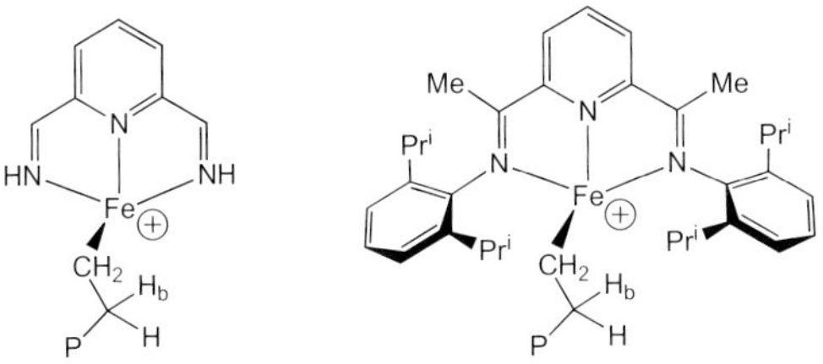

Figure 5.11 Cationic alkyl 'generic' model system **15** and 'real' system derived from **1a**.

(DFT) and DFT/molecular mechanics for an ethylene polymerization process initiated by cationic Fe(II) alkyl species based on the bis(imino)pyridine iron model system **15** and the iron-alkyl cation derived from **1a** (Figure 5.11) [133]. In this study they again consider a singlet electronic state for the Fe(II) and show that the rate-determining step for both termination and propagation is the capture of ethylene by the iron alkyl cation. The steric properties imposed by the N-(2,6-^{i}Pr$_2$C$_6$H$_3$) group were found to inhibit ethylene capture for the termination step and increase the rate of insertion. Using the cationic alkyl complex derived from **1a**, similar calculations reveal that the activity is inhibited by steric crowding [134]. However, Morokuma et al. [135], Zakharov et al. [122] and Budzelaar and coworkers [136] found, using a different DFT method, that the electronic configuration of Fe(II) in the propagation reaction corresponds to a high spin (i.e., quintet or triplet) rather than a low spin state. With regard to the Morokuma work the calculations revealed that for the ethylene insertion step to occur, the alkyl ligand in the cationic form, $[(L1)Fe\text{-}Me]^+$, should occupy the equatorial position in the plane of the trichelating bis(imino)pyridine ligand, trans to the pyridine nitrogen. From geometry optimizations, triplet states for the cationic form were selected as the most likely ones. On the other hand, and to add to the uncertainty about spin states, recent DFT calculations have shown an $[(L1)Fe(III)Me]^{2+}$ species to have a lower ethylene insertion barrier than $[(L1)Fe(II)Me]^+$ [125]. Clearly more work is needed in this challenging area.

For the cobalt catalyst the story appears somewhat clearer. Both experiment and theory have been shown to be in agreement and support a stepwise pathway for the reaction of the cobalt(I) alkyl complexes with 1-alkenes, reacting by β-hydride transfer via a cobalt hydride intermediate [135]. Such cobalt alkyls have also been shown to contain low-spin cobalt(II) antiferromagnetically coupled to a ligand radical anion. The lowest triplet state is thermally accessible and accounts for the observed ^{1}H NMR chemical shifts at room temperature [66].

5.4.3
Well-Defined Iron and Cobalt Alkyls

The alkylation of halide-containing precatalysts **1** and **2** to give well-defined alkyl complexes has been a key synthetic target since the initial catalyst discovery. It was viewed that these alkyl derivatives may serve not only as precursors for single component olefin polymerization catalysts but may also help shed some light on the nature of the propagating species (see Section 5.4.2). To this end alkylation chemistry of **1** and **2** has been the subject of numerous studies.

Interaction of **1a** or **2a** with LiMe results in reduction to give the alkylated M(I) species [2,6-(ArN=CMe)$_2$C$_5$H$_3$N]MCH$_3$ [**16** M = Fe; **17** M = Co; Ar = 2,6-^{i}Pr$_2$C$_6$H$_3$] (Scheme 5.11). Alternatively, **16** or **17** can be prepared from the corresponding monovalent metal halide species [2,6-(ArN=CMe)$_2$C$_5$H$_3$N]MCl (**18** M = Fe; **19** M = Co) on reaction with LiMe [49, 124, 128, 129, 137, 138]. Likewise, the Ph-ketimine cobalt(II) complex, [2,6-{ArN=CR}$_2$C$_5$H$_3$N]CoCl$_2$ (**20** R = Ph, Ar = 2,6-^{i}Pr$_2$C$_6$H$_3$), has been reduced to afford the cobalt(I) chloride [2,6-{(2,6-^{i}Pr$_2$C$_6$H$_3$)N=CPh}$_2$C$_5$H$_3$N]CoCl (**21**) which can then be alkylated with LiMe to afford the corresponding cobalt(I) methyl complex. Activation of the latter complex with Li[B(C$_6$F$_5$)$_4$] gave low activity ethylene polymerization catalysts [49]. Grignard reagents have also been used to generate reduced alkylated species of the form **17** [139].

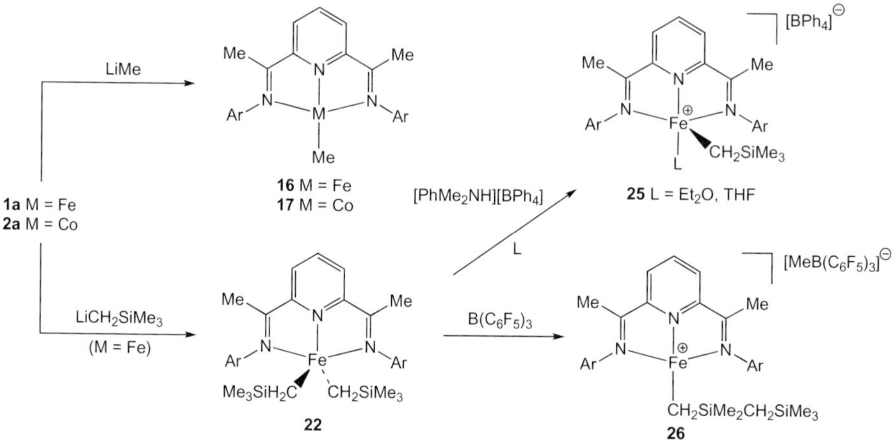

Scheme 5.11 Alkylation chemistry of **1a** and **2a** (Ar = 2,6-^{i}Pr$_2$C$_6$H$_3$).

Preservation of the ferrous oxidation state can be maintained if a more bulky alkyl group is employed. Thus, reaction of **1a** with LiCH$_2$SiMe$_3$ gave the crystallographically characterized dialkyl ferrous species [2,6-(ArN=CMe)$_2$C$_5$H$_3$N]Fe(CH$_2$SiMe$_3$)$_2$ (**22** Ar = 2,6-^{i}Pr$_2$C$_6$H$_3$) (Scheme 5.11) [138]. However, alkylation with LiCH$_2$CMe$_3$ or LiPh does result in reduction to give the monoalkylated derivatives [2,6-(ArN=CMe)$_2$C$_5$H$_3$N]FeR (**23** R = CH$_2$CMe$_3$; **24** R = Ph); the X-ray structure of **23** reveals a distorted square planar geometry [140]. Dialkyl iron(II) species of the type **22** can be more conveniently prepared from the reaction of **1** with the precursor complex [FeR$_2$(py)$_2$] (R = CH$_2$Ph, CH$_2$CMe$_2$Ph and CH$_2$SiMe$_3$) [141].

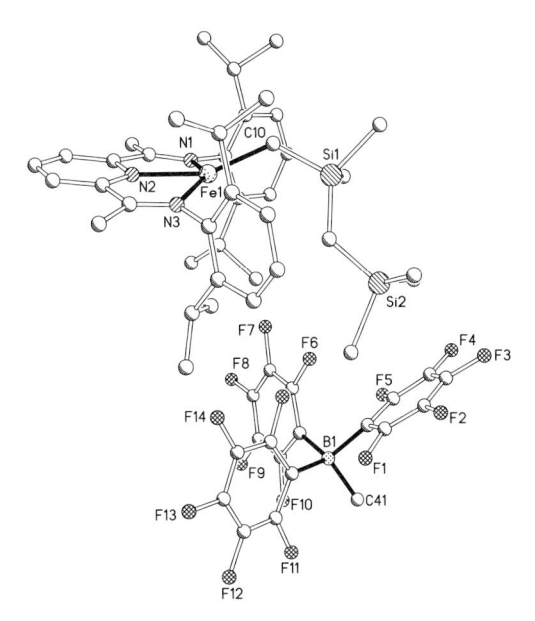

Figure 5.12 Molecular structure of cation-anion pair in active catalyst **26** (data from Ref. [129], with permission of the copyright holders, was used to generate the figure).

Despite a clear understanding of the nature of the active species for MAO-activated iron species (see Section 5.4.1.1), it has been demonstrated that iron(II) alkyl cations can perform as propagating species [142]. Thus, interaction of **22** with [PhMe$_2$NH][BPh$_4$] in the presence of diethyl ether or THF gives cationic [{2,6-(ArN=CMe)$_2$C$_5$H$_3$N}M(CH$_2$SiMe$_3$)(L)][BPh$_4$] (**25** L = Et$_2$O or THF, Ar = 2,6-^{i}Pr$_2$C$_6$H$_3$), respectively (Scheme 5.11); the base-free species [{2,6-(ArN=CMe)$_2$C$_5$H$_3$N}M(CH$_2$SiMe$_2$CH$_2$SiMe$_3$)][MeB(C$_6$F$_5$)$_3$] (**26** Ar = 2,6-^{i}Pr$_2$C$_6$H$_3$) is accessible on reaction of **22** with B(C$_6$F$_5$)$_3$. Although **25** gives very low activity for ethylene polymerization, the base-free species **26**, with the weaker coordinating anion, shows productivities approaching the MAO-activated catalyst **1a**; the molecular structure of the cation–anion pair in **26** is shown in Figure 5.12.

Curiously, the alkylation of **1a** with LiCH$_2$SiMe$_3$ has proved sensitive to the reaction conditions employed with a number of unusual minor side products isolable [136]. For example, reaction of *in situ* generated **1a** with 2 equiv of LiCH$_2$SiMe$_3$ gave as the major product the expected dialkyl complex **22** together with the minor products [{2,6-((2,6-^{i}Pr$_2$C$_6$H$_3$)N=CMe)$_2$-2-CH$_2$SiMe$_3$}C$_5$H$_3$N]Fe(CH$_2$SiMe$_3$) (**27**) and {2-{(2,6-^{i}Pr$_2$C$_6$H$_3$)N=CMe}-6-[{(2,6-^{i}Pr$_2$C$_6$H$_3$)NCMe(CH$_2$SiMe$_3$)}C$_5$H$_3$N]Fe(CH$_2$SiMe$_3$) (**28**), for which alkylation had occurred at either the pyridine ring 2-position or the imine C atom, respectively (Figure 5.13). On the other hand, use of analytically pure **1a** yielded **22** together with [2,6-{(2,6-^{i}Pr$_2$C$_6$H$_3$)NC=(CH$_2$)}$_2$C$_5$H$_3$N]Fe(μ-Cl)Li(THF)$_3$ (**29**) and the dinuclear iron(I) complex **30**, the result of reductive coupling of the two ligand frameworks through the methyl carbon wings (Figure 5.13). All the species (**27–30**) display, on

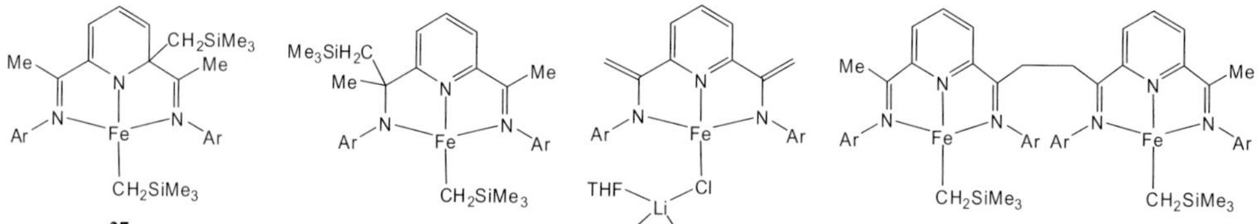

Figure 5.13 Side products isolable from the reaction of **1a** with LiCH$_2$SiMe$_3$ (Ar = 2,6-^{i}Pr$_2$C$_6$H$_3$).

activation with MAO, high activities for olefin polymerization, forming two types of polymer, the nature of which is related to the formal oxidation state of the metal center.

In order to mimic more effectively the propagating species in iron-based polymerization, a recent study has been conducted to prepare β-hydrogen-containing Fe(I) alkyl complexes, [2,6-{(2,6-^{i}Pr$_2$C$_6$H$_3$)N=CPh}$_2$C$_5$H$_3$N]FeR (R = Et, ^{i}Bu) [143]. It was found that the kinetic stability of these species is inversely related to the number of β-hydrogens present, with the iron ethyl complex undergoing clean loss of ethane over 3 h while the iron isobutyl compound had a half life of over 12 h. Deuterium labeling studies were performed and revealed a decomposition pathway in which β-H elimination was followed by cyclometallation of an isopropyl group (Scheme 5.12).

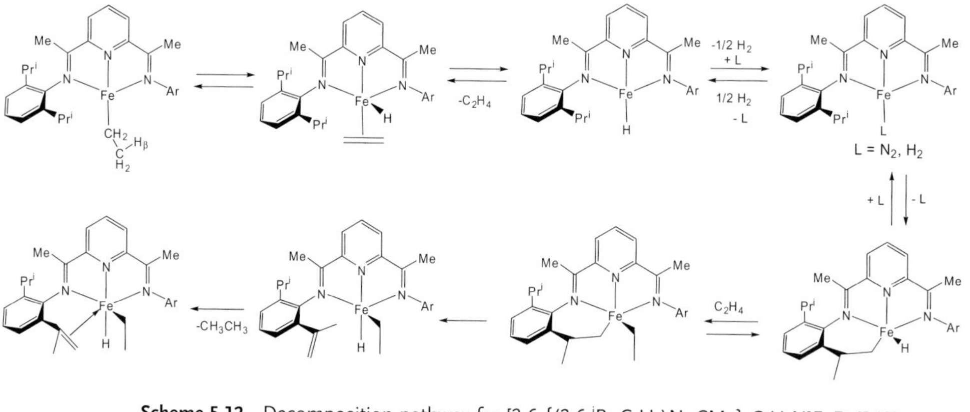

Scheme 5.12 Decomposition pathway for [2,6-{(2,6-^{i}Pr$_2$C$_6$H$_3$)N=CMe}$_2$C$_5$H$_3$N]FeEt [143].

It should be noted that, as with the free ligands, the bound ligand in **1** and **2** can be the site for a range of chemistries including deprotonation (see Chapter 4 and [135, 144]). In addition several studies have involved the reduction of **1** or **2** in the presence of nitrogen as a means of generating dinitrogen-containing complexes (see Chapter 4 and [53, 145–148]).

5.5
Other Applications

5.5.1
Immobilization

To circumvent problems such as reactor fouling and high polymerization exothermicities that have been encountered in continuous flow processes featuring iron and cobalt catalysts, considerable effort has been applied to anchoring (through covalent and non-covalent interactions) these homogeneous catalysts to various inorganic and organic media as supports. The reader is referred to a number of review articles that deal with this subject in more detail [13–15, 149]. With regard to covalent interactions a variety of inventive strategies have been employed and indeed most sites on the bis(imino)pyridine frame have been exploited as linker site to the support (Table 5.2).

One such example that has appeared recently features the central pyridine as the site for attachment. For example, Kim *et al.* have exploited the 4-substituted *O*-allyl group in L1 (Ar = 2,6-Me$_2$C$_6$H$_3$) as a means of generating silica-supported iron and cobalt systems (**31**) (Scheme 5.13) [58]. On activation with MAO these immobilized catalysts exhibited about 100-fold lower activity when compared to their homogeneous counterparts. This lowering in activity has been attributed to either diffusion limitation of monomer into the interior pores of the supported catalyst or to the result of reduced active sites present in the heterogeneous variant.

Table 5.2 Sites reported for tethering iron and cobalt polymerization catalysts.

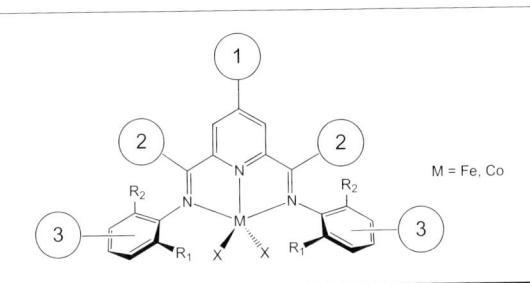

Site	Reference
1	[57, 58, 150]
2	[151]
3	[25, 137]

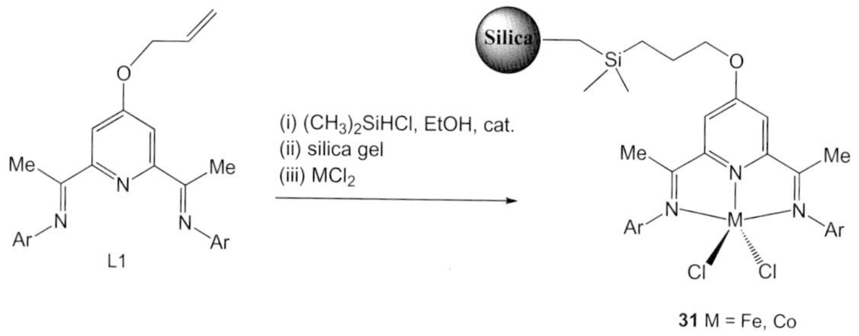

Scheme 5.13 Use of 4-pyridyl position as means of supporting **1** on silica.

5.5.2
Reactor Blending and Tandem Catalysis

As a means of controlling the properties of a polyolefin the use of two or more different polymerization catalysts in the same reactor, to allow in-reactor blending, has been extensively employed industrially to control molecular weight and molecular weight distribution. Bis(imino)pyridine iron and cobalt catalysts have found applications in the area, and a variety of reports have emerged on the use of bis(imino)pyridine iron/cobalt systems as one component of the process [150–160]. For example, by combining **1a**/MAO with an α-diimine nickel(II) halide/MAO catalyst, access to blended polymers consisting of a mixture of linear and branched polyethylenes is possible from a single ethylene feed [152]. By contrast, **1a**/MAO in combination with a metallocene can give rise to blends of polyethylenes with different molecular weights [152].

A more recent variation on reactor blending is to employ two catalysts that can operate in tandem. Such a process that lends itself to this approach is the commercially attractive formation of linear low density polyethylene (LLDPE) from a single ethylene feed. When an iron catalyst that forms α-olefins selectively is combined with a metallocene copolymerization catalyst, linear low density polyethylene (LLDPE) and, in some cases, ultra low density polyethylene (ULDPE) can be obtained from a single ethylene feedstock [155–159]. For example, the bis-indenyl *ansa*-metallocene [Me$_2$Si(Ind)$_2$]ZrCl$_2$/MAO has been used to copolymerize ethylene with the α-olefins obtained using [2,6-(ArN=CMe)$_2$C$_5$H$_3$N]FeCl$_2$ (**1** Ar = 2-EtC$_6$H$_4$)/MAO; branched polyethylenes with ethyl, butyl, and longer chain branches, have been detected (Scheme 5.14) [157]. Tandem polymerizations using bis(imino)pyridine cobalt catalysts have also been reported [160].

5.6
Conclusions and Outlook

Despite the discovery of iron and cobalt catalysts capable of ethylene polymerization occurring only ten years ago, the impact on polymerization catalysis has been

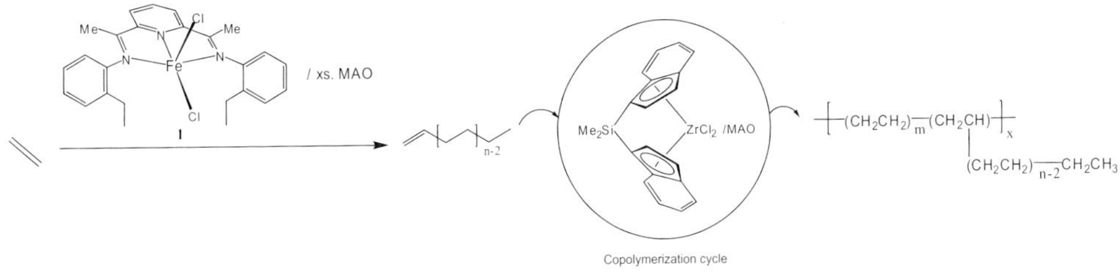

Scheme 5.14 Use of 1/MAO in tandem catalysis to form LLDPE.

truly remarkable, with research groups from around the world contributing to the field. This chapter has attempted to give a flavor of the more significant advances with special emphasis being placed on understanding the *modus operandi* of the catalyst and the role of the ligand set in supporting the catalytic transformations. The propensity of the precursor iron(II) and cobalt(II) halide species to undergo redox changes during alkylation, along with some conflicting spectroscopic studies, have helped to ignite the debate as to the true oxidation state and spin state of the active catalyst in an MAO-activated system. The potential onset of electron transfer processes within the metal–ligand unit of the active catalysts no doubt further complicates this exercise. The discovery of a well-defined single-component iron polymerization catalyst that approaches the activity of the MAO-initiated system represents a highlight and has, in turn, triggered a wealth of fascinating organometallic chemistry. From an industrial viewpoint these catalysts are highly versatile and not only allow access to HDPE and linear α-olefins (with Schulz–Flory distributions) but also can be adapted to facilitate the formation of LLDPE or to form linear α-olefins with a Poisson distribution. While considered beyond the scope of this chapter it should also be noted that bis(imino)pyridine iron complexes have been explored in recent years for a variety of other types of catalysis (see Chapter 4 and [161–165]). Likewise, the developments in the search for alternative ligand sets that can support active oligomerization/polymerization iron and cobalt catalysts have been the subject of considerable attention [13–15, 166] and help signpost further possibilities in this rich and fruitful area.

List of Abbreviations

acac	acetylacetonate
Ar	aryl
^{i}Bu	isobutyl
^{t}Bu	*tert*-butyl
Bn	benzyl
Cy	cyclohexyl
DFT	density functional theory
DEAC	diethylaluminium chloride

DRIFT	diffuse reflectance infrared Fourier transform
Et	ethyl
Et_2O	diethyl ether
EPR	electron paramagnetic resonance
ESI-MS	electrospray ionization tandem mass spectrometry
Fc	ferrocenyl
$g\,mmol^{-1}h^{-1}bar^{-1}$	grams per millimole (of precatalyst) per hour per bar
HDPE	high density polyethylene
IR	infrared
Pr^i	isopropyl
LLDPE	linear low density polyethylene
MAO	methylalumoxane
Mesityl or Mes	2,4,6-trimethylphenyl
MMAO	modified methylalumoxane, $AlMeO:Al^iBuO = 3:1$
Me	methyl
MeCN	acetonitrile
NMR	nuclear magnetic resonance
Ph	phenyl
PS	polystyrene
py	pyridine
TEA	triethylaluminum
THF	tetrahydrofuran
ULDPE	ultra low density polyethylene

References

1. Britovsek, G.J.P., Gibson, V.C., Kimberley, B.S., Maddox, P.J., McTavish, S.J., Solan, G.A., White, A.J.P., and Williams, D.J. (1998) *Chem. Commun.*, 849–850.
2. Small, B.L., and Brookhart, M. (1998) *J. Am. Chem. Soc.*, **120**, 7143–7144.
3. Small, B.L., Brookhart, M., and Bennett, A.M.A. (1998) *J. Am. Chem. Soc.*, **120**, 4049–4050.
4. Bennett, A.M.A. (1998) (E.I. du Pont de Nemours and Co, USA), PCT Int. Appl. WO 9827124 [*Chem. Abstr.*, 1998, **129**, 122973x].
5. Bennett, A.M.A. (1999) *Chemtech*, **29**, 24–28.
6. Britovsek, G.J.P., Bruce, M., Gibson, V.C., Kimberley, B.S., Maddox, P.J., Mastroianni, S., McTavish, S.J., Redshaw, C., Solan, G.A., Strömberg, S., White, A.J.P., and Williams, D.J. (1999) *J. Am. Chem. Soc.*, **121**, 8728–8740.
7. Britovsek, G.J.P., Dorer, B.A., Gibson, V.C., Kimberley, B.S., and Solan, G.A. (1999) (BP Chemicals Ltd, UK), PCT Int Appl WO 9912981 [*Chem. Abstr.*, 1999, **130**, 252793].
8. Gibson, V.C., and Spitzmesser, S.K. (2003) *Chem. Rev.*, **103**, 283–315.
9. Britovsek, G.J.P., Gibson, V.C., and Wass, D.F. (1999) *Angew. Chem. Int. Ed.*, **38**, 428–447.
10. Ittel, S.D., Johnson, L.K., and Brookhart, M. (2000) *Chem. Rev.*, **100**, 1169–1203.
11. Mecking, S. (2001) *Angew. Chem. Int. Ed.*, **40**, 534–540.
12. Park, S., Han, Y., Kim, S.K., Lee, J., Kim, H.K., and Do, Y. (2004) *J. Organomet. Chem.*, **689**, 4263–4276.
13. Gibson, V.C., Redshaw, C., and Solan, G.A. (2007) *Chem. Rev.*, **107**, 1745–1776.

14. Gibson, V.C., and Solan, G.A. (2009) *Topics in Organometallic Chemistry*, vol. 26 (ed. Z. Guan), Springer, Berlin/Heidelberg, Chapter 4, pp. 107–158.
15. Bianchini, C., Giambastiani, G., Rios, G.I., Mantovani, G., Meli, A., and Segarra, A.M. (2006) *Coord. Chem. Rev.*, **250**, 1391–1418.
16. Britovsek, G.J.P., Mastroianni, S., Solan, G.A., Baugh, S.P.D., Redshaw, C., Gibson, V.C., White, A.J.P., Williams, D.J., and Elsegood, M.R.J. (2000) *Chem. Eur. J.*, **6**, 2221–2231.
17. Granifo, J., Bird, S.J., Orrell, K.G., Osborne, A.G., and Sik, V. (1999) *Inorg. Chim. Acta*, **295**, 56–63.
18. Schmidt, R., Hammon, U., Gottfried, S., Welch, M.B., and Alt, H.G. (2003) *J. Appl. Polym. Sci.*, **88**, 476–482.
19. Schmidt, R., Welch, M.B., Knudsen, R.D., Gottfried, S., and Alt, H.G. (2004) *J. Mol. Catal. A: Chem.*, **222**, 9–15.
20. Cetinkaya, B., Cetinkaya, E., Brookhart, M., and White, P.S. (1999) *J. Mol. Catal. A: Chem.*, **142**, 101–112.
21. Kim, I., Han, B.H., Ha, Y.-S., Ha, C.-S., and Park, D.-W. (2004) *Catal. Today*, **93–95**, 281–285.
22. Gibson, V.C., Long, N.J., Oxford, P.J., White, A.J.P., and Williams, D.J. (2006) *Organometallics*, **25**, 1932–1939.
23. Schmidt, R., Welch, M.B., Palackal, S.J., and Alt, H.G. (2002) *J. Mol. Catal. A: Chem.*, **179**, 155–173.
24. Liu, J.-Y., Zheng, Y., Li, Y.-G., Pan, L., Li, Y.-S., and Hu, N.-H. (2005) *J. Organomet. Chem.*, **690**, 1233–1239.
25. Liu, C., and Jin, G. (2002) *New J. Chem.*, **26**, 1485–1489.
26. Small, B.L., and Brookhart, M. (1999) *Macromolecules*, **32**, 2120–2130.
27. Pelascini, F., Peruch, F., Lutz, P.J., Wesolek, M., and Kress, J. (2005) *Eur. Polym. J.*, **41**, 1288–1295.
28. Chen, J., Huang, Y., Li, Z., Zhang, Z., Wei, C., Lan, T., and Zhang, W. (2006) *J. Mol. Catal. A: Chem.*, **259**, 133–141.
29. Ivanchev, S.S., Yakimansky, A.V., and Rogozin, D.G. (2004) *Polymer*, **45**, 6453–6459.
30. Oleinik, I.I., Oleinik, I.V., Abdrakhmanov, I.B., Ivanchev, S.S., and Tolstikov, G.A. (2004) *Russ. J. Gen. Chem.*, **74**, 1575–1578.
31. Ivanchev, S.S., Tolstikov, G.A., Badaev, V.K., Oleinik, I.I., Ivancheva, N.I., Rogozin, D.G., Oleinik, I.V., and Myakin, S.V. (2004) *Kinet. Catal.*, **45**, 176–182.
32. Oleinik, I.I., Oleinik, I.V., Abdrakhmanov, I.B., Ivanchev, S.S., and Tolstikov, G.A. (2004) *Russ. J. Gen. Chem.*, **74**, 1423–1427.
33. Chen, Y., Chen, R., Qian, C., Dong, X., and Sun, J. (2003) *Organometallics*, **22**, 4312–4321.
34. Ionkin, A.S., Marshall, W.J., Adelman, D.J., Fones, B.B., Fish, B.M., Schiffhauer, M.F., Soper, P.D., Waterland, R.L., Spence, R.E., and Xie, T. (2007) *J. Polym. Sci. Part A: Polym. Chem.*, **46**, 585–611.
35. Zhang, Z., Zou, J., Cui, N., Ke, Y., and Hu, Y. (2004) *J. Mol. Catal. A: Chem.*, **219**, 249–254.
36. Zhang, Z., Chen, S., Zhang, X., Li, H., Ke, Y., Lu, Y., and Hu, Y. (2005) *J. Mol. Catal. A: Chem.*, **230**, 1–8.
37. Chen, Y., Qian, C., and Sun, J. (2003) *Organometallics*, **22**, 1231–1236.
38. Esteruelas, M.A., Lopez, A.M., Mendez, L., Olivan, M., and Onate, E. (2003) *Organometallics*, **22**, 395–406.
39. Tellmann, K.P., Gibson, V.C., White, A.J.P., and Williams, D.J. (2005) *Organometallics*, **24**, 280–286.
40. Bluhm, M.E., Folli, C., and Doring, M. (2004) *J. Mol. Catal. A: Chem.*, **212**, 13–18.
41. Paulino, I.S., and Schuchardt, U. (2004) *J. Mol. Catal. A: Chem.*, **211**, 55–58.
42. Ionkin, A.S., Marshall, W.J., Adelman, D.J., Fones, B.B., Fish, B.M., and Schiffhauer, M.F. (2006) *Organometallics*, **25**, 2978–2992.
43. Ionkin, A.S., Marshall, W.J., Adelman, D.J., Shoe, A.L., Spence, R.E., and Xie, T. (2006) *J. Polym. Sci. A, Polym. Chem.*, **44**, 2615–2635.
44. Adelman, D.J., and Ionkin, S.A. (2007) (E.I. du Pont de Nemours and Co, USA), PCT Int. Appl. WO 2007021955.
45. Abu-Surrah, A.S., Lappalainen, K., Piironen, P., Lehmus, P., Repo, T., and

Leskelä, M. (2002) *J. Organomet. Chem.*, **648**, 55–61.

46. Ma, Z., Qui, J., Xu, D., and Hu, Y. (2001) *Macromol. Rapid Commun.*, **22**, 1280–1283.
47. Kaul, F.A.R., Puchta, G.T., Frey, G.D., Herdtweck, E., and Herrmann, W.A. (2007) *Organometallics*, **26**, 988–999.
48. Small, B.L., and Marcucci, A.J. (2001) *Organometallics*, **20**, 5738–5744.
49. Kleigrewe, N., Steffen, W., Blőmker, T., Kehr, G., Fröhlich, R., Wibbeling, B., Erker, G., Wasilke, J.-C., Wu, G., and Bazan, G.C. (2005) *J. Am. Chem. Soc.*, **127**, 13955–13968.
50. Britovsek, G.J.P., Gibson, V.C., Hoarau, O.D., Spitzmesser, S.K., White, A.J.P., and Williams, D.J. (2003) *Inorg. Chem.*, **42**, 3454–3465.
51. McTavish, S., Britovsek, G.J.P., Smit, T.M., Gibson, V.C., White, A.J.P., and Williams, D.J. (2007) *J. Mol. Catal. A: Chem.*, **261**, 293–300.
52. Smit, T.M., Tomov, A.K., Gibson, V.C., White, A.J.P., and Williams, D.J. (2004) *Inorg. Chem.*, **43**, 6511–6512.
53. Archer, A.M., Bouwkamp, M.W., Cortez, M.P., Lobkovsky, E., and Chirik, P.J. (2006) *Organometallics*, **25**, 4269–4278.
54. Barbaro, P., Bianchini, C., Giambastiani, G., Rioa, I.G., Meli, A., Oberhauser, W., Segarra, A.M., Sorace, L., and Toti, A. (2007) *Organometallics*, **26**, 4639–4651.
55. Nückel, S., and Burger, P. (2001) *Organometallics*, **20**, 4345–4359.
56. Pelascini, F., Wesolek, M., Peruch, F., and Lutz, P.J. (2006) *Eur. J. Inorg. Chem.*, 4309–4316.
57. Seitz, M., Milius, W., and Alt, H.G. (2007) *J. Mol. Catal. A: Chem.*, **261**, 246–253.
58. Kim, I., Heui Han, B., Ha, C.-S., Kim, J.-K., and Suh, H. (2003) *Macromolecules*, **36**, 6689–6691.
59. Ionkin, A.S., Marshall, W.J., Adelman, D.J., Fones, B.B., Spence, R.E., and Xi, T. (2008) *Organometallics*, **27**, 1147–1156.
60. Bianchini, C., Giambastiani, G., Guerrero, I.R., Meli, A., Passaglia, E., and Gragnoli, T. (2004) *Organometallics*, **23**, 6087–6089.
61. Takeuchi, D., Anada, K., and Osakada, K. (2005) *Bull. Chem. Soc. Jpn.*, **78**, 1868–1878.
62. Small, B.L. (2003) *Organometallics*, **22**, 3178–3183.
63. Kim, I., Hwang, J.-M., Lee, J.K., Ha, C.S., and Woo, S.I. (2003) *Macromol. Rapid Commun.*, **24**, 508–511.
64. Edwards, D.A., Edwards, S.D., Pringle, T.J., and Thornton, P. (1992) *Polyhedron*, **11**, 1569–1573.
65. Lappalainen, K., Yliheikkilae, K., Abu-Surrah, A.S., Polamo, M., Lekellae, M., and Repo, T. (2005) *Z. Anorg. Allg. Chem.*, **631**, 763–768.
66. Knijnenburg, Q., Hetterschied, D., Kooistra, T.M., and Budzelaar, P.H.M. (2004) *Eur. J. Inorg. Chem.*, 1204–1211.
67. Kim, I., Byeong, H., Kim, J.S., and Ha, C.-S. (2005) *Catal. Lett.*, **101**, 249–253.
68. Qian, C., Gao, F., Chen, Y., and Gao, L. (2003) *Synlett*, 1419–1422.
69. Campora, J., Perez, C.M., Rodriguez-Delgado, A., Naz, A.M., Palma, P., and Alvarez, E. (2007) *Organometallics*, **26**, 1104–1107.
70. Campora, J., Naz, A.M., Palma, P., Rodriguez-Delgado, A., Alvarez, E., Tritto, I., and Boggioni, L. (2008) *Eur. J. Inorg. Chem.*, 1871–1879.
71. Small, B.L., Carney, M.J., Holman, D.M., O'Rourke, C.E., and Halfen, J.A. (2004) *Macromolecules*, **37**, 4375–4386.
72. Nakayama, Y., Sogo, K., Yasuda, H., and Shiono, T. (2005) *J. Polym. Sci., A, Polym. Chem.*, **43**, 3368–3375.
73. Lappalainen, K., Yliheikkilä, K., Abu-Surrah, A.S., Polamo, M., Leskelä, M., and Repo, T. (2005) *Z. Anorg. Allg. Chem.*, **631**, 763–768.
74. Britovsek, G.J.P., Gibson, V.C., Kimberley, B.S., Mastroianni, S., Redshaw, C., Solan, G.A., and White, A.J.P. (2001) *J. Chem. Soc., Dalton Trans.*, 1639–1644.
75. Moody, L.S., Mackenzie, P.B., Killian, C.M., Lavoie, G.G., Pansik, J.A., Jr., Barrett, A.G.M., Smith, T.W., and Pearson, J.C. (2000) (Eastman Chemical Company), WO 0050470, (*Chem. Abstr.*, 2000, **133**, 208316).
76. Amort, C., Malaun, M., Krajete, A., Kopacka, H., Wurst, K., Christ, M.,

Lilge, D., and Kristen, M.O. (2002) *Appl. Organomet. Chem.*, **16**, 506–516.

77. Bianchini, C., Mantovani, G., Meli, A., Migliacci, F., Zanobini, F., Laschi, F., and Sommazzi, A. (2003) *Eur. J. Inorg. Chem.*, 1620–1631.
78. Wallenhorst, C., Kehr, G., Luftmann, H., Frohlich, R., and Erker, G. (2008) *Organometallics*, **27**, 6547–6556.
79. Huang, H.Y.-B., Ma, M.X.L., Zheng, S.N., Chen, C.J.-X., and Wei, C.-X. (2006) *Acta Crystallogr. Sect. E*, **62**, o3044–o3045.
80. Vance, A.L., Alcock, N.W., Heppert, J.A., and Busch, D.H. (1998) *Inorg. Chem.*, **37**, 6912–6920.
81. Mentes, A., Fawcett, J., and Kemmitt, R.D.W. (2001) *Acta Crystallogr., Sect. E*, **57**, o424–o425.
82. Meehan, P.R., Alyea, E.C., and Ferguson, G. (1997) *Acta Crystallogr., Sect. C*, **C53**, 888–890.
83. Agrifoglio, G., Reyes, J., Atencio, R., and Briceno, A. (2008) *Acta Crystallogr., Sect. E*, **64**, o28–o29.
84. Sugiyama, H., Korobkov, I., Gambarotta, S., Möller, A., and Budzelaar, P.H.M. (2004) *Inorg. Chem.*, **43**, 5771–5779.
85. Sugiyama, H., Gambarotta, S., Yap, G.P.A., Wilson, D.R., and Thiele, S.K.-H. (2004) *Organometallics*, **23**, 5054–5061.
86. Clentsmith, G.K.B., Gibson, V.C., Hitchcock, P.B., Kimberley, B.S., and Rees, C.W. (2002) *Chem. Commun.*, 1498–1499.
87. Blackmore, I.J., Gibson, V.C., Hitchcock, P.B., Rees, C.W., Williams, D.J., and White, A.J.P. (2005) *J. Am. Chem. Soc.*, **127**, 6012–6020.
88. Khorobkov, I., Gambarotta, S., and Yap, G.P.A. (2002) *Organometallics*, **21**, 3088–3090.
89. Milione, S., Cavallo, G., Tedesco, C., and Grassi, A. (2002) *J. Chem. Soc., Dalton Trans.*, 1839–1846.
90. Bruce, M., Gibson, V.C., Redshaw, C., Solan, G.A., White, A.J.P., and Williams, D.J. (1998) *Chem. Commun.*, 2523–2524.
91. Knijnenburg, Q., Smits, J.M.M., and Budzelaar, P.H.M. (2004) *C. R. Chim.*, **7**, 865–869.
92. Knijnenburg, Q., Smits, J.M.M., and Budzelaar, P.H.M. (2006) *Organometallics*, **25**, 1036–1046.
93. Britovsek, G.J.P., Clentsmith, G.K.B., Gibson, V.C., Goodgame, D.M.L., McTavish, S.J., and Pankhurst, Q.A. (2002) *Catal. Commun.*, **3**, 207–211.
94. Enright, D., Gambarotta, S., Yap, P.A., and Budzelaar, P.H.M. (2002) *Angew. Chem. Int. Ed.*, **41**, 3873–3876.
95. Knijnenburg, Q., Gambarotta, S., and Budzelaar, P.H.M. (2006) *Dalton Trans.*, 5442–5448.
96. Rodriguez-Delgado, A., Campora, J., Naz, M.A., Palma, P., and Reyes, M.L. (2008) *Chem. Commun.*, 5230–5232.
97. CSD search December 2008.
98. Kaminsky, W. (2008) *Macromol. Chem. Phys.*, **209**, 459–466.
99. Talsi, E.P., Babushkin, D.E., Semikolenova, N.V., Zudin, V.N., Panchenko, V.N., and Zakharov, V.A. (2001) *Macromol. Chem. Phys.*, **202**, 2046–2051.
100. Bryliakov, K.P., Semikolenova, N.V., Zakharov, V.A., and Talsi, E.P. (2004) *Organometallics*, **23**, 5375–5378.
101. Kumar, K.R., and Sivaram, S. (2000) *Macromol. Chem. Phys.*, **201**, 1513–1520.
102. Semikolenova, N.V., Zakharov, V.A., Talsi, E.P., Babushkin, D.E., Sobolev, A.P., Echevskaya, L.G., and Khusniyarov, M.M. (2002) *J. Mol. Catal. A: Chem.*, **182**, 283–294.
103. Radhakrishnan, K., Cramail, H., Deffieux, A., Francois, P., and Momtaz, A. (2003) *Macromol. Rapid Commun.*, **24**, 251–254.
104. Wang, Q., Yang, H., and Fan, Z. (2002) *Macromol. Rapid Commun.*, **23**, 639–642.
105. Barabanov, A.A., Bukatov, G.D., Zakharov, V.A., Semikolenova, N.V., Echevskaja, L.G., and Matsko, M.A. (2005) *Macromol. Chem. Phys.*, **206**, 2292–2298.
106. Babik, S.T., and Fink, G. (2002) *J. Mol. Catal. A:Chem.*, **188**, 245–253.
107. Hanton, M.J., and Tenza, K. (2008) *Organometallics*, **27**, 5712–5716.
108. Al-Benna, S., Sarsfield, M.J., Thornton-Pett, M., Ormsby, D., Maddox, P.J., Bres, P., and Bochmann, M. (2000)

109. Kreisher, K., Kipke, J., Bauerfeind, M., and Sundermeyer, J. (2001) *Z. Anorg. Allg. Chem.*, **627**, 1023–1028.
110. Beaufort, L., Benvenuti, F., and Noels, A.F. (2006) *J. Mol. Catal. A: Chem.*, **260**, 210–214.
111. Beaufort, L., Benvenuti, F., and Noels, A.F. (2006) *J. Mol. Catal. A: Chem.*, **260**, 215–220.
112. Fernandes, S., M. Bellabarba, R., Ribeiro, D.F., Gomes, P.T., Ascenso, J.R., Mano, J.F., Dias, A.R., and Marques, M.M. (2002) *Polym. Int.*, **51**, 1301–1303.
113. Brookhart, M.S., and Small, B.L. (1998) (Dupont/UNC), WO 9830612 [*Chem. Abstr.*, 1998, **129**, 149375r].
114. Pellecchia, C., Mazzeo, M., and Pappalardo, D. (1998) *Macromol. Rapid Commun.*, **19**, 651–655.
115. Keim, W. (1990) *Angew. Chem. Int. Ed. Engl.*, **29**, 235–244.
116. Ionkin, A.S., Marshall, W.J., Adelman, D.J., Fones, B.B., Fish, B.M., and Schliffhauer, M.F. (2008) *Organometallics*, **27**, 1902–1911.
117. Flory, P.J. (1940) *J. Am. Chem. Soc.*, **62**, 1561–1565.
118. Schulz, G.V. (1939) *Z. Phys. Chem.*, **B43**, 25–46.
119. Britovsek, G.J.P., Cohen, S.A., Gibson, V.C., and van Meurs, M. (2004) *J. Am. Chem. Soc.*, **126**, 10701–10702.
120. Britovsek, G.J.P., Cohen, S.A., Gibson, V.C., Maddox, P.J., and Van Meurs, M. (2002) *Angew. Chem. Int. Ed.*, **41**, 489–491.
121. Van Meurs, M., Britovsek, G.J.P., Gibson, V.C., and Cohen, S.A. (2005) *J. Am. Chem. Soc.*, **127**, 9913–9923.
122. Bryliakov, K.P., Semikolenova, N.V., Zudin, V.N., Zakharov, V.A., and Talsi, E.P. (2004) *Catal. Commun.*, **5**, 45–48.
123. Talsi, E.P., Bryliakov, K.P., Semikolenova, N.V., Zakharov, V.A., and Bochmann, M. (2007) *Kinet. Catal.*, **48**, 490–504.
124. Scott, J., Gambarotta, S., Korobkov, I., and Budzelaar, P.H.M. (2005) *Organometallics*, **24**, 6298–6300.
125. Raucoules, R., de Bruin, T., Raybaud, P., and Adamo, C. (2008) *Organometallics*, **27**, 3368–3377.
126. Castro, P.M., Lahtinen, P., Axenov, K., Viidanoja, J., Kotiaho, T., Leskela, M., and Repo, T. (2005) *Organometallics*, **24**, 3664–3670.
127. Schmidt, R., Das, P.K., Welch, M.B., and Knudsen, R.D. (2004) *J. Mol. Catal. A. Chem.*, **222**, 27–45.
128. Kooistra, T.M., Knijnenburg, Q., Smits, J.M.M., Horton, A.D., Budzelaar, P.H.M., and Gal, A.W. (2001) *Angew. Chem. Int. Ed.*, **40**, 4719–4722.
129. Humphries, M.J., Tellmann, K.P., Gibson, V.C., White, A.J.P., and Williams, D.J. (2005) *Organometallics*, **24**, 2039–2050.
130. Cossee, P. (1964) *J. Catal.*, **3**, 80–88.
131. Tellmann, K.P., Humphries, M.J., Rzepa, H.S., and Gibson, V.C. (2004) *Organometallics*, **23**, 5503–5513.
132. Griffiths, E.A.H., Britovsek, G.J.P., Gibson, V.C., and Gould, I.R. (1999) *Chem. Commun.*, 1333–1334.
133. Deng, L., Margl, P., and Ziegler, T. (1999) *J. Am. Chem. Soc.*, **121**, 6479–6487.
134. Margl, P., Deng, L., and Ziegler, T. (1999) *Organometallics*, **18**, 5701–7708.
135. Khoroshun, D.V., Musaev, D.G., Vreven, T., and Morokuma, K. (2001) *Organometallics*, **20**, 2007–2026.
136. Scott, J., Gambarotta, S., Korobkov, I., and Budzelaar, P.H.M. (2005) *J. Am. Chem. Soc.*, **127**, 13019–13209.
137. Steffen, W., Blömker, T., Kleigrewe, N., Kehr, G., Fröhlich, R., and Erker, G. (2004) *Chem. Commun.*, 1188–1189.
138. Boukamp, M.W., Bart, S.C., Hawrelak, E.J., Trovitch, R.J., Lobkovsky, E., and Chirik, P.J. (2005) *Chem Commun.*, 3406–3408.
139. Gibson, V.C., Humphries, M.J., Tellmann, K.P., Wass, D.F., White, A.J.P., and Williams, D.J. (2001) *Chem. Commun.*, 2252–2253.
140. Trovitch, R.J., Lobkovsky, E., Bill, E., and Chirik, P.J. (2008) *Organometallics*, **27**, 109–118.
141. Cámpora, J., Naz, A.M., Palma, P., Álvarez, E., and Reyes, M.L. (2005) *Organometallics*, **24**, 4878–4881.

142. Boukamp, M.W., Lobkovsky, E., and Chirik, P.J. (2005) *J. Am. Chem. Soc.*, **127**, 9660–9661.
143. Trovitch, R.J., Lobkovsky, E., and Chirik, P.J. (2008) *J. Am. Chem. Soc.*, **130**, 11631–11640.
144. Bouwkamp, M.W., Lobkovsky, E., and Chirik, P.J. (2006) *Inorg. Chem.*, **45**, 2–4.
145. Bart, S.C., Lobkovsky, E., and Chirik, P.J. (2004) *J. Am. Chem. Soc.*, **126**, 13794–13807.
146. Bart, S.C., Lobkovsky, E., Bill, E., and Chirik, P.J. (2006) *J. Am. Chem. Soc.*, **128**, 5302–5303.
147. Scott, J., Vidyaratne, I., Korobkov, I., Gambarotta, S., and Budzelaar, P.H.M. (2008) *Inorg. Chem.*, **47**, 896–911.
148. Trovitch, R.J., Lobkovsky, E., Bouwkamp, M.W., and Chirik, P.J. (2008) *Organometallics*, **27**, 6264–6278.
149. Mastrorilli, P., and Nobile, C.F. (2004) *Coord. Chem. Rev.*, **248**, 377–395.
150. Goerl, C., and Alt, H.G. (2007) *J. Mol. Catal. A: Chem.*, **273**, 118–132.
151. Kaul, F.A.R., Puchta, G.T., Schneider, H., Bielert, F., Mihalios, D., and Hermann, W.A. (2002) *Organometallics*, **21**, 74–82.
152. Mecking, S. (1999) *Macromol. Rapid Commun.*, **20**, 139–143.
153. Bennett, A.M.A., Coughlin, E.B., Citron, J.D., and Wang, L. (1999) (E.I. Du Pont de Nemours and Co., US), WO 99/50318.
154. Pan, L., Zhang, K.Y., Li, Y.G., Bo, S.Q., and Li, Y.S. (2007) *J. Appl. Polym. Sci.*, **104**, 4188–4198.
155. Ivanchev, S.S., Badaev, V.K., Ivancheva, N.I., Sviridova, E.V., Rogozina, D.G., and Khaikin, S.Y. (2004) *Vysokomol. Soedin., Ser. A Ser. B*, **46**, 1959–1964.
156. Quijada, R., Rojas, R., Bazan, G., Komon, Z.J.A., Mauler, R.S., and Galland, G.B. (2001) *Macromolecules*, **34**, 2411–2417.
157. Galland, G.B., Quijada, R., Rojas, R., Bazan, G., and Komon, Z.J.A. (2002) *Macromolecules*, **35**, 339–345.
158. Lu, Z., Zhang, Z., Li, Y., Wu, C., and Hu, Y. (2006) *J. Appl. Polym. Sci.*, **99**, 2898–2903.
159. Wang, R., Zheng, Y., Cui, N., Zhang, Z., Ke, Y., and Hu, Y. (2005) *Gaofenzi Xuebao*, 132–136.
160. Wang, H., Ma, Z., Ke, Y., and Hu, Y. (2003) *Polym. Int.*, **52**, 1546–1552.
161. Trovitch, R.J., Lobkovsky, E., Bouwkamp, M.W., and Chirik, P.J. (2008) *Organometallics*, **27**, 6264–6278.
162. Archer, M.A., Bouwkamp, M.W., Cortez, M.-P., Lobkovsky, E., and Chirik, P.J. (2006) *Organometallics*, **25**, 4269–4278.
163. Bouwkamp, M.W., Bowman, A.C., Lobkovsky, E., and Chirik, P.J. (2006) *J. Am. Chem. Soc.*, **128**, 13901–13912.
164. Trovitch, R.J., Lobkovsky, E., Bill, E., and Chirik, P.J. (2008) *Organometallics*, **27**, 1470–1478.
165. Bedford, R.B., Betham, M., Bruce, D.W., Davis, S.A., Frost, R.M., and Hird, M. (2006) *Chem. Commun.*, 1398–1400.
166. Sun, W.-H., Zhang, S., and Zuo, W. (2008) *C. R. Chim.*, **11**, 307–316.

6
Cobalt and Nickel Catalyzed Reactions Involving C–H and C–N Activation Reactions

Renee Becker and William D. Jones

6.1
Introduction

Homogeneous transition metal catalyzed reactions have been very well studied over the past 50 years. The metal complexes used to accomplish these reactions have typically included noble transition metals or first row metals. Due to the cost and pathogenicity of these noble metals, the development of new catalytic routes using non-precious metals is an active area of research. In this chapter, a review of the use of selected cobalt and nickel species towards catalytic reactions analogous to those seen with noble metals is presented.

6.2
Catalysis with Cobalt

Transition metal catalyzed hetero-annulation provides a useful and convenient tool for the construction of N-heterocycles [1]. Quinolines are of special interest in that they display attractive applications in pharmaceuticals and are synthetic building blocks [2, 3]. Catalytic processes employing palladium [4–6], rhodium [7–9], ruthenium [10–14], and iron [15] have been studied and developed to synthesize quinoline skeletons. There are five common methods used to prepare substituted quinolines: the Skraup reaction [16], the Doebner–Von Miller reaction [17], the Conrad–Limpach reaction [18], the Friedlaender reaction [19, 20], and the Pfitzinger reaction [21, 22]. All five of the reactions require environmentally unfriendly acids or bases, high temperatures, or harsh conditions. Quinoline yields are usually low due to numerous side reactions. Even though much work has been done to find catalytic routes to quinolines, the use of non-precious metals remains an active area of research.

Recently, Jones *et al.* have published work on the conversion of diallylanilines and arylimines to quinolines [23, 24]. It was found that N-allylaniline, when heated in the presence of 10 mol% $Co_2(CO)_8$ and 1 atm of CO at 85 °C, leads to the selective formation of 2-ethyl-3-methylquinoline (Equation (6.1)). Aniline and propene

Catalysis Without Precious Metals. Edited by R. Morris Bullock
© 2010 WILEY-VCH Verlag GmbH & Co. KGaA, Weinheim
ISBN: 978-3-527-32354-8

Table 6.1 Isolated yields for the conversion of diallylanilines to quinolines.[a]

Entry	Diallylaniline	Product	Temp. (°C)	Isolated yield, (%)
1	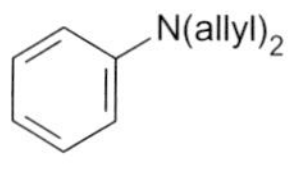 PhN(allyl)$_2$	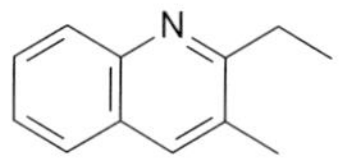 2-ethyl-3-methylquinoline	95	63
2	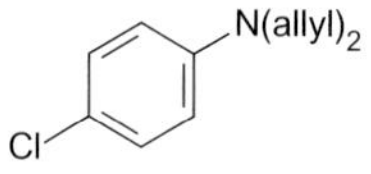 4-Cl-C$_6$H$_4$-N(allyl)$_2$	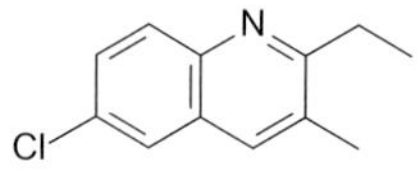 6-Cl quinoline	105	56
3	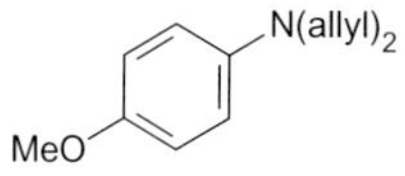 4-MeO-C$_6$H$_4$-N(allyl)$_2$	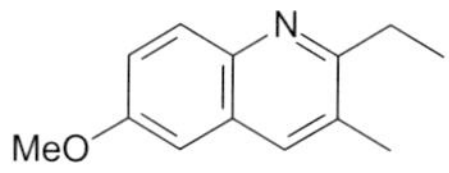 6-MeO quinoline	95	68
4	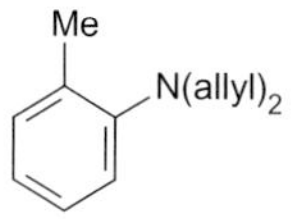 2-Me-C$_6$H$_4$-N(allyl)$_2$	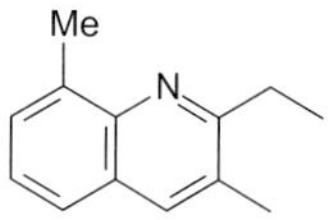 8-Me quinoline	95	56
5	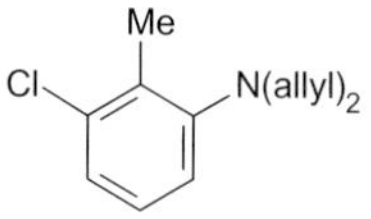 2-Me-6-Cl-C$_6$H$_3$-N(allyl)$_2$	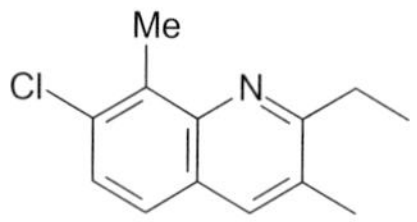 8-Me-5-Cl quinoline	105	42

a) Isolated yields at 100% conversion using 10% Co$_2$(CO)$_8$ in THF solvent under 1 atm CO. Typical reaction time is 36–48 h.

were also observed in the reaction. The constant presence of carbon monoxide was necessary to stabilize the Co$_2$(CO)$_8$ under the reaction conditions, without this reduced yields of products were obtained due to the instability of the Co$_2$(CO)$_8$ upon heating. Use of 4-methoxyallylaniline led to the corresponding 6-substituted quinoline in a 24% yield. Although the reaction was selective for quinoline formation, half of the starting material acted as a sacrificial allyl source, limiting the maximum yield to only 50%. A more atom economical approach was achieved by using diallylanilines. The diallylanilines formed the same products with higher yields and the same selectivity under the same reaction conditions (Equation (6.2)).

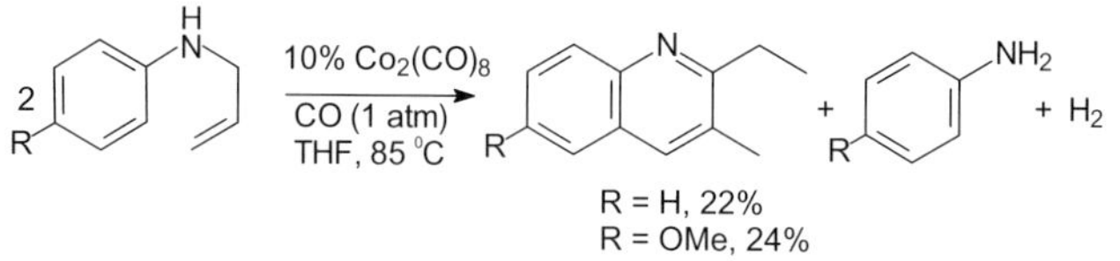

$$2 \text{ R-C}_6\text{H}_4\text{-NH-allyl} \xrightarrow[\text{THF, 85 °C}]{10\% \text{ Co}_2(\text{CO})_8,\ \text{CO (1 atm)}} \text{quinoline} + \text{R-C}_6\text{H}_4\text{-NH}_2 + \text{H}_2 \quad (6.1)$$

R = H, 22%
R = OMe, 24%

$$\text{PhN(allyl)}_2 \xrightarrow[\text{CO, THF, 85 °C}]{\text{cat. Co}_2(\text{CO})_8} \text{quinoline} + \text{H}_2 \quad (6.2)$$

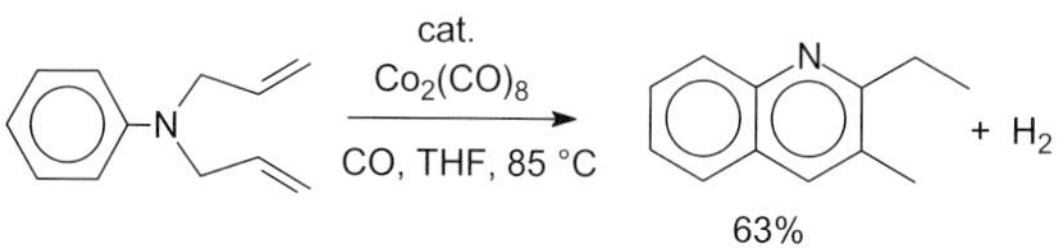

63%

This reaction can easily be extended to other diallylanilines to generate the corresponding quinolines (Table 6.1). It is unique in that it forms quinolines selectively and in good yields. The 2-ethyl-3-methyl substitution pattern is common to products derived from both mono-and di-allylanilines. The pathway initially suggested was that there was an initial cleavage of the allylic C–N bond by $Co_2(CO)_8$ to generate a π-allyl intermediate. Palladium and ruthenium have been shown to cleave allylic C–N bonds in a similar fashion [25, 26]. Further support of this initial cleavage of the allylic C–N bond is that allylcobalttricarbonyl was detected and isolated at intermediate times during the reaction. The formation of imine intermediates was suggested to occur by beta-elimination to give an enimine followed by olefin hydrogenation with $HCo(CO)_4$ or H_2. Quinoline could then form via a Schiff-base dimer, as seen earlier with ruthenium [12].

Further studies on regioselectivity, solvent effects, temperature effects, and catalyst loadings were also conducted [27]. A series of 2-ethyl-3-methylquinolines was obtained and substituents were added at the 5-, 6-, 7- and 8- positions. As seen in Table 6.2, introduction of an electron-donating group favored product formation, whereas electron-withdrawing groups inhibited product formation.

Reaction conditions were studied to maximize the yield of quinoline and to determine which conditions affect the reaction time (Table 6.3). As the concentration of catalyst was increased, the reaction time decreased but the yield remained the same. With regard to variation of the solvent, the reaction rate in hexane was much slower than in other solvents and the yield was lower due to solubility issues. In the non-polar solvent benzene, the reaction was as fast as in THF, and the yield was comparable to that obtained in THF [27].

Mechanistic studies using isotopic labeling have provided additional insights into the details of the mechanism. These studies allowed a complete mechanism to be proposed that accounts for the experimental observations and provides a rational route to the products using established methods [27]. Equation (6.3) shows how one deuterium on the aryl ring in diallylaniline-d_5 is selectively transferred to the α-methylene of the ethyl group.

$$(6.3)$$

The proposed mechanistic sequence is shown in Scheme 6.1. Because the reaction is carried out under an atmosphere of CO, carbonyl loss and addition are likely to be reversible in many of these species. For simplicity, deuterium is shown only in one ortho-aryl position so that its rearrangements can be followed. The proposed mechanism relies upon the isotopic preference for deuterium on carbon to account for the location of deuterium in the quinoline product. Commencing with the cleavage of the ortho-aryl C–D bond by $Co_2(CO)_7$ in an intramolecular complex, hydride is eliminated as $DCo(CO)_3$ which then coordinates to and inserts into an allyl C–C double bond. The preference for a primary Co–C bond over a secondary Co–C bond leads to deuterium incorporation selectively at the β-position of the alkyl group to form the intermediate **A**. This insertion is reversible, but the isotopic

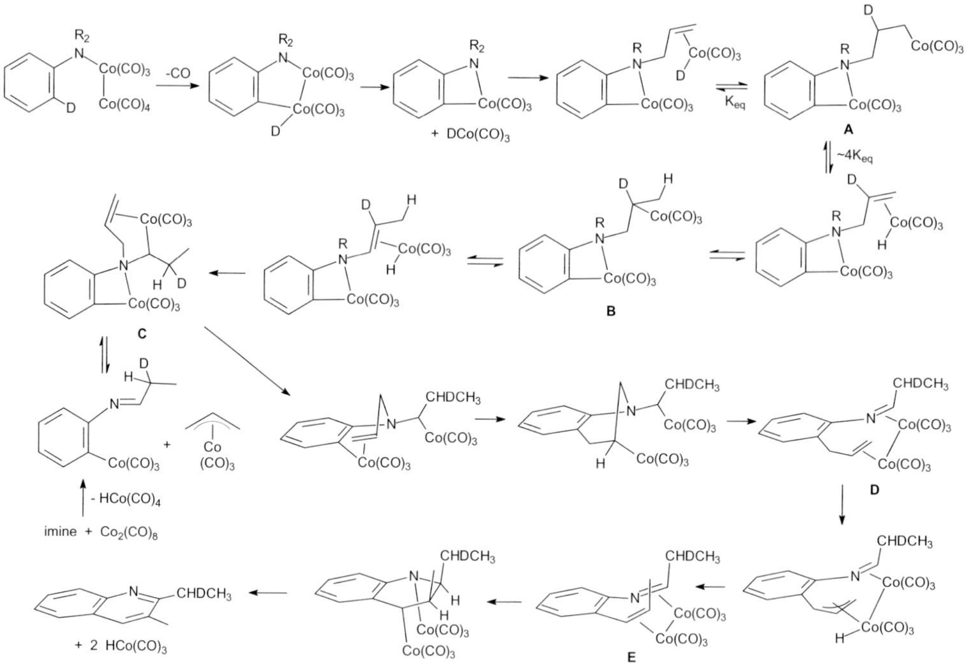

Scheme 6.1 Proposed mechanistic sequence.

Table 6.2 Yields for the conversion of diallylanilines to quinoline derivatives.[a]

Entry	Substrate	Product	T (°C)	Yield (%)[b]
1	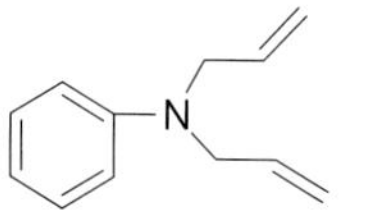	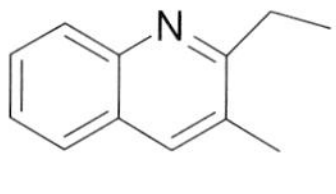	105	85(65)
2	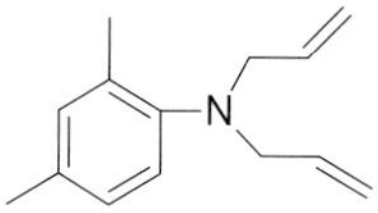	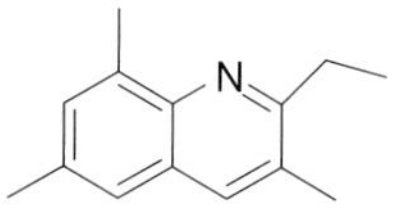	105	73(32)
3	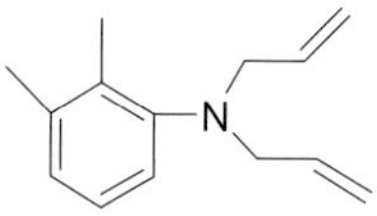	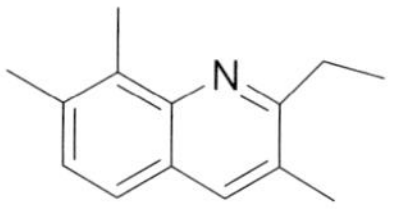	105	66(28)
4	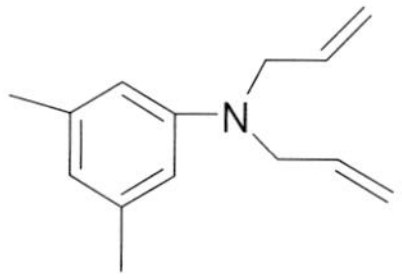	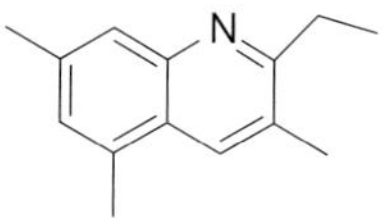	105	87(67)

6.2 Catalysis with Cobalt | 147

Table 6.2 Continued

Entry	Substrate	Product	T (°C)	Yield (%)[b]
5	3,5-(MeO)₂-N,N-diallylaniline	5,7-(MeO)₂-2-ethyl-3-methylquinoline	105	72(49)
6	2,5-dimethyl-N,N-diallylaniline	5,8-dimethyl-2-ethyl-3-methylquinoline	105	40(17)
7	2-methyl-N,N-diallylaniline	8-methyl-2-ethyl-3-methylquinoline	120	58(35)
8	4-methyl-N,N-diallylaniline	6-methyl-2-ethyl-3-methylquinoline	120	48(29)
9	4-MeO-N,N-diallylaniline	6-MeO-2-ethyl-3-methylquinoline	105	31(20)
10	3-MeO-6-methyl-N,N-diallylaniline	6-MeO-8-methyl-2-ethyl-3-methylquinoline	120	50(26)
11	4-CF₃-N,N-diallylaniline	6-CF₃-2-ethyl-3-methylquinoline	120	20(10)
12	3,5-(CF₃)₂-N,N-diallylaniline	—	120	—
13	2-OMe-N,N-diallylaniline	—	120	—
14	2-CN-N,N-diallylaniline	—	120	—

a) Yields at 100% conversion of the diallylaniline, using 10 mol% Co₂(CO)₈ in THF solvent under 1 atm CO in a closed ampule. Reaction time is 36–48 h.
b) NMR yield (isolated). The major byproduct is the corresponding aniline.

Table 6.3 Yields for the conversion of diallylaniline to 2-ethyl-3-methylquinoline with different catalyst loadings and solvents. T = 105 °C, 1 atm CO, [diallylaniline] = 0.2 M.

Entry	$Co_2(CO)_8$	Solvent	Yield (%)	Reaction time (h)
1	10%	THF	85(65)	~24
2	20%	THF	92(67)	8–24
3	50%	THF	87(63)	<2
4	10%	Toluene	79(53)	~38
5	10%	Benzene	82(58)	~21
6	10%	Hexane	60(35)	>55

a) NMR yield(isolated).

preference for deuterium on carbon leads to β-elimination of a C–H bond, which accounts for the observed 85% D in this position in the quinoline product. Insertion to give a secondary Co–C isomer **B** leads to β-elimination in the other direction to give an enamine, which upon rearrangement [28] to **C** can undergo β-carbon elimination of the allyl group to give imine plus π-allylcobalttricarbonyl. This pathway also permits the alternate use of imine as starting material. From the intermediate **C**, the allyl group can bridge to a second cobalt in the ortho-aryl position where olefin insertion is followed by another β-carbon elimination to regenerate the imine and a cobalt-olefin complex **D**. Double bond isomerization to a *cis*-olefin-imine complex **E** leads to reductive coupling for the formation of the second six-membered ring. This coupling also produces a product in which the two metal atoms are cis to adjacent hydrogens, permitting β-elimination to give the quinoline product and $HCo(CO)_3$. This mechanism, while still speculative, accounts for all intermediates and labeling results using known reactions of cobalt carbonyls.

With the use of $Co_2(CO)_8$ it has been shown that arylimines can undergo selective cross-coupling with diallylanilines to generate the corresponding quinolines [29]. The cross-coupling of phenyl benzaldimine with diallylaniline was studied in the presence of 10 mol% of $Co_2(CO)_8$. The cross-coupling quinoline product was isolated in 47% yield (Equation (6.4)). The minor products that were also formed were secondary amine (by reduction of the imine) and aniline (from diallylaniline after loss of two allyl groups). Using this cross-coupling reaction improved the scope of the reaction in that it allowed the use of a variety of substituents at the 2-position of the quinoline skeleton [24].

$$0.6\text{-}0.7\ \text{PhN(allyl)}_2 + \text{PhN=CHR} \xrightarrow[\text{CO, THF, 106 °C}]{\text{cat. } Co_2(CO)_8} \text{quinoline} + \text{Ph-NH-CH}_2\text{-R} + \text{PhNH}_2 \quad (6.4)$$

This cross-coupling reaction was used with a range of arylimines, as seen in Table 6.4. For these experiments, 0.6–0.7 equiv of diallylaniline was used as the allyl source. The reported yields are based on the quantity of starting imine. Some secondary amine formed by reduction of the imine was observed for all substrates. No cross-coupling product and only 2-ethyl-3-methylquinoline was observed for entry 6, and the formation of a stable chelate complex with $Co_2(CO)_8$ was suggested as the probable reason [24]. Carbonylation products accounted for 2–4% for entry 7. Without the introduction of diallylaniline while maintaining the same reaction conditions, the carbonylation product was the sole product in the system although the reaction was very slow (14 days at 120 °C). The carbonylation product was identified as a five-membered ring lactam [27]. Entries 9–14 were found to undergo an imine exchange reaction with the substrate (Equation (6.5)).

Table 6.4 Isolated yields of quinoline derivatives by cross coupling with diallylaniline.[a]

Entry	Imine	Product	Yield (%)
1			43
2			51
3			39
4			64[b]
5			50[c]

150 | *6 Cobalt and Nickel Catalyzed Reactions Involving C–H and C–N Activation Reactions*

Table 6.4 Continued

Entry	Imine	Product	Yield (%)
6	PhN=CH-(2-pyridyl)	2-(2-pyridyl)-3-methylquinoline	—
7	PhN=CH-C₆H₄-NMe₂	2-(4-NMe₂-C₆H₄)-3-methylquinoline	58
8	PhN=CH-(1-naphthyl)	2-(1-naphthyl)-3-methylquinoline	47
9	Me₂N-C₆H₄-N=CHPh	6-NMe₂-2-phenyl-3-methylquinoline	47
10	Ph-NH-C₆H₄-N=CHPh	6-(PhNH)-2-phenyl-3-methylquinoline	49
11	Cy-C₆H₄-N=CHPh	6-Cy-2-phenyl-3-methylquinoline	56
12	t-Bu-C₆H₄-N=CHPh	6-t-Bu-2-phenyl-3-methylquinoline	53
13	2-t-Bu-C₆H₄-N=CHPh	8-t-Bu-2-phenyl-3-methylquinoline	17[b]
14	2-Me-C₆H₄-N=CHPh	8-Me-2-phenyl-3-methylquinoline	29

a) 0.6–0.7 equiv diallylaniline was used as the allyl source in all the reactions: 10% $Co_2(CO)_8$ and 1 atm. of CO is used in all the reactions. Typical reaction time is 36–48 h at 100–120 °C.
b) NMR yield.
c) Small amounts (2–4%) of a carbonylation product are also observed in the reaction.

Table 6.5 Isolated yields of quinolines by cross-coupling of imines with diallylaniline catalyzed by $Co_2(CO)_8$.[a]

Entry	Imine	Diallyl aniline	Product	Temp. (°C)	Yield (%)
1	4-Me-C6H4-N=CH-Ph	4-Me-C6H4-N(allyl)2	6-Me-2-Ph-3-Me-quinoline	100	61
2	2-Me-C6H4-N=CH-Ph	2-Me-C6H4-N(allyl)2	8-Me-2-Ph-3-Me-quinoline	100	33
3	4-MeO-C6H4-N=CH-Ph	4-MeO-C6H4-N(allyl)2	6-MeO-2-Ph-3-Me-quinoline	105	60
4	3,5-Me2-C6H3-N=CH-Ph	3,5-Me2-C6H3-N(allyl)2	5,7-Me2-2-Ph-3-Me-quinoline	105	73
5	4-Cl-C6H4-N=CH-Ph	4-Cl-C6H4-N(allyl)2	6-Cl-2-Ph-3-Me-quinoline	105	55
6	4-Ph-C6H4-N=CH-Ph	4-Ph-C6H4-N(allyl)2	6-Ph-2-Ph-3-Me-quinoline	105	54

a) The yields reported are based on the starting imine; 10% catalyst loading and typical reaction time 36–48 h.

While there was no significant equilibration for imines derived from anilines bearing a para-substituent (entries 9–12), it was significant for imines derived from ortho-substituted anilines (entries 13 and 14). For entry 13, there was significant equilibration and the major product observed was 2-phenyl-3-methylquinoline, which can be explained by an aniline–imine exchange as in Equation (6.5). For entry 14, an approximately 1:1 mixture of the two quinolines was observed [24].

To counteract this problem, the cross-coupling was studied by using diallylanilines derived from the same aniline fragment as the imine. This modification significantly improved the yields of the cross-coupling reaction while avoiding any side reactions (Table 6.5). It can be seen from Tables 6.4 and 6.5 that arylimines with alkyl, alkoxy, dialkylamino, arylamino, and aryl substituents on either side of the ring undergo this cross-coupling efficiently [24].

The overall reaction is one in which heteroannulation of diallylaniline results in the corresponding quinoline with the elimination of hydrogen. The detection and isolation of allylcobalttricarbonyl suggests that the reaction is not intramolecular. Kinetic studies have shown a second order dependence of initial rate on the concentration of diallylaniline, further suggesting the possibility of allyl group transfer between substrates. To test this hypothesis the asymmetric substrate shown in Equation (6.6) was synthesized and its reactivity studied in the presence of $Co_2(CO)_8$ (Equation (6.6)). GC-MS analysis of the reaction showed the formation of three quinoline products, 2-ethyl-3-methyl, 2-phenethyl-3-methyl, and 2-phenethyl-3-benzylquinoline (ratio 1:4.8:5.8, respectively). If the reaction proceeded via an intramolecular process one would expect to see the 2-phenethyl-3-methyl isomer exclusively. The reaction of diallylaniline with dicinnamyl-aniline gave 2-phenethyl-3-methyl quinoline only, further supporting the concept of intermolecular allyl transfer.

$$(6.6)$$

Alper et al. have shown that cobalt carbonyl can catalyze the carbonylation of thiiranes to β-mercapto acids. This was accomplished by using methyl iodide or a benzylic bromide, 3 N potassium hydroxide, benzene as the organic phase, and polyethylene glycol as the phase-transfer agent [30].

Phase-transfer catalysis has been found to be a useful tool for promoting carbonylation reactions under mild conditions [31]. Examples of this are the cyanonickel(II) complex phase-transfer catalyzed carbonylation of allyl halides to acids [32], and the conversion of benzylic halides and methyl iodide to acids using cobalt carbonyl as the catalyst [33].

It has also been shown that a double carbonylation occurs when a styrene oxide is employed as the reactant, methyl iodide, carbon monoxide, and benzene as the organic phase, and aqueous sodium hydroxide, and cetyltrimethylammonium bromide as the phase-transfer agent (Equation (6.7)) [34].

$$(6.7)$$

There has not been much research in the area of metal-catalyzed reactions of thiiranes, which are sulfur analogues of epoxides. It has been shown that chlorodicarbonylrhodium(I) dimer can catalyze the homogeneous desulfurization of thiiranes to olefins [35]. This reaction is stereospecific and proceeds for thiiranes having aliphatic, aromatic, ether, and ester substituents (Equation (6.8)). This chemistry contrasts with that observed with cobalt. At room temperature, when 2-phenylthiirane, was treated with CO (1 atm), CH_3I, KOH, C_6H_6, tetrabutylammonium bromide, and a catalytic amount of cobalt carbonyl, β-mercapto acid, was isolated in 17% yield (Equation (6.9)). Using cetyltrimethylammonium bromide increased the yield to 39%, and substitution of polyethylene glycol for the quaternary ammonium salt increased the yield to 78% [30].

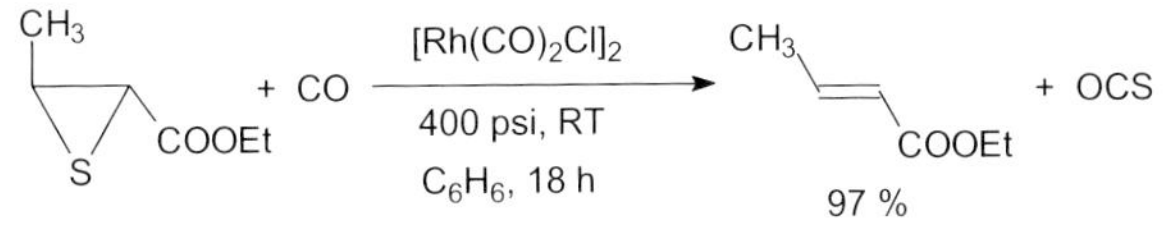

$$\text{thiirane(CH}_3\text{, COOEt)} + CO \xrightarrow[\substack{400 \text{ psi, RT} \\ C_6H_6, 18 \text{ h}}]{[Rh(CO)_2Cl]_2} \text{CH}_3\text{-CH=CH-COOEt} + OCS \quad (6.8)$$

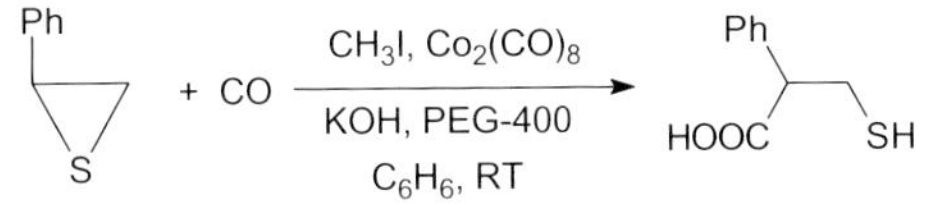

97 %

$$\text{thiirane(Ph)} + CO \xrightarrow[\substack{KOH, PEG-400 \\ C_6H_6, RT}]{CH_3I, Co_2(CO)_8} \text{HOOC-CH(Ph)-CH}_2\text{SH} \quad (6.9)$$

The proposed mechanism, which is based on the double carbonylation of styrene oxide, is shown in Scheme 6.2. The generation of an acylcobalt carbonyl complex from the reaction of cobalt tetracarbonyl anion with an alkyl halide is followed by reaction with a thiirane. This species can undergo carbonylation, the thioester function can undergo hydrolysis to reveal a sulfido nucleophile, and intramolecular cyclization then produces thietan-2-one. The thietan-2-one can undergo ring cleavage and the mercapto acid results by protonation.

The new route cited in this paper has several advantages to old routes, in that it is simple in execution and work-up, as well as regiospecific. This was the first

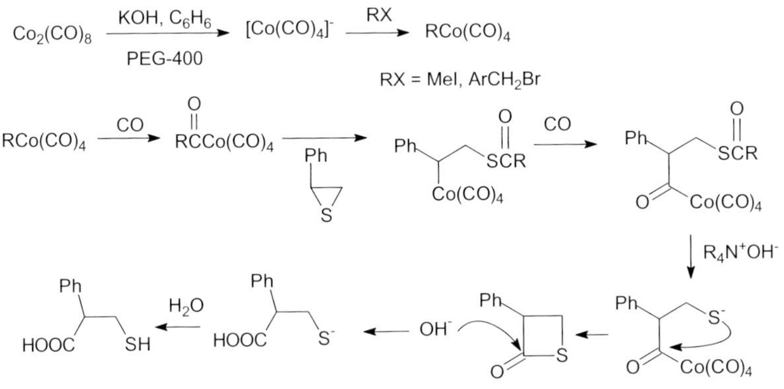

$$Co_2(CO)_8 \xrightarrow[\text{PEG-400}]{KOH, C_6H_6} [Co(CO)_4]^- \xrightarrow{RX} RCo(CO)_4$$

RX = MeI, ArCH$_2$Br

$$RCo(CO)_4 \xrightarrow{CO} RCCo(CO)_4 \xrightarrow{\text{Ph-thiirane}} \text{Ph-CH(Co(CO)}_4\text{)-CH}_2\text{-SCR} \xrightarrow{CO} \text{Ph-CH(OC-Co(CO)}_4\text{)-CH}_2\text{-SCR}$$

$$\downarrow R_4N^+OH^-$$

$$\text{HOOC-CH(Ph)-CH}_2\text{SH} \xleftarrow{H_2O} \text{HOOC-CH(Ph)-CH}_2\text{S}^- \xleftarrow{OH^-} \text{thietanone(Ph, S, O)} \xleftarrow{} \text{Ph-CH(OC-Co(CO)}_4\text{)-CH}_2\text{S}^-$$

Scheme 6.2

example of a metal complex catalyzed carbonylation reaction of a thiirane. In the past, reactions involving organosulfur reactants had been thought to poison the transition metal catalyst via the sulfur atom of the substrate. It has now been shown that this is not the case and that cobalt carbonyl is a valuable catalyst for the carbonylation of organic sulfur compounds. Other approaches to produce β-mercapto acids are known, including the reaction of unsaturated acids with benzyl mercaptan followed by debenzylation with sodium in liquid ammonia [36].

6.3
Catalysis with Nickel

It has been shown that the nickel complexes $Ni(PPh_3)_2X_2$ (X = Cl, Br) in the presence of oxygen can be used for the catalytic selective oxidative coupling of aryl thiols with secondary amines and carbon monoxide to give thiocarbamates [37]. A solution of benzylmethylamine and a slight excess of 4-chlorobenzenethiol in THF with 5% $(PPh_3)_2NiBr_2$ under 1 atm CO at room temperature gave 63% yield of the desired thiocarbamate (Equation (6.10)). Using N-benzylmethylamine as the excess reagent, a range of thiols with varying substituents at the ortho- and para-positions was investigated under the same conditions. The results of these experiments are shown in Table 6.6. The % yields indicate that the reaction is amenable to a range of aromatic thiols with a variety of substituents.

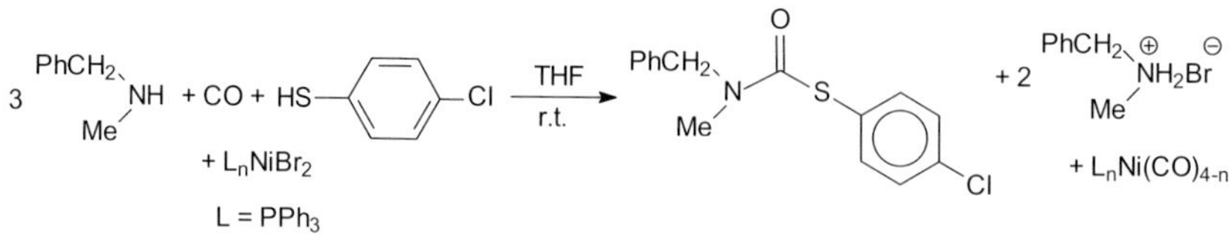

$$(6.10)$$

Various amines were studied using 4-chlorobenzenethiol as the thiol reactant, as summarized in Table 6.7. The reaction is very selective for secondary aliphatic amines. The primary aliphatic amines and anilines failed to undergo this reaction (entries 6–12).

A plausible mechanism for these reactions has been offered and can be seen in Scheme 6.3. There are two probable pathways, one in which complex 2 is initially formed and the other in which complex 4 is initially formed. If complex 2 is

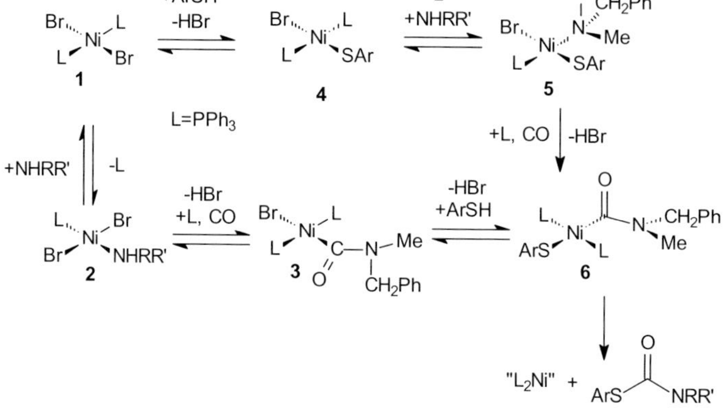

Scheme 6.3

6.3 Catalysis with Nickel

Table 6.6 Isolated yields (based on Ni) of thiocarbamate using N-benzylmethylamine.[a]

Entry	Thiol	Product	Yield (%)
1	HS–C₆H₄–OMe	PhCH₂(Me)N–C(O)–S–C₆H₄–OMe	67
2	HS–C₆H₄–Me	PhCH₂(Me)N–C(O)–S–C₆H₄–Me	55
3	HS–C₆H₄–Br	PhCH₂(Me)N–C(O)–S–C₆H₄–Br	65
4	HS–C₆H₄–Cl	PhCH₂(Me)N–C(O)–S–C₆H₄–Cl	63
5	HS–C₆H₄(o-Cl)	PhCH₂(Me)N–C(O)–S–C₆H₄(o-Cl)	59
6	HS–C₆H₃(2,6-Cl₂)	PhCH₂(Me)N–C(O)–S–C₆H₃(2,6-Cl₂)	62
7	HS–C₆H₄–F	PhCH₂(Me)N–C(O)–S–C₆H₄–F	38
8	HS–C₆H₄–NO₂	PhCH₂(Me)N–C(O)–S–C₆H₄–NO₂	53
9	HS–CH₂–Ph	PhCH₂(Me)N–C(O)–S–CH₂–Ph	18
10	o-MeC₆H₄–SH	—	0
11	HO₂C–C₆H₄–SH	—	0
12	CH₃CONH–C₆H₄–SH	—	0

a) Reaction conditions: (PPh₃)₂NiBr₂ (0.3 mmol), N-benzylmethylamine (4.8 mmol), benzenethiol (0.37 mmol), in THF (16 mL), 1 atm CO, reaction time 15 h.

Table 6.7 Isolated yields of thiocarbamate using 4-chlorobenzenethiol.[a]

Entry	Substrate	Product, yield (%)
1	pyrrolidine (NH)	pyrrolidine-C(O)-S-C$_6$H$_4$-Cl, 66
2	piperidine (NH)	piperidine-C(O)-S-C$_6$H$_4$-Cl, 50
3	morpholine (NH)	morpholine-C(O)-S-C$_6$H$_4$-Cl, 53
4	Et$_2$NH	Et$_2$N-C(O)-S-C$_6$H$_4$-Cl, 63
5	PhCH$_2$N(Me)H	PhCH$_2$N(Me)-C(O)-S-C$_6$H$_4$-Cl, 63
6	PhNH$_2$	—, 0
7	4-MeO-C$_6$H$_4$-NH$_2$	—, 0
8	Ph$_2$NH	—, 0
9	PhNHMe	—, 0
10	PhCH$_2$CH$_2$NH$_2$	—, 0
11	CyNH$_2$	—, 0
12	Cy$_2$NH	—, 0

a) Reaction conditions: (PPh$_3$)$_2$NiBr$_2$ (0.3 mmol), N-benzylmethylamine (4.8 mmol), benzenethiol (0.37 mmol), in THF (16 mL), 1 atm CO, reaction time 15 h.

originally formed, CO insertion to give the amido complex **3** with elimination of HBr and re-coordination of PPh$_3$ can take place. Complex **3** can generate complex **6** upon reaction with thiol, which then undergoes reductive elimination to generate a nickel(0) species and the desired thiocarbamate product. If the thiolatonickel complex **4** is initially formed, amine coordination and CO insertion can generate complex **6**. Complex **6** can then undergo reductive elimination to form the desired thiocarbamate product.

This system can be rendered catalytic by the addition of an appropriate oxidant to regenerate the Ni(II) from Ni(0). It would be beneficial to use oxygen as the oxidant, so that HX can be recycled to produce H$_2$O which would deter side-product formation (Equation (6.11)). To test this possibility, experiments were run using catalytic amounts of (PPh$_3$)$_2$NiBr$_2$ and air as the oxidant (Table 6.8). The catalytic formation of the thiocarbamate is accompanied by competitive

Table 6.8 Catalytic studies using $(Ph_3P)_2NiBr_2$ and air as oxidant.

Entry	Thiol	Amine	Product	GC yield[a]	GC yield of disulfide
1	HS–C₆H₄–Cl	pyrrolidine-NH	pyrrolidine-C(O)-S-C₆H₄-Cl	36(31)	61
2	HS–C₆H₄–Cl	PhCH₂N(Me)H	PhCH₂N(Me)-C(O)-S-C₆H₄-Cl	24(21)	58
3	HS–C₆H₄–Cl	piperidine-NH	piperidine-C(O)-S-C₆H₄-Cl	21(12)	59
4	HS–C₆H₄–OMe	PhCH₂N(Me)H	PhCH₂N(Me)-C(O)-S-C₆H₄-OMe	11(6)	86
5	HS–C₆H₄–Br	PhCH₂N(Me)H	PhCH₂N(Me)-C(O)-S-C₆H₄-Br	30(24)	60

a) Isolated yields are shown in parentheses.

oxidation of the thiol to the disulfide and presents a drawback to using O_2 as a reoxidant.

$$\text{pyrrolidine-NH} + CO + HS\text{-C}_6\text{H}_4\text{-Cl} + 1/2\, O_2 \xrightarrow[\text{L}_n\text{NiBr}_2]{\text{cat.}} \text{pyrrolidine-C(O)-S-C}_6\text{H}_4\text{-Cl} + H_2O \quad (6.11)$$

Another catalytic reaction that uses nickel is one in which (dippe)Ni(RC≡CR) has been shown to functionalize biphenylene with disubstituted alkynes to give the corresponding 9,10-disubstituted phenanthrenes [38]. Homogeneous C–C bond cleavage and functionalization by transition metal complexes is currently an active area of research in organometallic chemistry. Although catalytic C–C bond activation is at the forefront of innovative research, stoichiometric C–C bond activation has been reported [39]. These examples rely largely on relief of ring strain [40], attainment of aromaticity [41], or intramolecular addition in which the C–C bond is forced into close proximity to the metal [42].

In 1985 Eisch et al. formed 9,10-diphenylphenanthrene in greater than 50% yield by heating (PEt₃)₂Ni(2,2'-biphenyl) at 70 °C in the presence of diphenylacetylene [43]. 9,10-Diphenylphenanthrene was also produced in a 45% yield by Wakatsuki et al. by heating diphenylacetylene and CpCo(PPh₃)(2,2'-biphenyl) at 95 °C for 3 days [44]. Jones et al. obtained similar results when heating a mixture of diphenylacetylene (2 equiv) and (dippe)Ni(2,2'-biphenyl) at 70 °C (Equation (6.12)) [38].

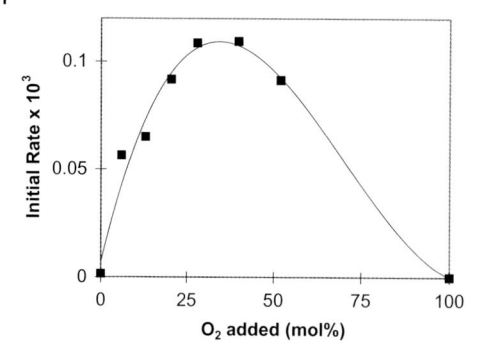

Figure 6.1 Initial rate of formation of 9,10-diphenylphenanthrene vs. mol % O_2 for the reaction of **1** (0.030 M) with diphenylacetylene (0.30 M) and biphenylene (0.30 M) at 60 °C. The initial rate ($M^{-1} s^{-1}$) was calculated from the slope of the concentration of 9,10-diphenylphenanthrene versus time during the first 10 min of reaction.

$$(6.12)$$

This reaction has been found to be catalytic upon addition of oxygen. Note that increasing the concentration of O_2 increased the initial rate of catalysis moderately up to 30–40 mol% O_2, after which point additional O_2 led to a decrease in the rate of catalysis (Figure 6.1). This observation led to the idea that the role of the O_2 was to remove the phosphine ligand from the metal center, resulting in a more active catalyst.

A proposed pathway was offered and is shown in Scheme 6.4. The addition of oxygen oxidizes a phosphine to its phosphine-oxide form, which allows cleavage of the strained C–C bond in biphenylene by the Ni(0) species. This results in the formation of intermediate **A**. Intermediate A can then coordinate a second acetylene molecule to create **B**. Rapid insertion into the Ni–C bond can occur which would produce **C**, which can reductively eliminate the phenanthrene and regenerate **A**.

The Ni(0) complexes ($^i Pr_2 PCH_2 CH_2 NMe_2$)Ni(η^2-PhC≡CR) (R = Ph, ^{t}Bu) are effective catalysts for the C–C bond activation in biphenylene and functionalization with PhC≡CPh or ^{t}BuC≡CPh to give 9,10-disubstituted phenanthrenes [45]. The proposed mechanism for the production of 9,10-diphenylphenanthrene can be seen in Scheme 6.5. The first step is assumed to be the dissociation of the NR_2 group from the nickel atom, creating a vacant coordination site at the metal center. The highly reactive metal center is able to cleave the C–C bond in biphenylene, resulting in the Ni(II) intermediate **I**. Alkyne insertion into the N–C bond leads to the Ni(II) species **II**. Reductive elimination of the

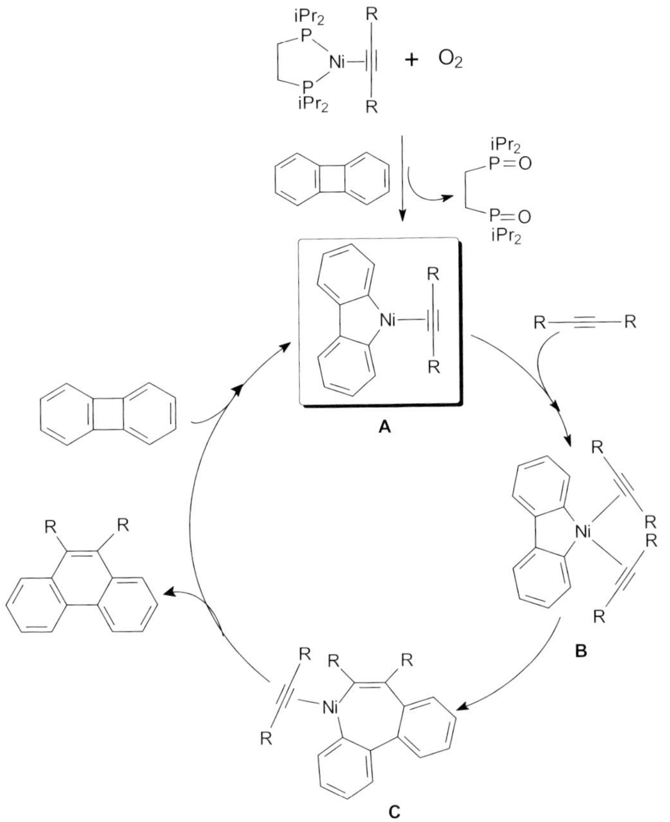

Scheme 6.4

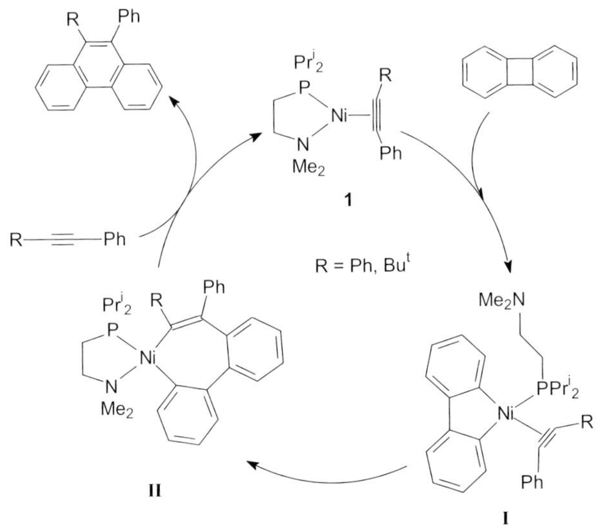

Scheme 6.5

9,10-diphenylphenanthrene can take place and a second alkyne insertion regenerates the Ni(0) complex **1**.

In the absence of biphenylene, hexaphenylbenzene is produced catalytically from diphenylacetylene in the presence of the Ni(0) complex **1** (Scheme 6.6). The lability of the Ni–NR$_2$ bond may facilitate the coordination of a second alkyne to generate the Ni(0) intermediate **Ia** [46]. According to the commonly accepted mechanism for the cyclotrimerization of alkynes [47], reductive coupling of the ligands leads to the formation of the donor-stabilized Ni(II) metallacyclopentadiene **IIa**. Dissociation of the NR$_2$ group provides a vacant coordination site for the third alkyne (**IIIa**), which then undergoes insertion into the Ni–C bond to form the Ni(II) intermediate **IVa**. Reductive elimination of the hexaphenylbenzene and complexation of an incoming alkyne regenerates the Ni(0) species **1**.

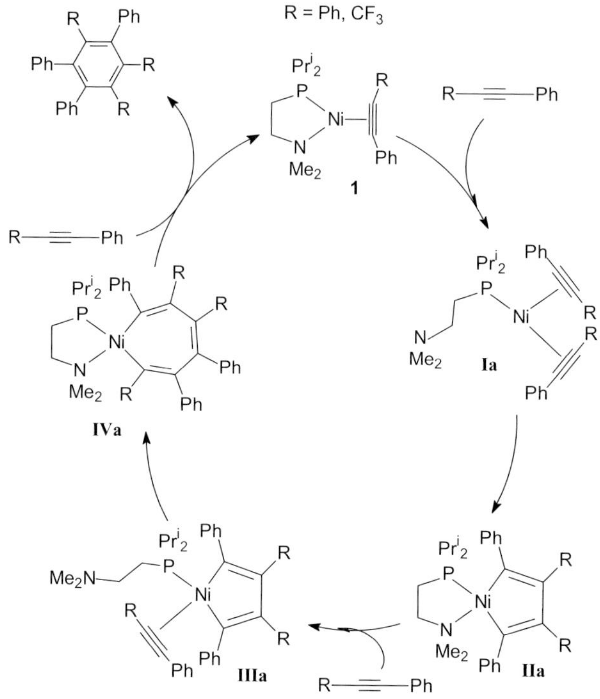

Scheme 6.6

Another catalytic reaction involving nickel involves the carbonylation of biphenylene [38]. Heating a THF-d_8 mixture of (dippe)Ni(CO)$_2$ at 95 °C with 5 equiv of biphenylene under 1 atm of CO resulted in the catalytic formation of fluorenone. The proposed mechanism can be seen in Scheme 6.7. The first step is the reversible loss of CO from (dippe)Ni(CO)$_2$. This gives the 14-electron complex (dippe)Ni(CO), to which C–C bond cleavage of biphenylene and CO insertion into the

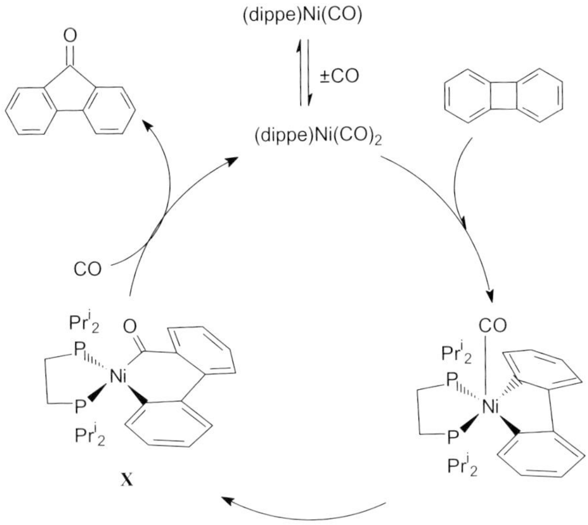

6.3 Catalysis with Nickel

Scheme 6.7

Ni–C bond of the coordinated biphenyl then occur. The regeneration of (dippe)Ni(CO)$_2$ is accomplished by the displacement of fluorenone by CO.

A recent publication has reported the coupling reaction between azolium salts and ethylene, catalyzed by zerovalent nickel complexes [48]. A combination of experiment and density functional theory (DFT) was used to show that the reaction proceeds via a redox mechanism. This redox mechanism involves the generation of Ni-carbene intermediates.

N-heterocyclic carbenes (NHC) are considered extremely effective ligands for homogeneous catalysis (Figure 6.2). These specific carbenes often lead to high efficiencies in metal-catalyzed reactions compared to traditional phosphines [49, 50]. NHC complexes are usually considered to be very stable, due to their electronic properties and the unusually high bond dissociation energies (BDE) associated with NHCs [51]. Previous work has shown that Ni-hydrocarbyl complexes of NHCs readily decompose by reductive elimination to yield the 2-substituted imidazolium salts [52, 53]. Later studies have shown that the "reverse" reaction, oxidative addition of imidazolium salts to zerovalent Group 10 metals, is feasible [53].

Cavell et al. have also combined these redox reactions to generate an atom-efficient coupling reaction between azolium salts and alkenes (Equation (6.13))

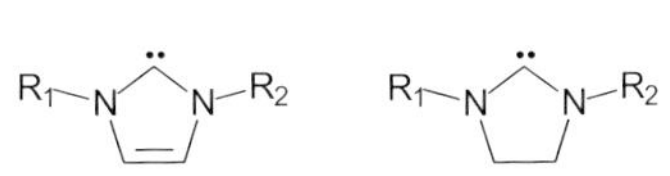

Figure 6.2 Saturated and unsaturated N-heterocyclic carbenes.

[54]. This reaction also demonstrates the *in situ* formation of reactive metal-hydride complexes.

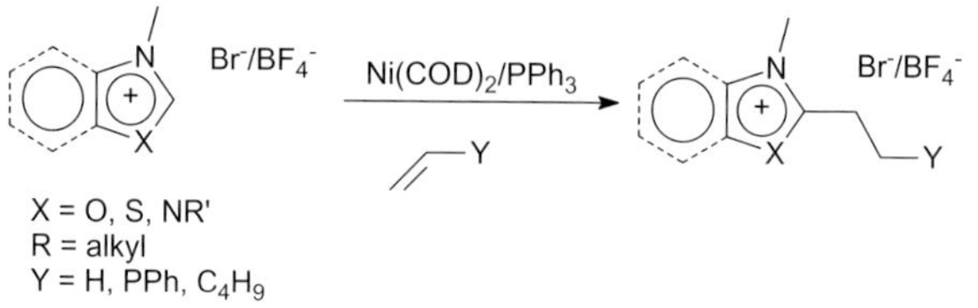

$$\text{(imidazolium)}^+ \text{Br}^-/\text{BF}_4^- \xrightarrow{\text{Ni(COD)}_2/\text{PPh}_3, \; \diagup\!\!\!\diagup Y} \text{(imidazolium-Y)}^+ \text{Br}^-/\text{BF}_4^- \quad (6.13)$$

X = O, S, NR'
R = alkyl
Y = H, PPh, C$_4$H$_9$

There is an interest in these types of ionic liquids as "green" media for chemical reactions, this reaction provides a useful approach to the synthesis of finely tuned ionic liquids, using imidazolium salts as building blocks. Functionalization of heterocyclic substrates via transition metals is an important extension of this work (Equation (6.14)). Bergman has shown that these species can be successfully reacted with olefins in the presence of a rhodium catalyst via an NHC intermediate [55, 56].

$$\text{(azole)} \xrightarrow{\text{ML}_n, \; \diagup\!\!\!\diagup R} \text{(azole-R)} \quad (6.14)$$

X = O, S, NR'

Cavell *et al.* previously proposed a catalytic cycle using a Ni(COD)$_2$/PPh$_3$ system (Scheme 6.8) [54]. The reaction is shown to proceed via oxidative addition of the

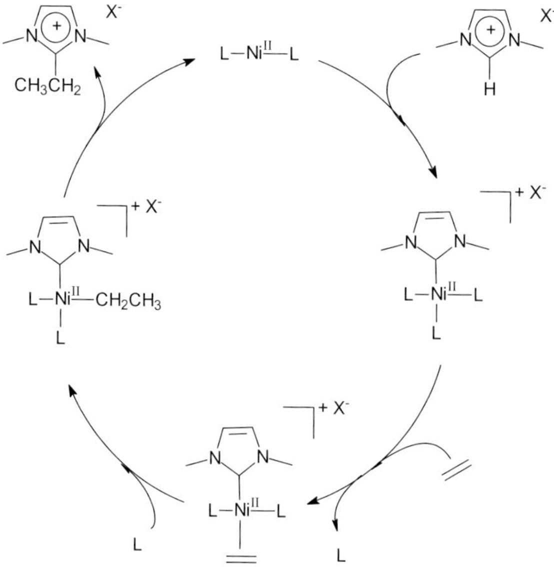

Scheme 6.8 Proposed mechanism for the reaction of olefins with imidazolium salts.

azolium salt and replacement of a weakly bound ligand by alkene. This is followed by insertion of the alkene into the metal-hydride bond, and, finally, reductive elimination of the product. Further experimental and computational studies have resulted in a detailed analysis of the full catalytic cycle that involves oxidative addition of the imidazolium salt, coordination of ethylene, ethylene insertion into the metal-hydride bond, and, finally, reductive elimination of the coupled product. The calculations show that the positive charge shuttles between the imidizolium ion and the metal during the catalysis.

References

1. Hegedus, L.S. (1998) *Angew. Chem. Int. Ed. Engl*, **27**, 1113.
2. Balasubramanian, M., and Keay, J.G. (1996) *Comprehensive Heterocyclic Chemistry II*, vol. **5** (eds A.R. Katritzky, C.W. Rees, and Scriven E.F.V.), Pergamon Press, Oxford, p. 254.
3. Jones, G. (1996) *Comprehensive Heterocyclic Chemistry II*, vol. **5** (eds A.R. Katritzky, C.W. Rees, and Scriven E.F.V.), Pergamon Press, Oxford, p. 167.
4. Hegedus, L.S., Allen, G.F., Bozell, J.J., and Waterman, E.L. (1978) *J. Am. Chem. Soc.*, **100**, 5800.
5. Larock, R.C., Hightower, T.R., Hasvold, L.A., and Peterson, K.P. (1996) *Org. Chem.*, **61**, 3584.
6. Larock, R.C., and Kuo, M.Y. (1991) *Tetrahedron Lett.*, **32**, 569.
7. Diamond, S.E., Szalkiewicz, A., and Mares, F.J. (1979) *J. Am. Chem Soc.*, **101**, 490.
8. Beller, M., Thiel, O.R., Trauthwein, H., and Hartung, C.G. (2000) *Chem. Eur. J.*, **6**, 2513.
9. Watanabe, Y., Suzuki, N., Tsuji, Y., Shim, S.C., and Mitsudo, T. (1982) *Bull. Chem. Soc. Jpn.*, **55**, 1116.
10. Watanabe, Y., Tsuji, Y., and Ohsugi, Y. (1981) *Tetrahedron Lett.*, **22**, 2667.
11. Cho, C.S., Oh, B.H., and Shim, S.C. (1999) *Tetrahedron Lett.*, **40**, 1499.
12. Watanabe, Y., Tsuji, Y., Ohsugi, Y., and Shida, J. (1983) *Bull. Chem. Soc. Jpn.*, **56**, 2452.
13. Wantanabe, Y., Tsuji, Y., and Shida, J. (1984) *Bull. Chem. Soc. Jpn.*, **57**, 435.
14. Cho, C.S., Kim, J.S., Oh, B.H., Kim, T.J., Shim, S.C., and Yoon, N.S. (2000) *Tetrahedron*, **56**, 7747.
15. Watanabe, Y., Takatsuki, K., Shim, S.C., Mitsudo, T., and Takegami, Y. (1978) *Bull. Chem. Soc. Jpn.*, **51**, 3397.
16. Kraup, S. (1888) *Berichte*, **21**, 1077.
17. Denmark, S.E., and Venkatraman, S. (2006) *J. Org. Chem.*, **71**, 1668.
18. Steck, E.A., Hallock, L.L., Holland, A.J., and Fletcher, L.T. (1948) *J. Am. Chem. Soc.*, **70**, 1012.
19. Yadav, J.S., Reddy, B.V.S., and Premalatha, K. (2004) *Synlett*, **6**, 963.
20. McNaughton, B.R., and Miller, B.L. (2003) *Org. Lett.*, **5**, 4257.
21. Henze, H.R., and Carroll, D.W. (1954) *J. Am. Chem. Soc.*, **76**, 4580.
22. Buu-Hoie, N.P., Royer, R., Xuong, N.D., and Jacquignon, P. (1953) *J. Am. Chem. Soc.*, **75**, 1209.
23. Josemon, J., Cavalier, C.M., Jones, W.D., Godleski, S.A., and Valente, R.R. (2002) *J. Mol. Catal.*, **182–183**, 565–570.
24. Josemon, J., and Jones, W.D. (2003) *J. Org. Chem.*, **68**, 3563–3568.
25. Hiraki, K., Matsunaga, T., and Kawano, H. (1994) *Organometallics*, **13**, 1878.
26. Moberg, C., and Antonsson, T. (1985) *Organometallics*, **4**, 1083.
27. Li, L., and Jones, W.D. (2007) *J. Am. Chem. Soc.*, **129**, 10707–10713.
28. Kumobayashi, H., Akutagawa, S., and Otsuka, S. (1978) *J. Am. Chem. Soc.*, **100**, 3949.
29. Bruce, D.W., and Liu, X.H. (1994) *J. Chem. Soc., Chem. Commun.*, 729.
30. Calet, S., and Alper, H. (1987) *Organometallics*, **6**, 1625–1628.

31. Alper, H. (1984) *Fundam. Res. Homogeneous Catal.*, **4**, 79.
32. Joo, F., and Alper, H. (1985) *Organometallics*, **4**, 1775.
33. Alper, H., and Abbayes, H. (1977) *J. Organomet. Chem.*, **134**, C11.
34. Alper, H., Arzoumanian, H., Petrignani, J.F., and Saldana-Maldonado, M. (1985) *J. Chem. Soc., Chem. Commun.*, 340.
35. Calet, S., and Alper, H. (1986) *Tetrahedron Lett.*, **27**, 3573.
36. Field, L., and Giles, P.M. (1970) *J. Med. Chem*, **13**, 317.
37. Jacob, J., Reynolds, K.A., and Jones, W.D. (2001) *Organometallics*, **20**, 1028–1031.
38. Edelbach, B.L., Lachicotte, R.J., and Jones, W.D. (1999) *Organometallics*, **18**, 4040–4049.
39. Murakami, M., and Yoshihiko, I. (1999) *Topics in Organometallic Chemistry. Activation of Unreactive Bonds and Organic Synthesis*, vol. **3** (ed. S. Murai), Springer, New York.
40. For example: Adams, D.M., Chatt, J., Guy, R.G., Sheppard, N. (1961) *J. Chem. Soc.*, 738–742.
41. An example: Kang, J.W., Moseley, K., Maitlis, P.J. (1969) *J. Am. Chem. Soc.*, **91**, 5970–5977.
42. An example: Suggs, J.W., Jun, C.H. (1985) *J. Chem. Soc., Chem. Commun.*, 92–93.
43. Eisch, J.J., Piotrowski, A.M., Han, K.I., and Kruger, C. (1988) *Angew. Chem. Int. Ed. Engl.*, **27**, 833.
44. Wakatsuki, Y., Nomura, O., Tone, H., and Yamazaki, H. (1980) *J. Chem. Soc., Perkin Trans.2*, 1344.
45. Muller, C., Lachicotte, R.J., and Jones, W.D. (2002) *Organometallics*, **21**, 1975–1981.
46. Rosenthal, U., Pulst, S., Kempe, R., Porschke, K.R., Goddard, R., and Proft, B. (1998) *Tetrahedron*, **54**, 1277.
47. Rosenthal, U., and Schulz, W. (1987) *J. Organomet. Chem.*, **321**, 103.
48. Normand, A.T., Hawkes, K.J., Clement, N.D., Cavell, K.J., and Yates, B.F. (2007) *Organometallics*, **26**, 5353–5363.
49. Bourissou, D., Guerret, O., Gabbai, F.P., and Bertrand, G. (2000) *Chem Rev.*, **100**, 39–91.
50. Herrmnn, W.A. (2002) *Angew. Chem. Int. Ed.*, **41**, 1290–1309.
51. Crudden, C.M., and Allen, D.P. (2004) *Coord. Chem Rev.*, **248**, 2247–2273.
52. Graham, D.C., Cavell, K.J., and Yates, B.F. (2005) *Dalton Trans.*, 1093–1100.
53. Bacciu, D., Cavell, K.J., Fallis, I.A., and Ooi, L. (2005) *Angew. Chem. Int. Ed.*, **44**, 5282–5284.
54. Clement, N.D., and Cavell, K.J. (2004) *Angew. Chem. Int. Ed.*, **43**, 3845–3847.
55. Tan, K.L., Bergman, R.G., and Ellman, J.A. (2002) *J. Am. Chem. Soc.*, **124**, 3202–3203.
56. Wiedemann, S.H., Lewis, J.C., Ellman, J.A., and Bergman, R.G. (2006) *J. Am. Chem. Soc.*, **128**, 2452–2462.

7
A Modular Approach to the Development of Molecular Electrocatalysts for H_2 Oxidation and Production Based on Inexpensive Metals

M. Rakowski DuBois and Daniel L. DuBois

7.1
Introduction

The development of inexpensive electrocatalysts for the production and oxidation of hydrogen will play a vital role in future energy storage and delivery systems. The generation of hydrogen from non-fossil energy sources such as solar, wind, geothermal, and nuclear energy is one approach being considered for storing the electrical energy generated by these sources for transportation and other uses that are not temporally matched to electrical energy production. In the reverse process, in which fuels are used to produce electricity, it is recognized that fuel cells have significant thermodynamic advantages in terms of energy efficiency compared to internal combustion engines and other Carnot processes. Both fuel generation and fuel utilization require electrocatalysts for efficient interconversion of electrical energy and chemical energy.

The best catalysts for the electrochemical oxidation and production of hydrogen are platinum metal and the hydrogenase enzymes. Both catalyze the reaction of two protons with two electrons to form H_2, as shown in Equation 7.1. Because of its superior catalytic rates and overpotentials compared to other metals and because of its high stability compared to hydrogenase enzymes, platinum is currently used as the catalyst for both half reactions (the oxidation of H_2 and the reduction of O_2) in polymer electrolyte membrane (PEM) fuel cells, which have been proposed for automotive transportation [1]. However, the high cost of platinum provides a strong impetus for developing less expensive alternatives.

$$2H^+ + 2e^- \Leftrightarrow H_2 \qquad (7.1)$$

The high activity of the hydrogenase enzymes containing iron or nickel, clearly demonstrates that catalytic activity for H^+/H_2 interconversion is not unique to platinum [2–5]. This observation has provided much of the impetus for attempting to develop small single site catalysts for hydrogen oxidation and production based on metals much less expensive than platinum. Artero and Fontecave have summarized these efforts in an excellent review [6]. Some notable developments since the time of that review have included progress on cobalt complexes containing

Catalysis Without Precious Metals. Edited by R. Morris Bullock
© 2010 WILEY-VCH Verlag GmbH & Co. KGaA, Weinheim
ISBN: 978-3-527-32354-8

glyoxime ligands (**1**) [7–11], molybdenum sulfur dimers (**2**) [12], and nickel and cobalt complexes containing diphosphine ligands with pendant nitrogen bases (**3**) [13–17]. All of these complexes are capable of catalyzing the reduction of protons to H_2 with low overpotentials. Although these results are promising, achieving molecular catalysts that simultaneously exhibit overpotentials of less than 100 mV with catalytic rates equivalent to or surpassing those of the enzymes has not yet been achieved.

7.2
Concepts in Catalyst Design Based on Structural Studies of Hydrogenase Enzymes

Structural studies of the [FeFe] and [NiFe] hydrogenase enzymes have provided important information on the geometry and composition of the active sites [2–5]. Some of these features are depicted in structures **4** and **5**. Notable features in both structures are the presence of two metal atoms bridged by thiolate ligands and the presence of both CO and CN^- ligands bound to iron. In addition it has been suggested that the central atom of the dithiolate ligand of [FeFe] hydrogenase is a nitrogen atom. Although the structural data alone are not sufficient to resolve the latter point, theoretical studies [18] and studies of simple model complexes containing similar pendant amines [13–16, 19, 20] suggest that this is likely.

This understanding of the structure and composition of the enzyme active sites has resulted in considerable efforts to synthesize structural analogs and explore the catalytic activity of these analogs. This effort has been summarized in recent thematic issues on hydrogen and bio-inspired hydrogen production and uptake catalysts [21, 22]. The most recent studies in the area of [FeFe] molecules have focused on favoring so-called "rotated" or "entatic" structures in which the site for protonation and H_2 activation is endo with respect to the chelate ring of the

7.2 Concepts in Catalyst Design Based on Structural Studies of Hydrogenase Enzymes | 167

dithiolate ligand [23–26]. An example is shown in Equation 7.2 [23], in which protonation of complex **6** at low temperature results in the formation of a terminal hydride that is properly positioned to interact with the pendant nitrogen of the dithiolate bridge (if present). Unfortunately, these terminal hydride complexes rearrange to form bridging hydrides, **8**, that cannot interact with a pendant base in an azadithiolate ligand.

$$(7.2)$$

X-ray diffraction studies of the hydrogenase enzymes not only provide detailed information on the structure of the active site, they also suggest a number of other concepts that may be useful in catalyst design. Figure 7.1 shows a schematic depiction of [FeFe] hydrogenase. The active site of this enzyme is located at the intersection of an electron-transport chain, hydrophobic channels for H_2 diffusion, and a proton-transport chain. The electron transport chain consists of three [Fe$_4$S$_4$] clusters spaced about 12 Å apart [2–5]. This electron transport chain serves as a molecular wire to move electrons between the active site and the exterior of the protein. The outermost [Fe$_4$S$_4$] clusters are embedded in recognition sites that allow attachment of only specific redox partners [27]. This arrangement allows regulation of the overall process. The distance between the [Fe$_4$S$_4$] clusters would

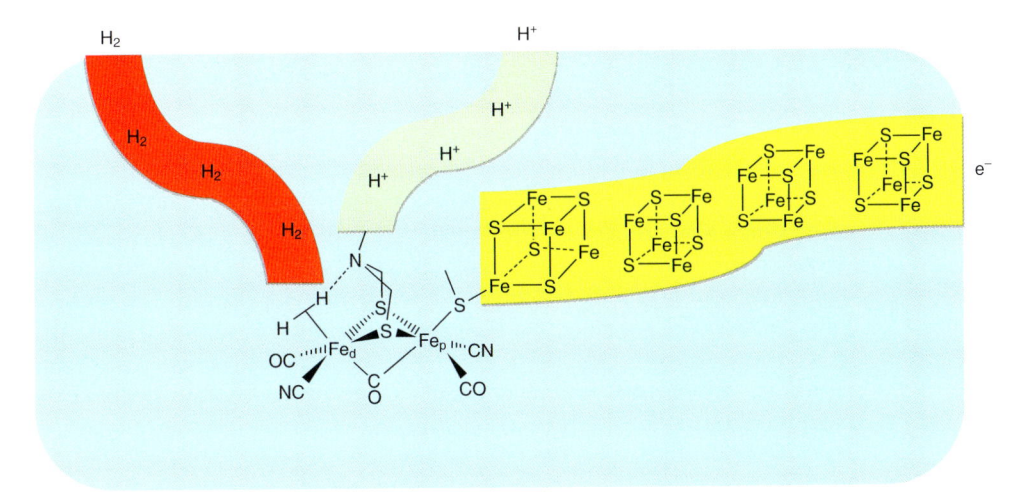

Figure 7.1 Depiction of hydrogenase enzyme emphasizing channels for H_2, proton, and electron transport.

suggest that intimate contact is not required for efficient electron transport, that is, electron transfer over several angstroms is efficient [28, 29]. On the other hand, the [Fe$_4$S$_4$] cluster proximal to the active site of the [FeFe] hydrogenase enzyme is intimately connected to the active site, suggesting it may serve as an electron reservoir. This is not observed for the [NiFe] hydrogenase [27].

The proton transport chain is composed of acidic or basic groups that allow facile transport between the exterior of the enzyme and the active site. The precise pathway in many instances is a subject of debate [27]. For the [FeFe] hydrogenase enzymes the pendant base shown in structure **4** is capable of approaching the Fe atom closely, allowing proton transfer. This pendant base is adjacent to a proton transfer channel that achieves proton transport via a series of proton transfer steps over short distances (less than 1 Å) between acidic and basic functional groups followed by rotation and translation of these groups [30]. Thus the movement of protons involves a continuous pathway from the iron atom to the exterior of the molecule. This observation suggests that similar control may be important for synthetic catalysts. It is clear that the distances between the individual groups of the *proton* transport chain during the actual proton transfer event are much shorter than the distances between the *electron* transport clusters [2]. This difference in distances is reasonable when one considers the difference in mass of these two particles. This suggests that much more attention should be paid to this particular aspect in the design of synthetic catalysts for H$_2$ production and oxidation. We will return to this point later.

A second important observation is that although water may be involved in the proton transport chain, it is protons that are transported and not water. Indeed, the pocket immediately surrounding the active site is often described as hydrophobic [2, 31]. One of the reasons for placing the active site in the interior of the enzyme instead of on its surface may be to provide a nonaqueous environment. This possibility calls into question the commonly expressed view that hydrogen oxidation and production catalysts should be designed to operate in aqueous conditions. It would seem that if there were strong advantages to operating in an aqueous environment, hydrogenase enzymes would have evolved to position the active site in contact with the aqueous environment of the cell.

It is also thought that hydrophobic tunnels provide a pathway for hydrogen diffusion that connects the distal iron atom of the active site in the [FeFe] hydrogenase enzyme with the surface of the protein [27]. A similar channel is proposed to exist in the [NiFe] hydrogenase, and this tunnel appears to connect the nickel atom of the active site with the surface of the enzyme [32]. From the existence of these channels, we can infer that the protein structure has evolved to control or assist the movement of all three substrates to the catalytically active site. In this respect the hydrogenase enzyme is a molecular entity that incorporates many of the features of a fuel cell. Both fuel cells and the enzymes are designed to deliver H$_2$ to the catalytically active site, remove protons after oxidation (either via a proton-conducting membrane in a fuel cell or the proton transport pathway in the enzyme), and conduct electrons away from the active site (via a carbon conductor in a fuel cell or the electron transport chain in the enzyme). A central message to

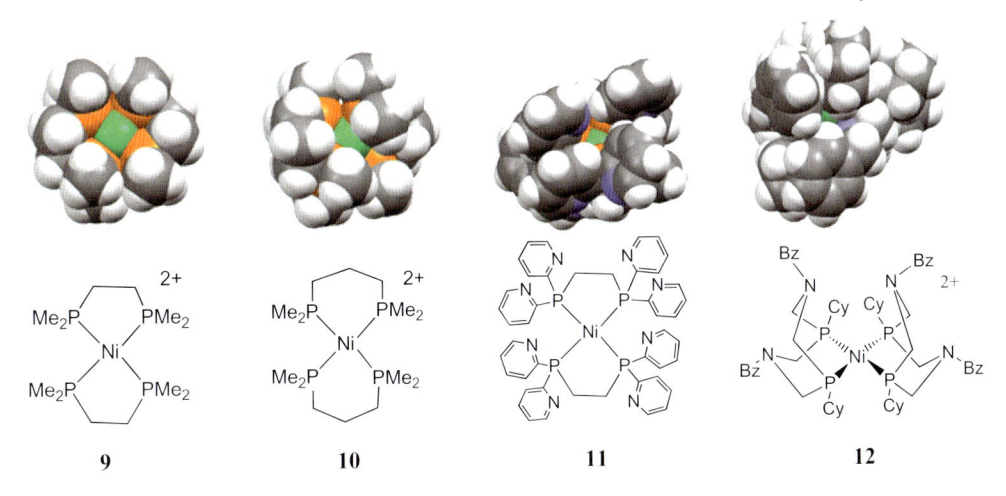

Figure 7.2 Space filling models of some typical small inorganic complexes and their corresponding structural representations, **9–12** [14, 19, 33].

be derived from these structural considerations of the entire enzyme is that the design of efficient catalysts must consider the approach and departure of all reactants and products, as well as the complex chemistry occurring at the catalytically active site.

Because typical synthetic organometallic molecules have much lower molecular weights than enzymes, is it reasonable to think that the same design principles will apply? Figure 7.2 shows space filling models of some typical small inorganic complexes and their corresponding structural representations, **9–12** [14, 19, 33]. It can be seen that only for the structure with the least sterically demanding ligand set, **9**, is the metal atom readily accessible to molecules of intermediate size, for example an organic acid or base. However, close approach of these reactants is required for facile proton transfer. If organic acids or bases are to transfer protons directly to the metal in structures **10–12**, significant structural reorganization of the complex must occur to allow close approach of these molecules. These rearrangements are likely to be associated with large energy barriers and slow rates. As discussed below, such energy barriers can be avoided by incorporation of proton relays (analogous to those used in the hydrogenase enzymes) within the second coordination sphere of these complexes. Small hydrophobic molecules like hydrogen should be able to access the metal centers of **9–12**, and electron transfer between molecules at the surface of the catalyst and the metal center, or between an electrode and the metal center, should still occur, as the radius of the largest complex, **12**, is approximately 8 Å. Of the three reactants and products that must be able to access or leave the active metal site, protons are the most likely to need assistance. As a result, management of proton transfer will be a central issue in the design of even relatively small synthetic catalysts for multi-electron/multi-proton processes.

7.3
A Layered or Modular Approach to Catalyst Design

Our approach to addressing the catalytic issues that are so effectively accomplished by the hydrogenase enzymes is to divide the design of catalysts into coordination spheres and to try to understand the roles of each coordination sphere. The first coordination sphere consists of those atoms directly bound to the metal of the catalytically active site. In Figure 7.3, the first coordination sphere of the nickel atom consists of nickel and the four phosphorus atoms shown in blue. The substituents on the phosphorus atoms and the P–Ni–P angles play dominant roles in controlling H_2 binding, pK_a values, hydride donor/acceptor properties, homolytic bond energies, and redox potentials [34–36]. Thus the first coordination sphere plays a primary role in determining the relative free energies of all catalytic intermediates observed during a catalytic cycle.

The second coordination sphere refers to atoms or functional groups that are in close proximity to the metal site, but that are not directly bound to the metal (shown in yellow in Figure 7.3). As a result, direct interactions of the second coordination sphere may occur with substrate molecules bound to the metal (e.g., interaction of a pendant amine with an H_2 or hydride ligand), but not with the metal center itself. This second coordination sphere can modify some of the energies of the catalytic intermediates, but an additional important role is the ability of the second coordination sphere to facilitate the movement of protons between the active metal site and species in solution. In this role the second coordination sphere acts as a relay for proton transfer.

In hydrogenase enzymes, there is an outer coordination sphere that does not involve direct interactions with the metal center of the active site or the coordinated substrate. However, this outer coordination sphere is important because it controls interactions between the catalytic core and the external environment. For example,

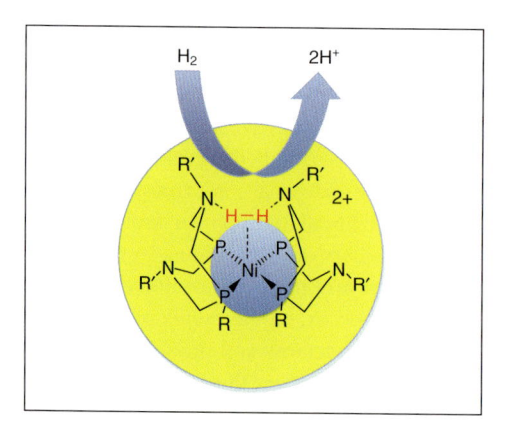

Figure 7.3 Ni catalyst for oxidation of hydrogen, showing the first coordination sphere in blue, and the second coordination sphere in yellow.

the outer coordination sphere can provide a mechanism for regulating catalyst activity or can provide a link between the electrocatalyst and an electrode or semiconductor surface. The optimization of the properties of each coordination sphere will be extremely important in the design of highly active molecular catalysts that are integrated with an external environment. At this stage of catalyst development, we have only begun to explore the role of the outer coordination sphere. As a result, only the first two coordination spheres will be discussed in this chapter.

7.4 Using the First Coordination Sphere to Control the Energies of Catalytic Intermediates

In the active site of [FeFe] hydrogenase, shown by structure **4**, the rapid and efficient catalysis of H_2 production and oxidation requires that the hydride acceptor ability of the metal match the proton acceptor ability of the base incorporated into the dithiolate ligand. The energies associated with H^- addition to the metal and H^+ addition to this pendant base should ideally be balanced so that the free energy associated with the heterolytic cleavage or formation of H_2 is close to $0\,\text{kcal}\,\text{mol}^{-1}$. High energy intermediates would lead to a large barrier for this step and result in either slow rates and/or large overpotentials. A highly exoergic step in a mechanism, leading to a low energy intermediate, will result in a high barrier for a subsequent step. Thus the energies of the proton acceptor and the hydride acceptor must be matched for optimum performance.

Similar considerations are important for developing *synthetic* catalysts for hydrogen production or oxidation. To achieve a better understanding of the factors controlling the hydride acceptor abilities of transition metal complexes, a series of studies of the hydride acceptor/donor abilities of a variety of $[M(\text{diphosphine})_2]^{2+}$ complexes (where M = Ni, Pd, and Pt) and other transition metal complexes was undertaken. It was determined that the hydride acceptor abilities of $[M(\text{diphosphine})_2]^{2+}$ complexes are determined by three factors: the metal, the electron donor ability of the ligand substituents, and the dihedral angle between the two planes defined by the two phosphorus atoms of each diphosphine ligand and the metal. The hydride acceptor ability follows the order Ni > Pt > Pd [35]. For example, the free energy associated with Equation 7.3 is $-66\,\text{kcal}\,\text{mol}^{-1}$ for Ni, $-55\,\text{kcal}\,\text{mol}^{-1}$ for Pt, and $-51\,\text{kcal}\,\text{mol}^{-1}$ for Pd (where PNP is $Et_2PCH_2NMeCH_2PEt_2$).

$$[M(PNP)_2]^{2+} + H^- = [HM(PNP)_2]^+ \tag{7.3}$$

Thus, nickel complexes are approximately $15\,\text{kcal}\,\text{mol}^{-1}$ better hydride acceptors than the corresponding palladium complexes. The hydride acceptor ability of $[M(\text{diphosphine})_2]^{2+}$ complexes is also a function of the substituents on the diphosphine ligands. The hydride acceptor abilities of $[Ni(dppe)_2]^{2+}$ and $[Ni(dmpe)_2]^{2+}$ are -63 and $-51\,\text{kcal}\,\text{mol}^{-1}$, respectively (where dppe is bis(diphenylphosphino)ethane and dmpe is bis(dimethylphosphino)ethane) [33, 36]. Thus, replacement of the methyl substituents with phenyl substituents increases the hydride acceptor ability by 12 kcal mol^{-1}. Increasing the dihedral angle between the two diphosphine ligands also increases the hydride acceptor ability of [M(diphosphine)$_2$]$^{2+}$ complexes. [Ni(dmpe)$_2$]$^{2+}$ has a dihedral angle of 3° and a hydride acceptor ability of −51 kcal mol^{-1}, whereas [Ni(dmpp)$_2$]$^{2+}$ (where dmpp is bis(dimethylphosphino) propane) has a dihedral angle of 44° and a hydride acceptor ability of −61 kcal mol^{-1} [33]. Even larger effects (more than 25 kcal mol^{-1}) have been observed for a series of [Pd(diphosphine)$_2$]$^{2+}$ and [Pt(diphosphine)$_2$]$^{2+}$ complexes [34, 37]. These large changes in hydride acceptor abilities with dihedral angle can be understood in terms of simple overlap arguments for the lowest occupied molecular orbitals of the [M(diphosphine)$_2$]$^{2+}$ complexes [38, 39]. An understanding of the factors controlling the hydride acceptor ability of these complexes allows this property to be tuned to match the proton acceptor ability of a pendant base.

Similar energy matching strategies are also important for other steps in the catalytic cycle including H$_2$ binding, transfer of a second proton between the metal and the pendant base following or preceding oxidation, and so on. As a result, a deeper and more general understanding of the thermodynamic relationships relevant to catalysis is desirable. Using free energy relationships and other empirical relationships, a thermodynamic model has been developed for [Ni(diphosphine)$_2$]$^{2+}$ complexes that allows us to predict thermodynamic schemes (e.g., Figure 7.4a for the Ni(depp)$_2$ system, depp = bis(diethylphosphino)propane) [16]. Such schemes can be used to construct free energy maps or landscapes, Figure 7.4b, depicting the relative free energies of possible catalytic intermediates. These free energy maps can, in turn, be used to develop reaction profiles (Figure 7.4c) that illustrate the relative energies of the reaction intermediates (but not the transition states) of the catalytic cycle for H$_2$ oxidation. These free energy landscapes and reaction profiles can be determined for different conditions (such as pH values and hydrogen pressures) and for catalysts with different substituents and different dihedral angles. This quantitative model of how the free energy

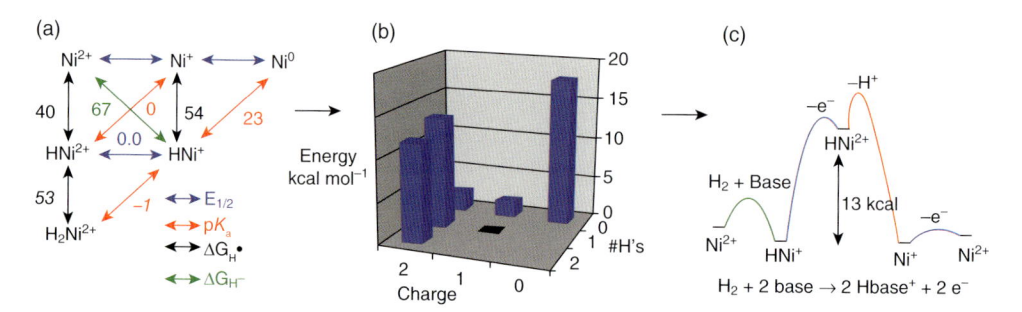

Figure 7.4 Thermodynamic scheme (a) for the Ni(depp)$_2$ system, free energy landscape (b), H$_2$ pressure = 1 atm, pH = 8.5, energy in kcal mol^{-1}), reaction profile for H$_2$ oxidation (c), same conditions).

surfaces and reaction profiles of catalytic cycles respond to changes in reaction conditions and to catalyst structure is extremely useful in catalyst design for this class of nickel complexes [13, 14]. These free energy landscapes can also be usefully applied to the development of molybdenum sulfur complexes such as the one shown by structure **2** [40].

7.5 Using the Second Coordination Sphere to Control the Movement of Protons between the Metal and the Exterior of the Molecular Catalyst

As shown in structure **4**, it is thought that the pendant amine in the dithiolate bridge of the [FeFe] hydrogenase acts as a relay to facilitate the movement of protons between the distal iron atom and the proton conduction channel [5]. In addition, transition metal hydride and dihydrogen complexes with pendant bases in the second coordination sphere are known to play a role in ligand exchange reactions of the hydride ligand [41–48]. This led us to undertake a series of studies of nickel and iron complexes in which we probed the roles of proton relays. The iron complex, $[HFe(PNP)(dmpm)(CH_3CN)]^+$ (**13**) (where dmpm is $Me_2PCH_2PMe_2$) was synthesized because the PNP ligand should provide a good model for the azadithiolate ligand thought to be present in the [FeFe] hydrogenase enzyme. In addition, iron complexes of this type are known to form stable dihydrogen complexes [49–53]. Complex **13** can be protonated at low temperature to form $[HFe(PNHP)(dmpm)(CH_3CN)]^{2+}$ (**14**; Equation 7.4) [20]. Variable-temperature NMR studies of **14** revealed a fast intramolecular exchange of the NH proton and the hydride ligand, which is thought to proceed through the dihydrogen intermediate **15** in Equation 7.4. Rotation of the dihydrogen ligand followed by cleavage of the H–H bond exchanges the original hydride and NH hydrogen atoms. These studies strongly support the role of the bridging amine in assisting the heterolytic cleavage of H_2, as proposed for the [FeFe] hydrogenase enzyme. The rapid exchange of the NH proton of **13** with protons in solution was also shown to occur. The intramolecular exchange shown in Equation 7.4, coupled with the intermolecular exchange of the protonated ligand with protons in solution, provides a facile pathway for the exchange of the hydride ligand of **13** with external protons. Similar processes are likely involved in the H/D exchange between D_2 gas and water observed for hydrogenase enzymes [2].

$$\text{13} \xrightarrow{H^+} \text{14} \rightleftharpoons \text{15} \tag{7.4}$$

In contrast, studies of cis-[HFe(PNHP)$_2$(CH$_3$CN)]$^{3+}$ (**16**) indicated that intramolecular exchange between the hydride ligand and the protons bound to N is much slower than is observed for **13** [54]. The origin of this difference was attributed to steric interactions between the two PNHP$^+$ ligands, which make the formation of the boat conformation of the ligand required for FeH/NH exchange, as shown for structures **13**–**15**, energetically unfavorable. These results suggest that although a pendant base can facilitate intramolecular exchange, steric interactions can prevent the exchange from occurring. In addition, when the acetonitrile ligand in **13** is replaced by CO, no NH/FeH exchange is observed [20]. This is attributed to the much higher acidity of a dihydrogen ligand trans to CO compared to one trans to acetonitrile. As a result, the transfer of a proton from nitrogen to the hydride ligand trans to CO is energetically unfavorable, and the intramolecular exchange is effectively turned off. From these results it is clear that although pendant bases or proton relays can facilitate the movement of protons between the metal center and solution, both steric *and* electronic energy barriers can prevent this movement from occurring. Both steric considerations and energy matching must be taken into account when designing proton relays.

Cyclic voltammetry studies reveal striking differences between complex **13** and the analogous complex [HFe(depp)(dmpm)(CH$_3$CN)]$^+$ (**17**) in which the NMe group of **13** has been replaced by a methylene group. At normal scan rates the Fe$^{III/II}$ couple is reversible for complex **17**, but irreversible for **13**. Scan rate dependence measurements and potential step experiments indicated that this difference in behavior arises from a rapid transfer of the proton of the Fe^{III} hydride to the N atom of the pendant base with a rate constant of $1.1 \times 10^2 \, \text{s}^{-1}$ at room temperature. This proton transfer results in an irreversible Fe$^{(III/II)}$ couple at low scan rates. A similar process cannot occur for **17**, and the Fe$^{(III/II)}$ couple remains reversible, even at slow scan rates in the presence of an external base. These results indicate that pendant bases in the second coordination sphere can facilitate the coupling of electron and proton transfer reactions.

The studies of iron complexes containing ligands with pendant nitrogen bases in the second coordination sphere demonstrated that pendant bases assist heterolytic cleavage of H$_2$, resulting in intramolecular proton exchange between hydride ligands and NH protons. Because the latter exchange rapidly with protons in solution, they provide a facile pathway for intermolecular exchange of protons in solution with the hydride ligand bound to iron. Pendant bases can also facilitate the coupling of electron and proton transfer reactions. All of these properties justify describing these pendant bases as proton relays that assist in the movement of protons between the interior and exterior of the molecule.

7.6
Integration of the First and Second Coordination Spheres

Our understanding of the factors controlling metal hydride acidity and hydride acceptor abilities for nickel diphosphine complexes, together with the potential

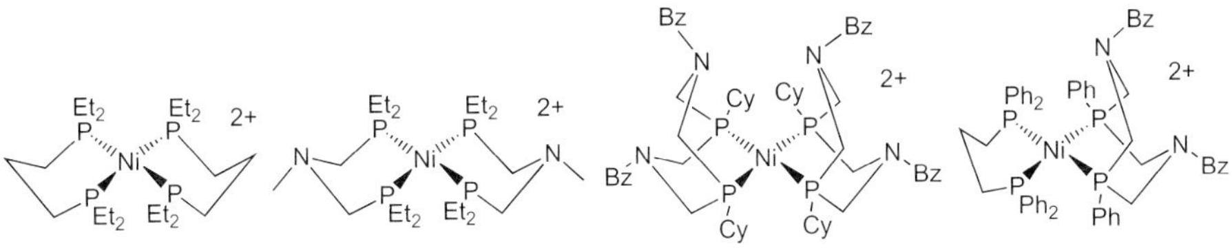

7.6 Integration of the First and Second Coordination Spheres

	18	19	20	28
rate of H_2 addition, s^{-1}		<0.2	10	0.4
$\Delta G°$, kcal mol^{-1} (H_2 addition)		-6.0	-3.1	-4.0

Figure 7.5 The catalytic rates and the thermodynamic driving forces for the formation of the products of H_2 cleavage for **19**, **20**, and **28** are shown below their respective structures [55, 56]

role of proton relays in the second coordination sphere, led us to design molecular electrocatalysts for H_2 oxidation and production by appropriately matching the thermodynamic properties of the metal and the nitrogen-based relays. The energy profile for the catalytic oxidation of H_2 by [Ni(depp)$_2$]$^{2+}$, structure **18**, is shown in Figure 7.4c. Replacement of the central CH_2 group of the $(CH_2)_3$ backbone of the diphosphine ligand in **18** with an NCH_3 group produced the nickel complex, [Ni(PNP)$_2$]$^{2+}$, **19** (see Figure 7.5) [19]. Both complexes are catalysts for H_2 oxidation, but the incorporation of the proton relay results in a 0.6–0.7 V decrease in the overpotentials for **19** relative to **18**. The substantial decrease in overpotential (down to approximately 100 mV) is attributed to the ability of the pendant nitrogen base to act as a relay to transfer protons from the nickel atom to solution without major structural reorganization of the first coordination sphere. NMR and electrochemical studies of these compounds strongly support the role of these ligands in promoting fast proton transfer steps. However, the catalytic rate for [Ni(PNP)$_2$]$^{2+}$ is slow, <0.2 s^{-1}, because H_2 activation is slow.

To overcome the problem of slow H_2 activation, complexes with *positioned* pendant bases (proton relays) were investigated. The proposed structure of the active site of the Fe-only hydrogenase active site (structure **4**) includes a pendant base incorporated in a six-membered ring that is forced to be in a boat conformation by steric interactions with the proximal Fe atom and its ligands. This boat conformation results in a positioned relay that assists in the binding and heterolytic cleavage of H_2. This structural feature is also incorporated into the nickel complex shown in structure **20** by the introduction of a second six-membered ring system in the chelating diphosphine ligands. The enforced boat conformations for at least two of the rings of these complexes avoid the activation energies associated with converting from the more stable chair conformation observed in the resting state of the [Ni(PNP)$_2$]$^{2+}$ complex, **19**, to the boat conformation required for catalysis. Complexes having the structure shown by **20** are indeed very active catalysts for either H_2 oxidation (turnover frequency of 10 s^{-1} for R = cyclohexyl, R' = benzyl) or H_2 production (turnover frequency of 350 s^{-1} for R = R' = phenyl) [13, 14]. These results clearly document the importance of these pendant bases in H_2 activation, *in addition* to their role as proton relays in these nickel complexes.

Scheme 7.1 shows a proposed catalytic cycle for the oxidation (clockwise direction) or production (counterclockwise direction) of H_2. This cycle is based on a combination of spectroscopic and electrochemical measurements and theoretical calculations for $[Ni(P_2^{Cy}N_2^{Bz})_2]^{2+}$ and related complexes. Reaction of H_2 with $[Ni(P_2^{Cy}N_2^{Bz})_2]^{2+}$, **20**, results in the formation of complex **21** which has not yet been observed experimentally. However, on the basis of theoretical calculations and kinetic studies, **21** is thought to be in equilibrium with **20**, with the equilibrium lying far to the left. For some complexes, a dihydride intermediate, **22**, has also been observed. It is step 2 that is postulated as the rate-determining step for the overall catalytic reaction. The *observed* catalytic rate is determined by the rate of oxidative addition of H_2 to the metal (step 2) and the value of the equilibrium constant for step 1. Sequential or concerted migration of the two hydride ligands of **22**, as protons, to nitrogen atoms of the diphosphine ligands results in the formation of **23**. This reaction is fast. Deprotonation of **23** is observed to produce a nickel hydride shown by structure **25**. However, the isomeric complex **24**, in which the proton is bound to N is probably only slightly uphill. This is supported by the observation of this isomer in a similar complex, $[Ni(dppp)(P_2^{Ph}N_2^{Bz}HN^{Bz})]^+$ (where dppp is bis(diphenylphosphino)propane). Oxidation of **24** produces **26**,

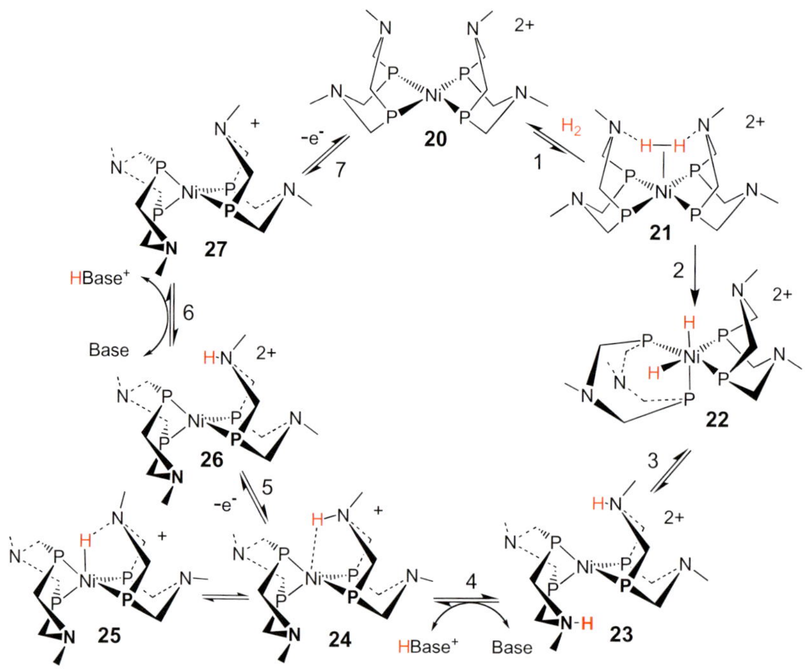

Scheme 7.1

which is deprotonated to form the $Ni^{(I)}$ complex, **27**. Oxidation of the latter species regenerates the original catalyst.

As discussed, for these Ni catalysts the rate of H_2 oxidation or production is determined by the equilibrium associated with step 1 and the rate constant for step 2. The ability of $[NiP_2^{Cy}N_2^{Bz})_2]^{2+}$, **20**, to position two pendant bases so that they stabilize the dihydrogen intermediate **21** thus facilitates the rate of cleavage of H_2 to form **23**, even when the thermodynamic driving force is relatively low. In Figure 7.5 the catalytic rates and the thermodynamic driving forces for the formation of the products of H_2 cleavage complexes for **19**, **20**, and **28** are shown below their respective structures [55, 56]. The slowest rate is observed for complex **19**, which has the largest driving force for H_2 addition under 1 atm H_2 at 22 °C. However, complex **19** has no positioned bases. Complex **28**, which has a smaller driving force but a larger rate, has one positioned base. Complex **20** with two positioned bases has the highest rate of H_2 addition, but the lowest driving force. These results are consistent with an increased stability for an H_2 adduct arising from interactions with two positioned pendant bases, as suggested by theoretical studies [13, 14].

If the rate-determining step is the formation of the dihydride intermediate from the dihydrogen complex in Scheme 7.1, then stabilizing this dihydride intermediate should increase the rate of H_2 oxidation. In previous studies of $[Rh(diphosphine)_2]^+$ complexes, the stability of the dihydride complex resulting from H_2 addition, $[(H)_2Rh(diphosphine)_2]^+$, was shown to be enhanced by electron-donating substituents and an increase in the natural bite angle of the diphosphine ligand with an associated increase in the dihedral angle between the two diphosphine ligands [57]. If this is true, then the same dihedral twist or tetrahedral distortion should facilitate H_2 addition to $[Ni(diphosphine)_2]^{2+}$ complexes to form the corresponding dihydrides. It is worth noting that large tetrahedral distortions are also observed for the Ni atom in [NiFe] hydrogenases. These distortions may play the same role in facilitating H_2 activation by these enzymes.

Another consideration for H_2 oxidation catalysts is their inhibition by CO. It is well known that CO poisons both Pt catalysts in fuel cells and hydrogenase enzymes. As a result, when platinum is used as a catalyst, it is necessary to keep CO levels in the hydrogen gas below 100 ppm. In contrast, studies of complex **20** using H_2 gas containing 5% CO shows that it is not significantly inhibited by CO, even at these very high CO concentrations [14]. Although CO binding does occur to this catalyst, H_2 binds 20 times more strongly [58]. As a result CO does not significantly inhibit this synthetic catalyst for H_2 oxidation at concentrations 100 times higher than those at which platinum is inhibited, and any inhibition caused by even pure CO gas is reversible. Structural studies of the CO adduct of **20** support a model for CO binding that involves stabilizing interactions between two pendant amines of the ligands and the electropositive carbon of the coordinated CO ligand [57].

The role of pendant bases in catalytic H_2 production is not unique to the nickel complexes described in the preceding paragraphs. Recent studies of cobalt complexes have shown that complexes with a single cyclic diphosphine ligand containing a pendant base, such as $[Co(P_2^{Ph}N_2^{Ph})(CH_3CN)_3](BF_4)_2$, are electrocatalysts for H_2 production with turnover frequencies and overpotentials very similar to those of $[Ni(P_2^{Ph}N_2^{Ph})_2(CH_3CN)_3](BF_4)_2$ [17]. Comparison of the catalytic behavior of $[Co(P_2^{Ph}N_2^{Ph})_2(CH_3CN)_3](BF_4)_2$ with that of $[Co(dppp)(CH_3CN)_3](BF_4)_2$ (where dppp is bis(diphenylphosphino)propane), for which no catalytic activity is observed, suggests that the pendant amine is playing an important role in the catalysis of H_2 production, similar to the nickel complexes. However, two amine bases near the metal ion are not required for high activity in the cobalt complex.

7.7
Summary

Simple molecular catalysts for hydrogen oxidation and production that are based on inexpensive first row transition metal complexes have been developed. Our approach has been to divide the design of catalysts into distinct coordination spheres and to try to understand the roles of each coordination sphere. Our studies have provided an understanding of those structural and electronic features of both the ligands and the metal (first coordination sphere) that control hydride acceptor abilities, pK_a values, and redox potentials of $[M(diphosphine)_2]^{2+}$ complexes (where M = Ni, Pd, and Pt). From this thermodynamic information, free energy maps and reaction profiles can be constructed that indicate the relative energies of intermediates in potential catalytic cycles. How these free energy maps and reaction profiles change as a function of the ligand and metal can be used to guide catalyst development. In an effort to lower the activation barriers between these intermediates, pendant bases were incorporated into the second coordination sphere and shown to facilitate H_2 binding, intramolecular heterolytic cleavage of H_2, and proton-coupled electron-transfer reactions. The structure, number, and positioning of the proton relays each play an important role in their ability to facilitate the overall catalytic process. It is also important that energy matching between the first and second coordination sphere be achieved. This approach has led to the development of highly active H_2 production and oxidation catalysts based on both nickel and cobalt. It is hoped that in the future some of the insights developed in these studies will contribute to the design and discovery of molecular catalysts based on inexpensive metals that can replace platinum with superior performance characteristics.

In a broader sense, the tools developed in the course of these studies may find applications in a range of multi-proton and multi-electron transfer processes. The understanding of free energy landscapes and how the first coordination sphere of the metal controls them provides a rational basis for initial catalyst design by providing the information needed to avoid high-energy intermediates. The incorporation of proton relays into the second coordination sphere can provide low energy pathways for controlling the movement of protons between the metal and solution. This requires energy matching of the proton relays in the second coordination sphere with hydrides in the first coordination sphere for efficient catalysis. This simple modular approach can be useful in catalyst development for many important redox reactions and provide insights into more complex biological systems.

Acknowledgements

Our research is supported by the Division of Chemical Sciences, Biosciences and Geosciences of the Office of Basic Energy Sciences of the Department of Energy. Pacific Northwest National Laboratory is operated by Battelle for the U.S. Department of Energy.

References

1. http://en.wikipedia.org/wiki/Fuel_cell, last access 12/31/2009.
2. Frey, M. (2002) *Chembiochem*, **3**, 153–160.
3. Peters, J.W. (1999) *Curr. Opin. Struct. Biol.*, **9**, 670–676.
4. Peters, J.W., Lanzilotta, W.N., Lemon, B.J., and Seefeldt, L.C. (1998) *Science*, **282**, 1853–1858.
5. Nicolet, Y., de Lacey, A.L., Vernède, X., Fernandez, V.M., Hatchikian, E.C., and Fontecilla-Camps, J.C. (2001) *J. Am. Chem. Soc.*, **123**, 1596–1601.
6. Artero, V., and Fontecave, M. (2005) *Coord. Chem. Rev.*, **249**, 1518–1535.
7. Hu, X., Brunschwig, B.S., and Peters, J.C. (2007) *J. Am. Chem. Soc.*, **129**, 8988–8998.
8. Hu, X.L., Cossairt, B.M., Brunschwig, B.S., Lewis, N.S., and Peters, J.C. (2005) *Chem. Commun.*, 4723–4725.
9. Connolly, P., and Espenson, J.H. (1986) *Inorg. Chem.*, **25**, 2684–2688.
10. Razavet, M., Artero, V., and Fontecave, M. (2005) *Inorg. Chem.*, **44**, 4786-4795.
11. Baffert, C., Artero, V., and Fontecave, M. (2007) *Inorg. Chem.*, **46**, 1817–1824.
12. Appel, A.M., DuBois, D.L., and Rakowski DuBois, M. (2005) *J. Am. Chem. Soc.*, **127**, 12717–12726.
13. Wilson, A.D., Shoemaker, R.K., Meidaner, A., Muckerman, J.T., Rakowski DuBois, M., and DuBois, D.L. (2007) *Proc. Natl. Acad. Sci. USA*, **104**, 6951–6956.
14. Wilson, A.D., Newell, R.H., McNevin, M.J., Muckerman, J.T., Rakowski DuBois, M., and DuBois, D.L. (2006) *J. Am. Chem. Soc.*, **128**, 358–366.
15. Rakowski DuBois, M., and DuBois, D.L. (2008) *Compt. Rend. Chim.*, **11**, 805–817.
16. Rakowski DuBois, M., and DuBois, D.L. (2009) *Chem. Soc. Rev.*, **38**, 62–72.
17. Jacobsen, G.M., Yang, J.Y., Twamley, B., Wilson, A.D., Bullock, R.M., Rakowski DuBois, M., and DuBois, D.L. (2008) *Energy Environ. Sci.*, **1**, 167–174.
18. Fan, H.-J., and Hall, M.B. (2001) *J. Am. Chem. Soc.*, **123**, 3828–3829.
19. Curtis, C.J., Miedaner, A., Ciancanelli, R.F., Ellis, W.W., Noll, B.C., Rakowski DuBois, M., and DuBois, D.L. (2003) *Inorg. Chem.*, **42**, 216–227.
20. Henry, R.M., Shoemaker, R.K., DuBois, D.L., and Rakowski DuBois, M. (2006) *J. Am. Chem. Soc.*, **128**, 3002–3010.
21. Thematic issue on bio-inspired hydrogen production/uptake catalysis. (2008) *Compt. Rend. Chim.*, **11** (8), 789–944.
22. Special Issue on hydrogenases (2005) *Coord. Chem. Rev.*, **249** (15–16), 1517–1690. devoted to hydrogenases
23. Barton, B.E., and Rauchfuss, T.B. (2008) *Inorg. Chem.*, **47**, 2261–2263.
24. Singleton, M.L., Jenkins, R.M., Klemashevich, C.L., and Darensbourg, M.Y. (2008) *Compt. Rend. Chim.*, **11**, 861–874.
25. Ezzaher, S., Capon, J.-F., Gloaguen, F., Kervarec, N., Petillon, F.Y., Pichon, R., Schollhammer, P., and Talamarin, J. (2008) *Compt. Rend. Chim.*, **11**, 906–914.

26. Adam, F.I., Hogarth, G., Kabir, S.E., and Richards, I. (2008) *Compt. Rend. Chim.*, **11**, 890–905.
27. Fontecilla-Camps, J.C., Volbeda, A., Cavazza, C., and Nicolet, Y. (2007) *Chem. Rev.*, **107**, 4273–4303.
28. Page, C.C., Moser, C.C., Chen, X., and Dutton, P.L. (1999) *Nature*, **402**, 47–52.
29. (a) Onuchic, J.N., Beratan, D.N., Winkler, J.R., and Gray, H.B. (1992) *Annu. Rev. Biophys. Biomol. Struct.*, **21**, 349–377; (b) Axup, A.W., Albin, M., Mayo, S.L., Crutchley, R.J., and Gray, H.B. (1988) *J. Am. Chem. Soc.*, **110**, 435–439.
30. Williams, R.J.P. (1995) *Nature*, **376**, 643.
31. Peters, J.W., Lanzilotta, W.N., Lemon, B.J., and Seefeldt, L.C. (1998) *Science*, **282**, 1853.
32. Montet, Y., Amara, P., Volbeda, A., Vernède, X., Hatchikian, E.C., Field, M.J., Frey, M., and Fontecilla-Camps, J.C. (1997) *Nat. Struct. Biol.*, **4**, 523.
33. Berning, D.E., Noll, B.C., and DuBois, D.L. (1999) *J. Am. Chem. Soc.*, **121**, 11432–11447.
34. Raebiger, J.W., Miedaner, A., Curtis, C.J., Miller, S.M., and DuBois, D.L. (2004) *J. Am. Chem. Soc.*, **126**, 5502–5514.
35. Curtis, C.J., Miedaner, A., Raebiger, J.W., and DuBois, D.L. (2004) *Organometallics*, **23**, 511–516.
36. Berning, D.E., Miedaner, A., Curtis, C.J., Noll, B.C., Rakowski DuBois, M., and DuBois, D.L. (2001) *Organometallics*, **20**, 1832–1839.
37. Miedaner, A., Raebiger, J.W., Curtis, C.J., Miller, S.M., and DuBois, D.L. (2004) *Organometallics*, **23**, 2670–2679.
38. Nimlos, M.R., Chang, C.H., Curtis, C.J., Miedaner, A., Pilath, H.M., and DuBois, D.L. (2008) *Organometallics*, **27**, 2715–2722.
39. Miedaner, A., Haltiwanger, R.C., and DuBois, D.L. (1991) *Inorg. Chem.*, **30**, 417–427.
40. Appel, A.M., Lee, S.-J., Franz, J.A., DuBois, D.L., Rakowski DuBois, M., and Twamley, B. (2009) *J. Am. Chem. Soc.*, **131**, 5224–5232.
41. Lough, A.J., Park, S., Ramachandran, R., and Morris, R.H. (1994) *J. Am. Chem. Soc.*, **116**, 8356–8357.
42. Park, S., Lough, A.J., and Morris, R.H. (1996) *Inorg. Chem.*, **35**, 3001–3006.
43. Xu, W., Lough, A.J., and Morris, R.H. (1996) *Inorg. Chem.*, **35**, 1549–1555.
44. Lee, D.-H., Patel, B.P., Clot, E., Eisenstein, O., and Crabtree, R.H. (1999) *Chem. Commun.*, 297–298.
45. Crabtree, R.H., Siegbahn, P.E.M., Eisenstein, O., Rheingold, A.L., and Koetzle, T.F. (1996) *Acc. Chem. Res.*, **29**, 348–354.
46. Chu, H.S., Lau, C.P., Wong, K.Y., and Wong, W.T. (1998) *Organometallics*, **17**, 2768–2777.
47. Ayllon, J.A., Sayers, S.F., Sabo-Etienne, S., Donnadieu, B., Chaudret, B., and Clot, E. (1999) *Organometallics*, **18**, 3981–3990.
48. Custelcean, R., and Jackson, J.E. (2001) *Chem. Rev.*, **101**, 1963–1980.
49. Kubas, G.J. (2001) *Metal Dihydrogen and σ-Bond Complexes*, Kluwer Academic/Plenum, NewYork, p. 159.
50. Hills, A., Hughes, D.L., Jimenez-Tenorio, M., and Leigh, G.J. (1993) *J. Chem. Soc., Dalton Trans.*, 3041–3049.
51. Morris, R.H., Sawyer, J.F., Shiralian, M., and Zubkowski, J.D. (1985) *J. Am. Chem. Soc.*, **107**, 5581–5582.
52. Ricci, J.S., Koetzle, T.F., Bautista, M.T., Hofstede, T.M., Morris, R.H., and Sawyer, J.F. (1989) *J. Am. Chem. Soc.*, **111**, 8823–8827.
53. Gilbertson, J.D., Szymczak, N.K., and Tyler, D.R. (2004) *Inorg. Chem.*, **43**, 3341–3343.
54. Jacobsen, G.M., Shoemaker, R.K., Rakowski DuBois, M., and DuBois, D.L. (2007) *Organometallics*, **26**, 4964–4971.
55. Fraze, K., Wilson, A.D., Appel, A.M., Rakowski DuBois, M., and DuBois, D.L. (2007) *Organometallics*, **26**, 3918–3924.
56. Yang, J.Y., Bullock, R.M., Shaw, W.J., Twamley, B., Fraze, K., Rakowski DuBois, M., and DuBois, D.L. (2009) *J. Am. Chem. Soc.*, **131**, 5935–5945.
57. DuBois, D.L., Blake, D.M., Miedaner, A., Curtis, C.J., DuBois, M.R., Franz, J.A., and Linehan, J.C. (2006) *Organometallics*, **25**, 4414–4419.
58. Wilson, A.D., Fraze, K., Twamley, B., Miller, S.M., DuBois, D.L., and DuBois, M.R. (2008) *J. Am. Chem. Soc.*, **130**, 1061–1068.

8
Nickel-Catalyzed Reductive Couplings and Cyclizations

Hasnain A. Malik, Ryan D. Baxter, and John Montgomery

8.1
Introduction

Many classes of standard organic transformations involve the coupling or cycloaddition of two different π-components to assemble a more functionalized product. Many such processes are amenable to the union of two complex fragments, thus allowing convergent approaches to complex organic molecules. Diels–Alder cycloadditions, Prins addition reactions, and ene reactions are examples of broadly useful processes that involve the union of two π-components in a selective manner. As a complement to these classical transformations, the *reductive* coupling of two π-components provides an alternative strategy for assembly of complex fragments from the same types of starting materials [1–8]. Whereas the starting components resemble those required for the types of standard organic transformations noted above, the products obtained differ structurally and are obtained in a reduced oxidation state based on the action of the reducing agent employed.

A number of transition metal catalyst/reducing agent combinations are known to promote various reductive coupling processes. In addition to the nickel-catalyzed variants that are the subject of this chapter, important advances have been made with titanium- [9–13], iridium- [14], and rhodium-catalyzed variants [15]. Many reducing agents have been employed, and the most widely used classes include silanes, organozincs, organoboranes, molecular hydrogen, and alcohols. The specific focus of this chapter will be the development of nickel-catalyzed reductive couplings and cyclizations between a polar π-component and a non-polar π-component. This strategy allows the two π-components to be differentiated in catalytic reactions, and thus avoids homocoupling processes that are often problematic in transformations of this type. The use of either aldehydes or α,β-unsaturated carbonyls as the polar component, and alkynes as the non-polar component has been a primary focus of our laboratories and will be the subject of this chapter. Extensive related developments involving dienes as the non-polar component reported by the laboratories of Mori [16], Sato [17], and Tamaru [18] have been reviewed elsewhere [3] and will not be extensively discussed in this

Catalysis Without Precious Metals. Edited by R. Morris Bullock
© 2010 WILEY-VCH Verlag GmbH & Co. KGaA, Weinheim
ISBN: 978-3-527-32354-8

chapter. In addition to advances from our laboratory, seminal contributions from the groups of Ikeda and Jamison are described.

8.2
Couplings of Alkynes with α,β-Unsaturated Carbonyls

The synthesis of stereochemically defined tri- or tetra-substituted alkenes is a challenging problem, one that has inspired the development of numerous methodologies. The nickel-catalyzed coupling of alkynes and α,β-unsaturated carbonyls offers a direct process to access complex olefin substitution patterns. Both intra- and inter-molecular variants of the process are known, and, depending on the reaction conditions, the processes may proceed by either alkylative (transferring a carbon substituent) or reductive (transferring a hydrogen atom substituent) coupling. 1,5-Cyclooctadiene (COD)-stabilized zero-valent nickel, monodentate phosphine ligands, and a reducing agent such as a silane, organozinc, organoborane, or vinylzirconium reagent are typically used. Generally, in the absence of monodentate phosphine ligands, the reaction undergoes an alkylative coupling pathway. However, in the presence of both a monodentate phosphine ligand and a reducing agent possessing a β-hydrogen, nearly complete crossover to a reductive coupling mechanism can be achieved. The mechanism of the nickel-catalyzed coupling of alkynes and α,β-unsaturated carbonyls has been studied in detail, and the intermediacy of a nickel metallacycle has been proposed based on both experimental and theoretical considerations (see Section 8.2.3). This methodology has been applied in the total syntheses of several natural products, highlighting the synthetic utility of this process (see Section 8.2.4).

8.2.1
Three-Component Couplings via Alkyl Group Transfer – Methods Development

The nickel-catalyzed alkylative coupling of alkynes and α,β-unsaturated carbonyls has been studied in a variety of contexts and can utilize numerous classes of organometallic reducing agents. The reactions are generally highly stereoselective and provide direct access to functionalized γ,δ-unsaturated carbonyls containing stereodefined tri- or tetra-substituted alkenes. The intermolecular coupling of alkynes and α,β-unsaturated carbonyls was first reported using alkynyltin reducing agents to generate conjugated enynes (Scheme 8.1) [19]. This methodology was effective for α,β-unsaturated enones as well as α,β-unsaturated enals and tolerated both terminal and internal alkynes.

The methodology was later expanded to include the introduction of sp^3-hybridized alkyl groups via organozinc reducing agents; this experimental modification had little effect on the generality of the reaction and was similarly tolerant of substrate variation to the method described previously [20]. The introduction of sp^2-hybridized alkyl groups was later achieved via the use of vinylzirconium reduc-

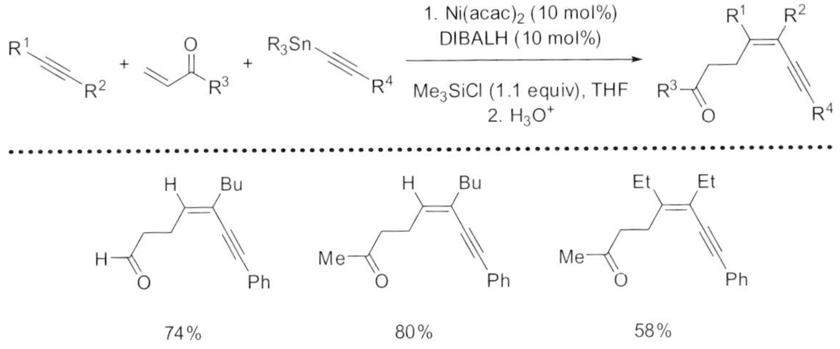

8.2 Couplings of Alkynes with α,β-Unsaturated Carbonyls

Scheme 8.1

ing agents [21]. However, this method was limited to intramolecular couplings of α,β-unsaturated enones and alkynes.

The intramolecular variant of the nickel-catalyzed alkylative coupling is especially robust and has been extensively developed for the cyclization of alkynes tethered to a variety of α,β-unsaturated systems in the presence of organozinc and organozirconium reducing agents (Scheme 8.2) [22]. Alkylative cyclizations were effective for both terminal and internal alkynes with high levels of chemo- and stereo-selectivity. The scope and tolerance of the electron-deficient alkene component was broad as enones, alkylidene malonates, nitroalkenes, and unsaturated imides were all tolerated (Scheme 8.3) [23].

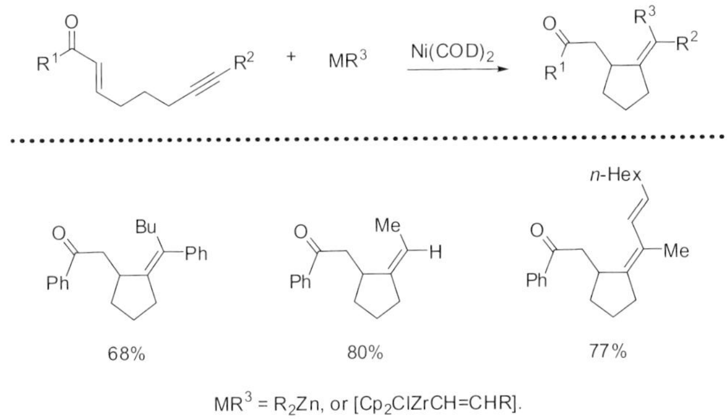

Scheme 8.2

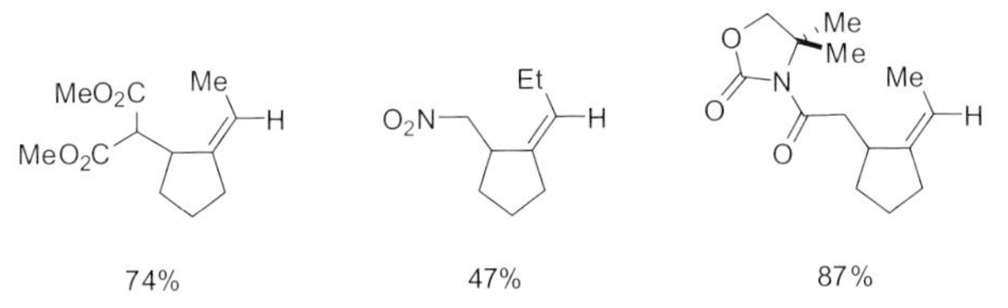

Scheme 8.3

Several insights on the mechanism of this process, from both experimental and theoretical studies, will be addressed in future sections (Section 8.2.3). Ligand effects are especially interesting features of this methodology as the alkylative coupling manifold can be tuned to favor reductive coupling in the presence of monodentate phosphine ligands and organozinc reducing agents bearing a β-hydrogen (Section 8.2.2). This alkylative coupling reaction has been shown to be an effective method, and several total synthesis applications have been demonstrated (Section 8.2.4).

8.2.2
Reductive Couplings via Hydrogen Atom Transfer – Methods Development

During the development of the nickel-catalyzed cyclization of tethered alkynyl enones, an interesting ligand effect was observed in which organozincs bearing a β-hydrogen provided hydrogen atom transfer to the distal carbon of the resulting exocyclic olefin (Scheme 8.4) [22]. This mechanistic change from alkylative to reductive coupling was seen only when Ni(COD)$_2$ was pretreated with 4 equiv of tributylphosphine prior to substrate addition in the presence of organozincs bearing β-hydrogens. Reducing agents bearing no β-hydrogens (such as Me$_2$Zn or t-Bu$_2$Zn) do not undergo the reductive coupling process, even in the presence of tributylphosphine.

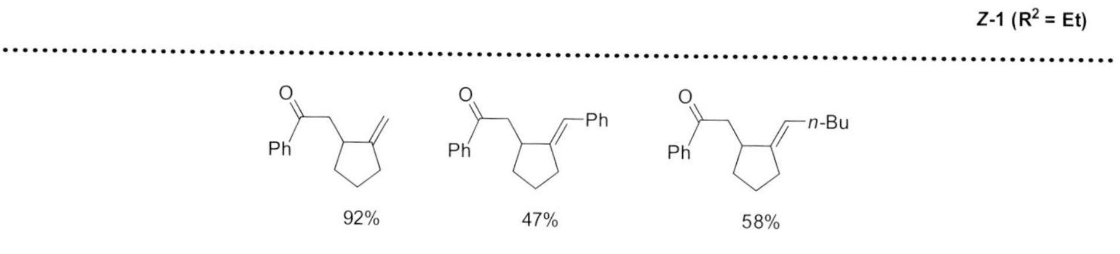

^{a} 25 mol% of PPh$_3$ is used for reductive coupling.

Scheme 8.4

The ability to promote reductive coupling provides a powerful complement to the alkylative coupling methodology discussed previously (Section 8.2.1), as either E- or Z-isomers of the exocyclic olefin **1** can be stereoselectively produced in an

8.2 Couplings of Alkynes with α,β-Unsaturated Carbonyls

efficient and direct manner (Scheme 8.4). Both aliphatic and aromatic alkynyl enones can undergo cyclization, and the reductive cyclization is also tolerant of unsaturated imides[1]. A change in reaction mechanism from alkylative to reductive coupling is dependent on the electronic environment of the nickel center. The σ-donation from the phosphine ligands leads to an electron-rich metal center that likely suppresses the carbon–carbon bond forming reductive elimination pathway.

Reaction conditions amenable to nickel-catalyzed reductive cyclization of alkynyl enones were ineffective in the intermolecular variant. In the course of developing catalytic [3 + 2] cycloadditions of alkynes and enals [24], it was discovered that by simply using an enone instead of an enal electrophile, catalytic intermolecular reductive coupling could be achieved in a Ni(COD)$_2$/Bu$_3$P catalyst system using Et$_3$B as a reducing agent in a MeOH/THF (8:1) solvent system (Scheme 8.5) [25]. A wide range of enones were shown to couple with an extensive variety of alkynes. It is important to note that both aliphatic and aromatic vinyl ketones were viable substrates in which β-alkyl and β-aryl substitutions were tolerated. Diaryl alkynes, mono-aryl internal alkynes, nonaromatic internal alkynes, terminal alkynes, and free hydroxyl-bearing alkynes were all efficient participants. It should be mentioned that while ynoates were successful nucleophiles, enoate electrophiles were found to be poor participants in nickel-catalyzed couplings.

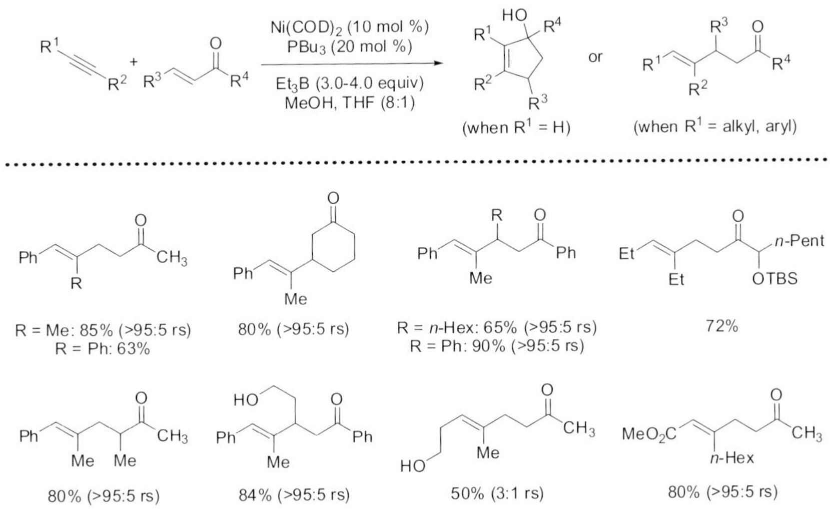

Scheme 8.5

The use of enals in nickel-promoted intermolecular couplings was initially limited to stoichiometric [3 + 2]/[2 + 1] cycloadditions [26]. This limitation was overcome with the development of a Ni(COD)$_2$/PCy$_3$/R$_3$SiH/THF reaction system, whereby the reductive coupling of alkynes and enals was achieved to afford a highly chemo- and stereo-selective synthesis of Z-enol silanes (Scheme 8.6) [27].

1) Intermolecular reductive couplings of alkynes and enoates are not yet achievable.

Both β-alkyl and β-aryl substituted enals were viable coupling partners with a variety of alkynes. Ynoates, diaryl alkynes, mono-aryl internal alkynes, non-aromatic internal alkynes, terminal alkynes, free hydroxyl-bearing alkynes, basic secondary amine-bearing alkynes, and even alkynes with tethered aldehydes or ketones were all efficient participants. The Z-enol silanes produced by this protocol are largely inaccessible by other synthetic methods and this process complements the powerful Trost and Jamison procedures that are E-selective [28, 29]. The utility of these substrates is evident by the emergence of a range of Mukaiyama aldol reaction methodologies. The high Z-isomer stereoselectivity (>98:2) observed in this coupling offers strong empirical mechanistic evidence of a highly ordered, stereochemically defined η^1, O-bound metallacyclic nickel enolate intermediate (further discussed in Section 8.2.3).

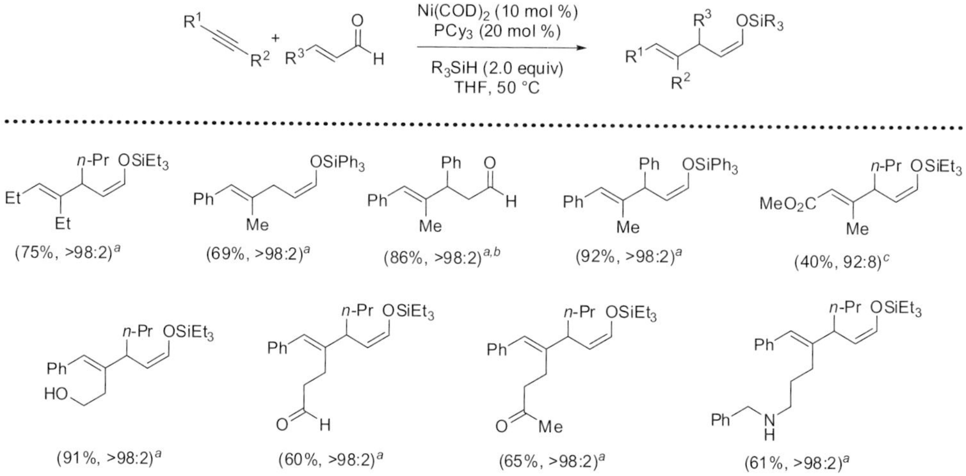

[a] Ratio refers to both stereo- and regioselectivity. [b] Me$_2$PhSiH was used as a reducing agent. [c] Isomer ratio refers to the regiochemistry of alkyne insertion.

Scheme 8.6

8.2.3
Mechanistic Insights

There are several possible mechanisms that have been proposed since the first report of nickel-catalyzed coupling of alkynes and α,β-unsaturated carbonyls. This section will briefly cover the metallacycle-based mechanism as it is generally considered in both the alkylative and reductive coupling manifolds. However, it should be emphasized that different variants may proceed via differing pathways.

8.2.3.1 Metallacycle-Based Mechanistic Pathway
The intermediacy of a nickel-metallacycle is likely involved in many classes of nickel-catalyzed couplings of π-systems. The highly ordered nature of the metal-

8.2 Couplings of Alkynes with α,β-Unsaturated Carbonyls

lacycle resulting from the oxidative cyclization of an alkyne and an α,β-unsaturated carbonyl would lead to products with high levels of stereo- and regioselectivity (Scheme 8.7) [26]. There are several examples that illustrate the feasibility of a metallacycle intermediate in this class of reactions and a select few will be highlighted in this section.

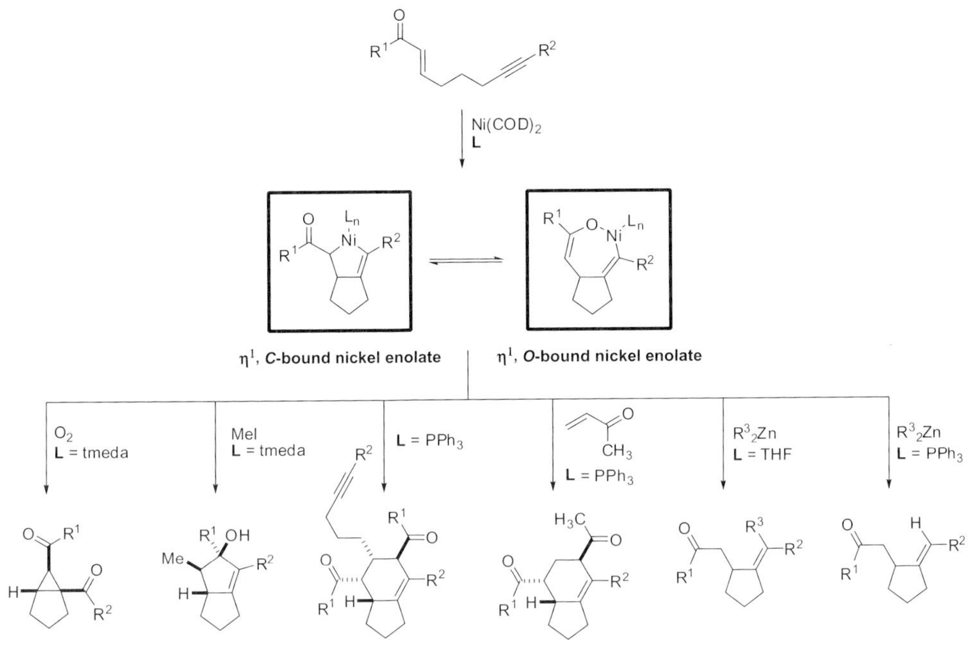

Scheme 8.7

A combined experimental and theoretical study was undertaken in which a well defined η^1, O-bound nickel enolate metallacycle **3** was isolated and crystal structure data were obtained (Figure 8.1, Scheme 8.8) [30]. The metallacycle isolated was not kinetically competent due to the stabilizing ligands required for isolation, but it clearly demonstrates the ability to access metallacycles via oxidative cyclization of a Ni(0) catalyst with alkynes and enals. Moreover, complex **3** displayed reactivity which paralleled the nickel-catalyzed methodology of alkynyl enal **2**. For example, metallacycle **3** was subjected to a reducing agent (Me$_2$Zn) and afforded aldehyde **4** in direct analogy to precedented catalytic alkyne/enal couplings. Furthermore, alkylative cycloadditions also proceeded in direct analogy to published accounts. In a series of DFT calculations it was found that, in the presence of dimethylzinc, a π-allyl intermediate would likely undergo conversion to a metallacycle with stereodefined olefin geometry via alkyne insertion. In the absence of dimethylzinc, the η^1, O-bound nickel enolate was found to be lower in energy than the η^1, C-bound metallacycle [31].

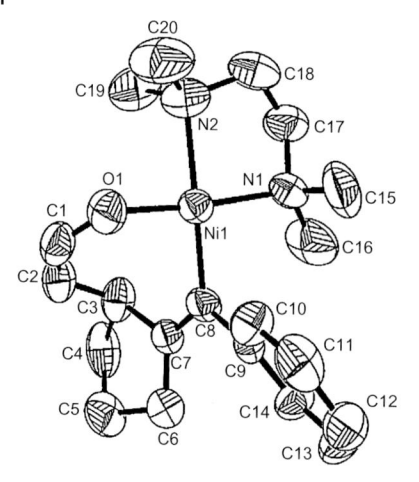

Figure 8.1 X-ray crystal structure of metallacycle **3**.

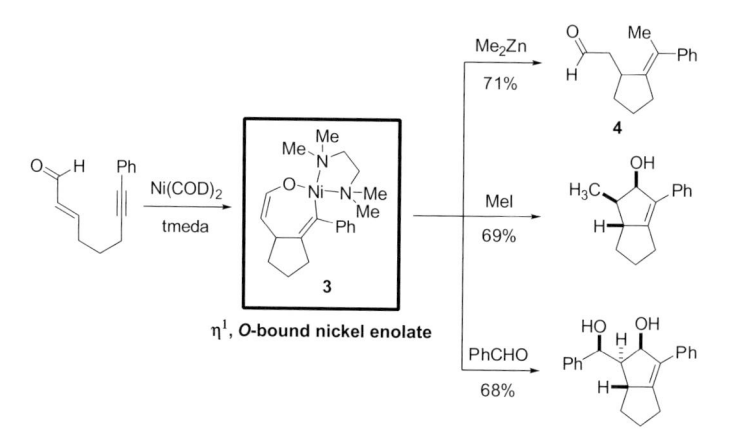

Scheme 8.8

The clearest empirical evidence for the productive involvement of an η^1, O-bound nickel enolate comes from the intermolecular reductive coupling of alkynes and enals (Scheme 8.9) [27]. The extremely high levels of Z-isomer stereoselectivity (>98:2) can best be rationalized via the metallacycle intermediate **5** which undergoes σ-bond metathesis to afford nickel hydride **6**, followed by reductive elimination to yield the Z-selective enol silane product **7**. A mechanism consisting of a nickel π-allyl species would not be expected to lead to high selectivities of Z-enol silanes, and has been implicated in reactions leading to the selective production of E-enol silanes [28].

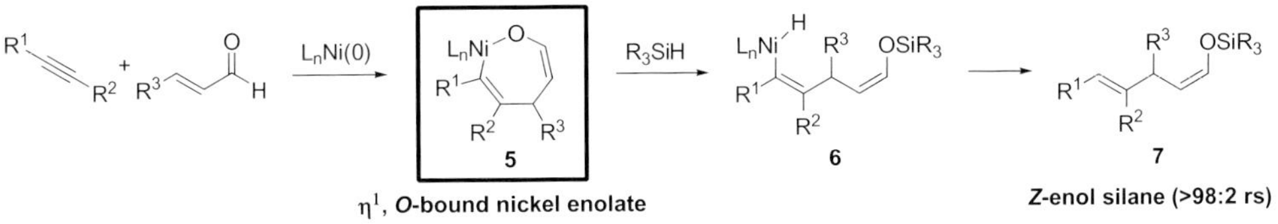

Scheme 8.9

8.2.4
Use in Natural Product Synthesis

Nickel-catalyzed couplings of alkynes and α,β-unsaturated carbonyls have been applied in the total synthesis of several complex natural products. For many of these examples, the use of nickel-catalyzed alkylative coupling offers a straightforward means of synthesis. For each of the following examples the general synthetic plan is briefly described, followed by the key nickel-catalyzed coupling step that is critical for the construction of the overall structural motif of the natural product in question.

The kainoid amino acid family of natural products has been the target of a number of synthetic efforts involving nickel-catalyzed alkylative couplings as the key synthetic step. Using structurally related alkyne- or allene-containing substrates, a versatile approach to both (−)-α-kainic acid and (+)-α-allokainic acid was developed [32, 33]. The synthesis of (−)-α-kainic acid involved an unsaturated imide/allene alkylative coupling with dimethylzinc to produce 8 (Scheme 8.10) [32]. This process led to the diastereoselective synthesis of the core five-membered ring in 57% yield, followed by hydrolysis of the protecting groups to complete the total synthesis of (−)-α-kainic acid.

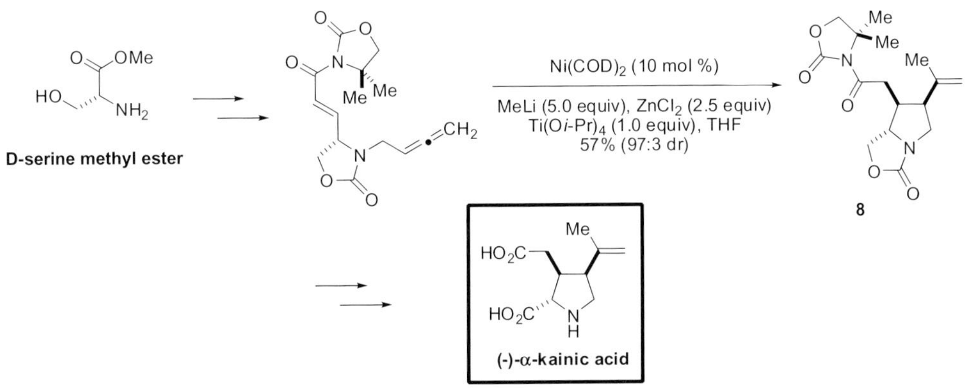

Scheme 8.10

The epimeric (+)-α-allokainic acid was constructed by the unsaturated imide/alkyne alkylative coupling of **9** with trimethylaluminum in 73% yield and 97:3 diastereoselectivity (Scheme 8.11) [33]. This was followed by silyl to carbonate protecting group transposition, stereoselective allylic reduction, and removal of protecting groups to afford (+)-α-allokainic acid. The complementary nature of these two stereodivergent approaches allowed access to both epimeric natural products.

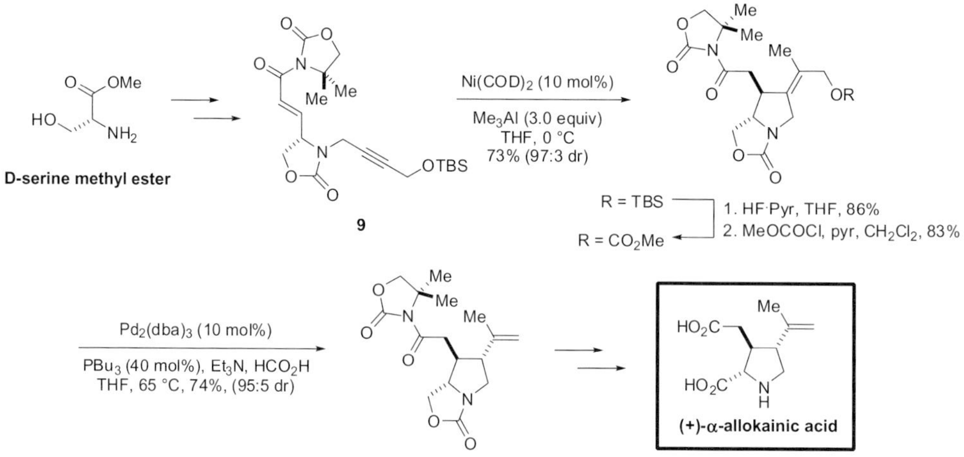

Scheme 8.11

A more complex member of the kainoid amino acid class of natural products is isodomoic acid G. This structure lacks the C_4 stereocenter but instead possesses an exocyclic tetrasubstituted alkene. The core structure was constructed in a single step by nickel-catalyzed cyclization of **10** with the vinylzirconium alkylating reagent **11** in 74% yield (Scheme 8.12) [34]. Notably, the pyrrolidine five-membered ring, C_2/C_3 relative stereochemistry, and 1,3-diene motif were assembled in a completely stereoselective fashion in a single operation. Sequential deprotection and oxidation steps afforded isodomoic acid G.

The core skeleton of geissoschizine, an important biosynthetic precursor to numerous polycyclic indole scaffolds, was the target of a nickel-catalyzed alkylative coupling strategy. Cyclization precursor **13** was prepared by ozonolysis and double reductive amination of cyclopentene **12** (Scheme 8.13) [35]. Nickel-catalyzed alkylative cyclization afforded **14** in 84% yield and 95:5 diastereoselectivity. A deprotection/oxidation sequence followed, and chromatography led to complete inversion of the C_3 stereocenter. A Fisher indole synthesis followed to afford (±)-deformyl-isogeissoschizine, the core skeleton of geissoschizine.

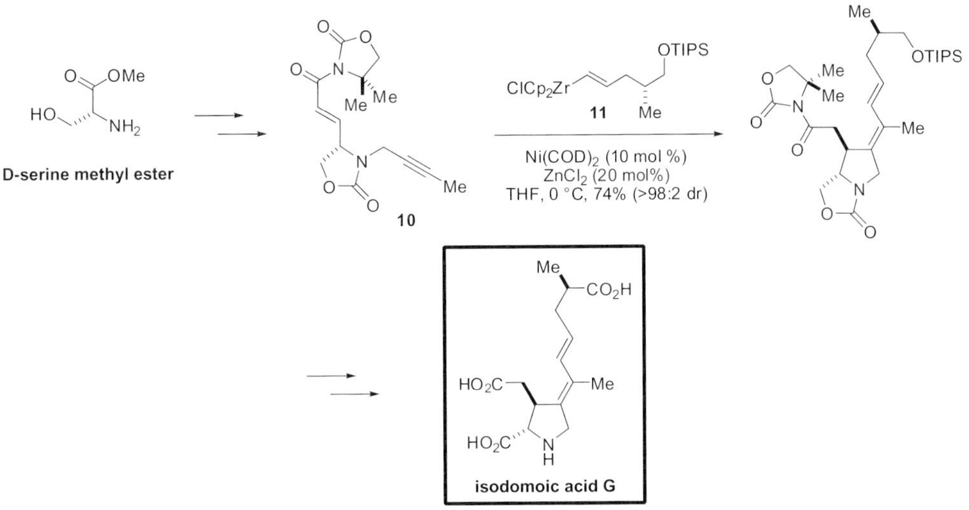

Scheme 8.12

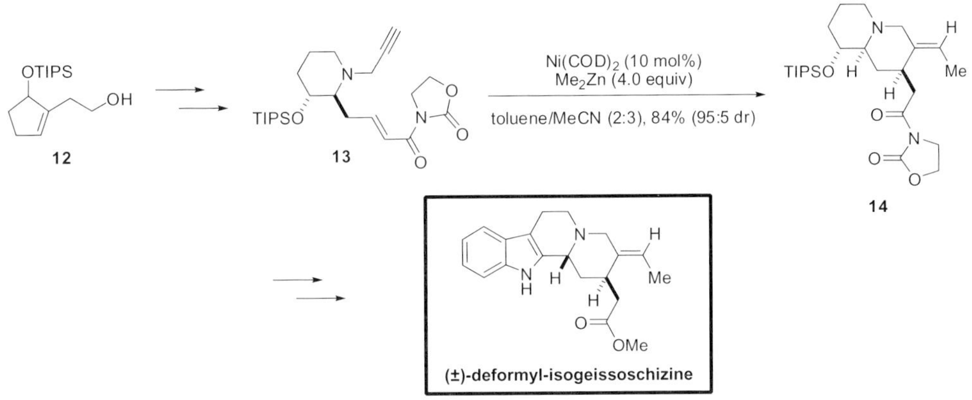

Scheme 8.13

8.3
Couplings of Alkynes with Aldehydes

Since the first reported example of nickel-catalyzed cyclization of ynals in 1997, many variants now exist that undergo either alkylative (transferring a carbon substituent) or reductive (transferring a hydrogen atom substituent) coupling. The methodology has been applied to both inter- and intra-molecular couplings, allowing access to acyclic products and small or macrocyclic ring systems. In this nickel-catalyzed process, COD-stabilized zero-valent nickel, monodentate phosphine or *N*-heterocyclic carbene (NHC) ligands, and a reducing agent such as a silane, organozinc, organoborane, or vinylzirconium reagent are typically used. There are several mechanistic insights that have emerged that shed light on the subtle aspects that govern the reactivity and outcome of various nickel-catalyzed processes (see Section 8.3.3). These protocols have been applied in complex settings, illustrating their utility in the total synthesis of various natural products (see Section 8.3.5).

8.3.1
Three-Component Couplings via Alkyl Group Transfer – Method Development

The nickel-catalyzed alkylative coupling of aldehydes and alkynes was first described using organozinc reducing agents [36]. Using methodology previously developed for the intramolecular coupling of alkynyl enones (Section 8.2.1), derivatives of 5-hexynal were cyclized in the presence of $Ni(COD)_2$ and various organozinc reducing agents to afford cyclic allylic alcohols with stereodefined exocyclic tri- or tetra-substituted olefins. The methodology was tolerant of a variety of alkyne substitutions as well as various organozinc reducing agents (Scheme 8.14).

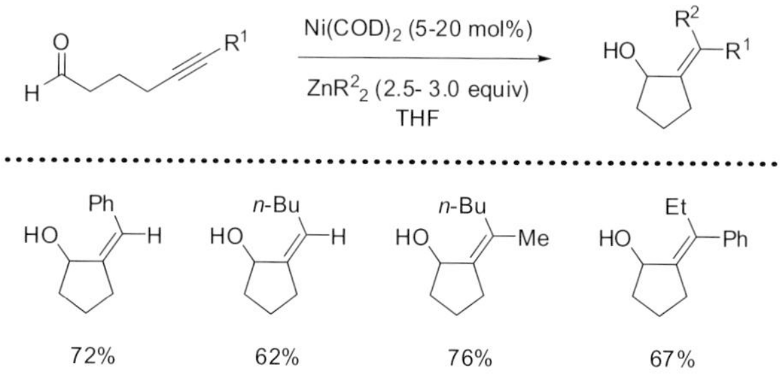

Scheme 8.14

Intermolecular couplings involving aldehydes, terminal alkynes, and organozincs also proceeded with high levels of chemo- and regioselectivity (Scheme 8.15). However, unlike intramolecular couplings, direct addition of more reactive organozincs to the aldehyde component competed with the desired three-

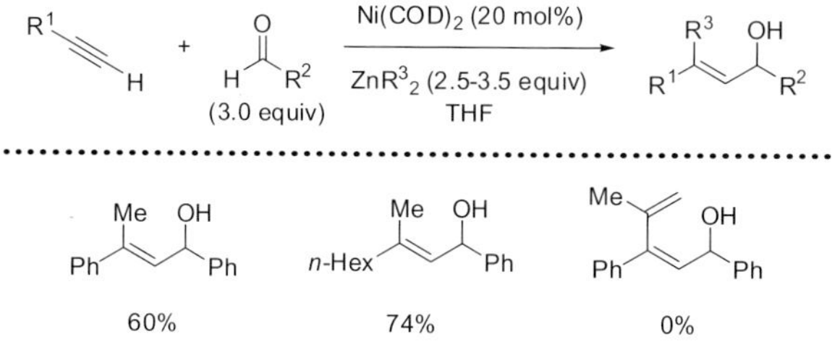

Scheme 8.15

component coupling. As a result, the scope of the intermolecular variant is limited to the use of sp^3-hybridized organozinc reagents, as sp^2-hybridized organozincs add to the aldehyde at a much faster rate than the three-component couplings.

The studies above illustrate the limitations associated with organozinc reducing agents in the intermolecular alkylative coupling of aldehydes and alkynes. However, alkenylzirconium reducing agents, derived from hydrozirconation of alkynes, participated in both inter- and intra-molecular coupling reactions without suffering competing addition to the carbonyl (Scheme 8.16) [21]. This methodology provides a facile entry to functionalized cyclic or acyclic 1,3-dienes, and expands the generality of the multicomponent coupling to include the introduction of sp^2-hybridized centers from the reducing agent. While aromatic, aliphatic, and terminal alkynes were tolerated in the intramolecular coupling, more highly functionalized ynals were also good substrates, as evidenced by the efficient cyclization of substrate **15** with the vinylzirconium reagent **16** (Scheme 8.17).

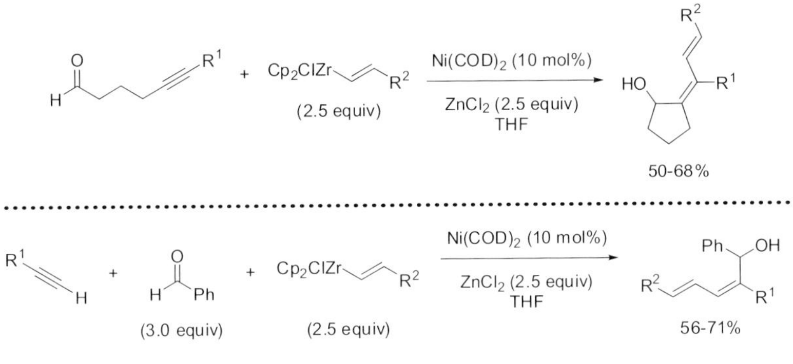

Scheme 8.16

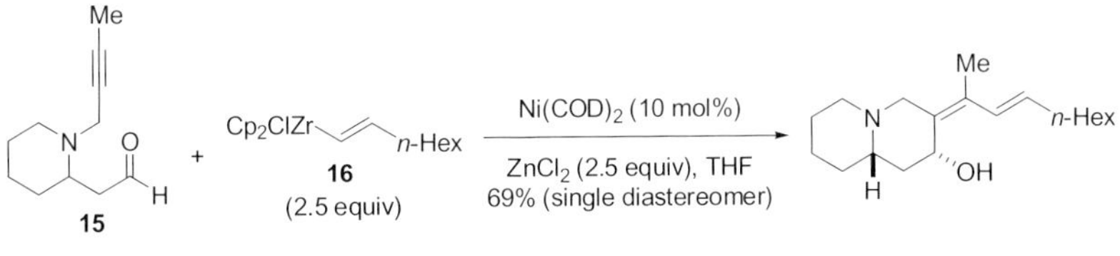

Scheme 8.17

8.3.2
Reductive Couplings via Hydrogen Atom Transfer – Method Development

The field of nickel-catalyzed reductive couplings has seen substantial interest in the past decade and numerous advances have been made in the area. There now exist several protocols, including simple aldehyde/alkyne coupling, diastereoselective variants using transfer of chirality, asymmetric variants using either chiral monodentate phosphine or NHC ligands, and macrocyclization methodology for which there are both diastereoselective and asymmetric variants.

8.3.2.1 Simple Aldehyde and Alkyne Reductive Couplings

The nickel-catalyzed aldehyde–alkyne reductive coupling was first reported using a Ni(COD)$_2$/PBu$_3$ catalyst system with ZnEt$_2$ as the reducing agent (the mechanistic rationale for crossover from alkylative to reductive coupling is analogous to that discussed in Section 8.2) [36]. Various ynals were cyclized to produce allylic alcohols in good yield with complete hydrogen atom transfer (Scheme 8.18). It is important to state that organozinc-mediated reductive coupling can only be achieved when zero-valent nickel is pretreated with tributylphosphine and used with organozinc reducing agents bearing β-hydrogens. However, under these specific reaction conditions, the methodology is limited to intramolecular couplings only, and requires the use of highly reactive reducing agents such as ZnEt$_2$.

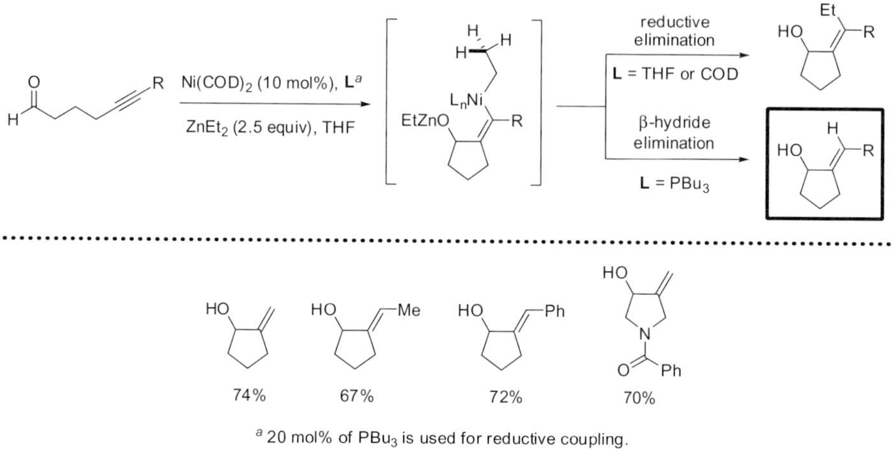

Scheme 8.18

^{a} 20 mol% of PBu$_3$ is used for reductive coupling.

The combination of monodentate phosphine ligands and silane reducing agents proved to be a much more effective process in intramolecular couplings, and involves the use of a mild, bench-stable reducing agent [37, 38]. This methodology was developed for the total synthesis of (+)-allopumiliotoxin 267A, (+)-allopumiliotoxin 339A, and (+)-allopumiliotoxin 339B (total synthesis discussed in Section 8.3.5). High yields and excellent diastereoselectivities were observed in the synthesis of bicyclic substrates (Scheme 8.19). However, the PBu$_3$/Et$_3$SiH combination was not amenable for intermolecular reductive couplings, for which there are generally two reaction systems conducive for coupling, namely either Ni(COD)$_2$/PR$_3$/BEt$_3$ or Ni(COD)$_2$/NHC/Et$_3$SiH reagent combinations.

The first nickel-catalyzed intermolecular reductive coupling utilized the combination of COD-stabilized zero-valent nickel, monodentate phosphine ligands, and

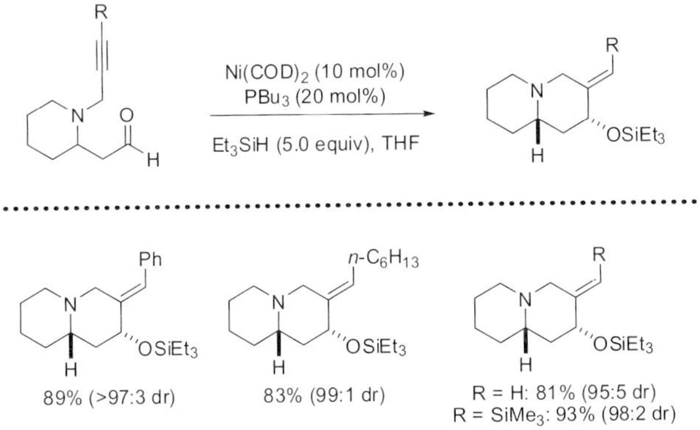

8.3 Couplings of Alkynes with Aldehydes

Scheme 8.19

a borane reducing agent (Scheme 8.20) [39]. The process works best for aromatic alkynes but is quite tolerant of various aldehydes.

Another effective catalyst system for intermolecular reductive coupling is the combination of Ni(COD)$_2$/NHC with Et$_3$SiH as a stoichiometric reducing agent (Scheme 8.21) [40]. The NHC catalyst is more reactive than analogous phosphine-based catalysts and a broad range of alkynes and aldehydes undergo reductive coupling under these conditions.

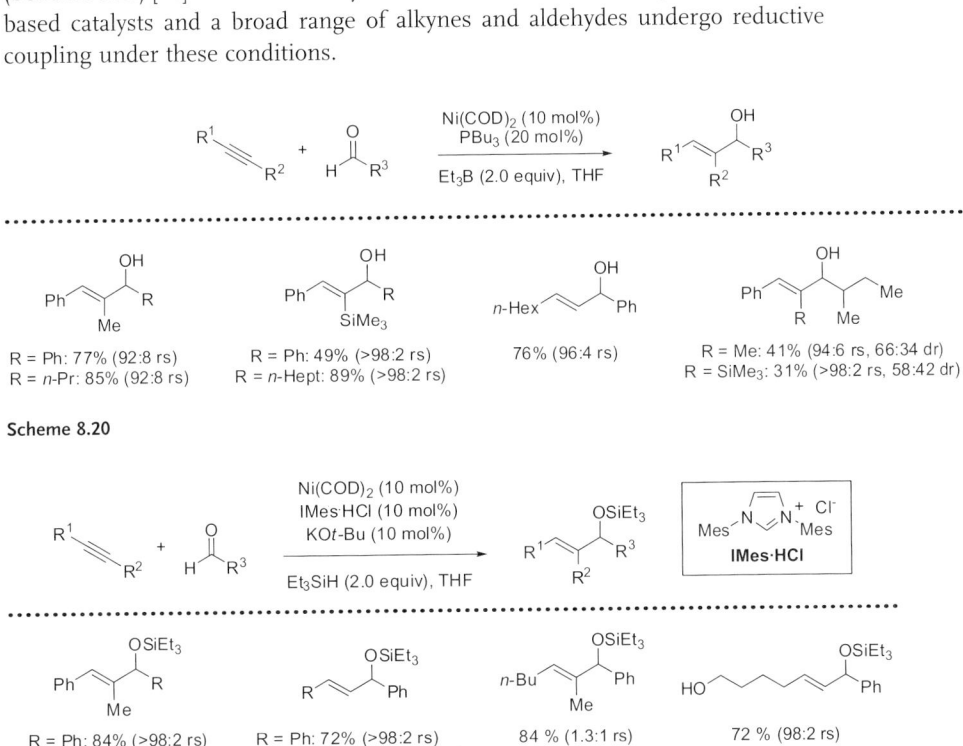

Scheme 8.20

Scheme 8.21

8 Nickel-Catalyzed Reductive Couplings and Cyclizations

Macrocyclization via nickel-catalyzed reductive coupling has been shown to be a powerful tool in the synthesis of several natural products, including amphidinolide T1 [41], amphidinolide T4 [41], (−)-terpestacin [42], and aigialomycin D [43]. A general protocol for nickel-catalyzed macrocyclizations was published using ligand-based control of regioselectivity (Scheme 8.22) [44]. Ni(COD)$_2$/PBu$_3$/BEt$_3$ and Ni(COD)$_2$/NHC/Et$_3$SiH combinations are both effective and complementary in reductive macrocyclizations. Ynal **17** was cyclized to afford the endocyclic product **19** in 89% yield with a 4.5:1 regioselectivity using a Ni(COD)$_2$/PBu$_3$ catalyst system and Et$_3$B reducing agent. However, with the Ni(COD)$_2$/NHC catalyst system and Et$_3$SiH reducing agent, the regioselectivity is reversed to favor the exocyclic product **18** which was formed in 93% yield and a 5:1 regioselectivity.

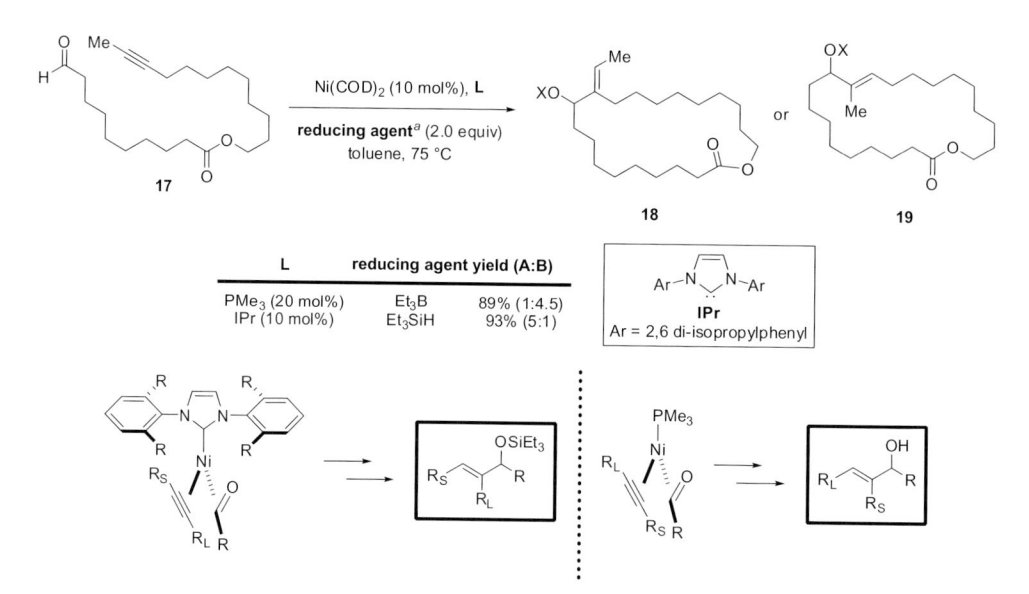

Scheme 8.22

8.3.2.2 Directed Processes

The reductive coupling of 1,3-enynes and aldehydes has been shown to be an extremely regioselective reaction, yielding high levels of regiocontrol for both the NHC/Et$_3$SiH and PBu$_3$/Et$_3$B combinations (Scheme 8.23) [40, 45]. The conjugated olefin of the 1,3-diene directs coupling to take place at the alkyne terminus distal to the alkene moiety. Olefin-directed couplings are also shown to be exceptionally effective in 1,6-enyne/aldehyde couplings, providing a high level of regiocontrol dependent on ligand structure and stoichiometry (Scheme 8.24) [46–48]. Mechanistic considerations for such control of regioselectivity for both 1,3- and 1,6-enynes will be discussed in a later section (Section 8.3.3).

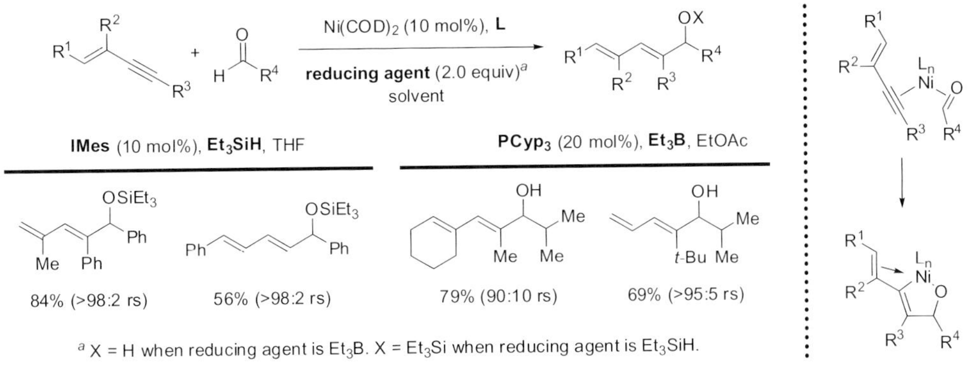

8.3 Couplings of Alkynes with Aldehydes

Scheme 8.23

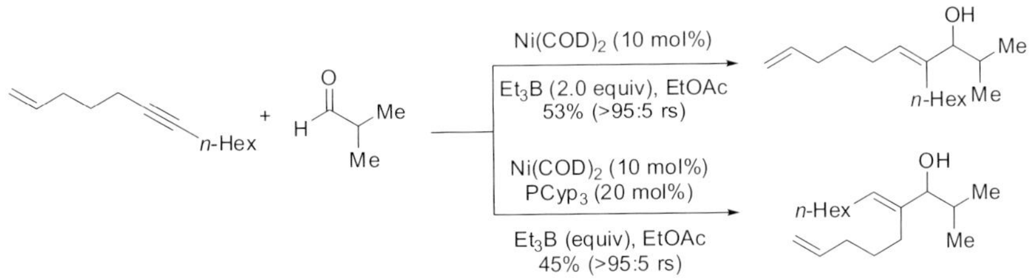

Scheme 8.24

8.3.2.3 Diastereoselective Variants: Transfer of Chirality

Reductive cyclizations have been shown to be diastereoselective in the synthesis of bicyclic products as well as various macrocycles. There are also instances of chirality transfer in intermolecular reductive couplings. The synthesis of *anti*-1,2-diols has been demonstrated using α-alkoxyaldehydes with a methoxymethyl ether (MOM) protecting group and mono-aryl internal alkynes (Scheme 8.25) [49]. Diastereoselectivities are high for the formation of *anti*-1,2-diols in cases where the aldehyde has a branched sp³-β-carbon.

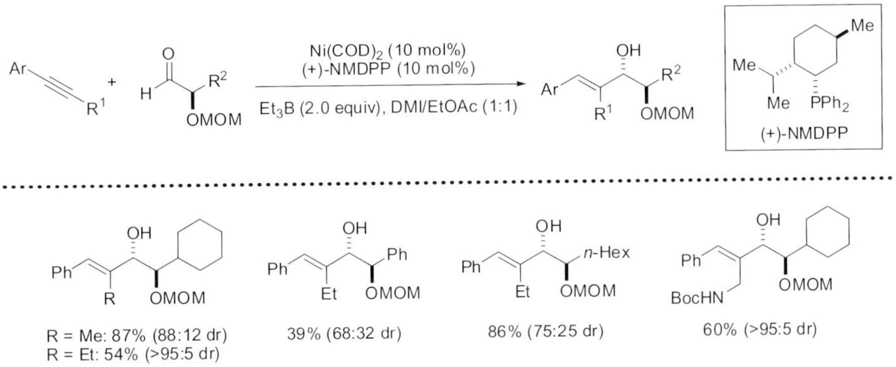

Scheme 8.25

A similar protocol employs α-silyloxyaldehydes with a *t*-butyldimethylsilyl (TBDMS) protecting group and TMS-protected alkynes [50]. In this protocol, β-unbranched aldehydes afforded the highest level of diastereoselectivity (Scheme 8.26). The method is general for both aromatic and non-aromatic alkynes and complements the prior report in terms of scope.

The transfer of chirality is not limited to chiral aldehydes. It has been shown that in 1,6-enyne/aldehyde couplings, a chiral moiety on the 1,6-enyne could make the process diastereoselective (Scheme 8.27) [47, 48]. If no phosphine ligands are employed, **20** was formed in >95:5 regioselectivity and 95:5 diastereoselectivity. However, the use of tricyclopentylphosphine as ligand led to reversal of regioselectivity with no chiral induction in the resulting allylic alcohol subunit of **21**. The mechanistic implications of this process will be discussed in a later section (Section 8.3.3).

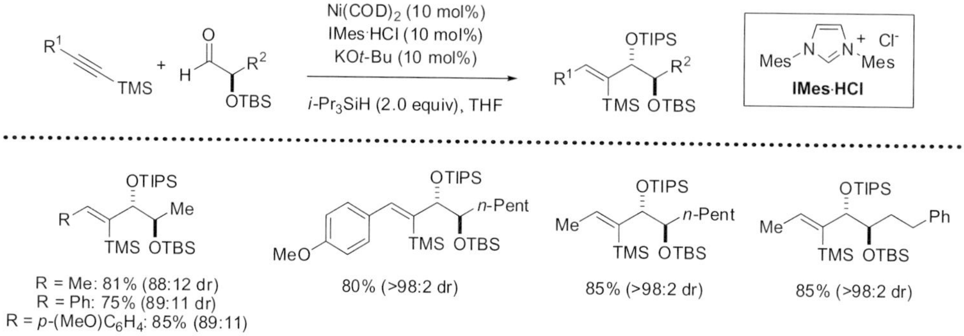

Scheme 8.26

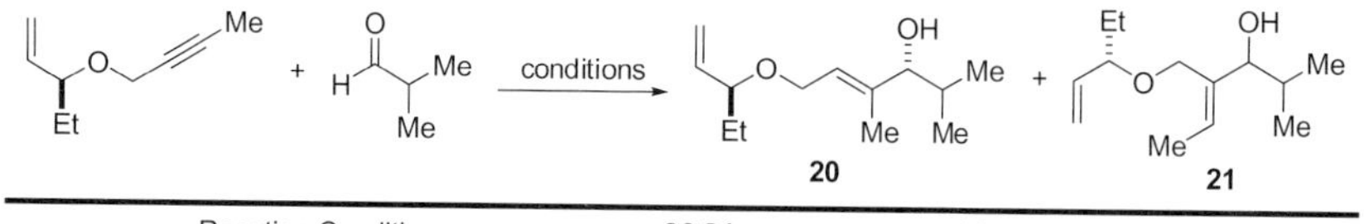

Scheme 8.27

Reaction Conditions	20:21	dr 20	dr 21
Ni(COD)$_2$ (10 mol%) Et$_3$B (2 equiv)	> 95:5	95:5	—
Ni(COD)$_2$ (10 mol%) PCyp$_3$ (20 mol%) Et$_3$B (2 equiv)	< 5:95	—	45:55

8.3.2.4 Asymmetric Variants

Two general protocols exist for nickel-catalyzed reductive couplings of aldehydes and alkynes. The first employs (+)-NMDPP as a chiral monodentate phosphine ligand. Internal aromatic alkynes were shown to undergo reductive coupling in good yields with enantioselectivities ranging from 42% to 96% (Scheme 8.28) [51]. The scope of this asymmetric reaction is limited to aryl-substituted alkynes, and α-branched aldehydes provide the highest levels of asymmetric induction.

R = i-Pr: 95% (90% ee)
R = n-Pr: 82% (65% ee)
R = c-Hex: 97% (90% ee)
R = Ph: 79% (91:9 rs, 73% ee)

R = c-Hex: 78% (89% ee)
R = i-Pr: 95% (90% ee)

74% (92% ee)

35% (42% ee)

Scheme 8.28

The second protocol for asymmetric nickel-catalyzed reductive coupling utilizes chiral NHC ligands (Scheme 8.29) [52]. The scope was general for alkynes and aldehydes, and enantioselectivities ranged from 65% to 85%. Both internal and terminal aromatic or non-aromatic alkynes were effective participants, as were aromatic and aliphatic aldehydes.

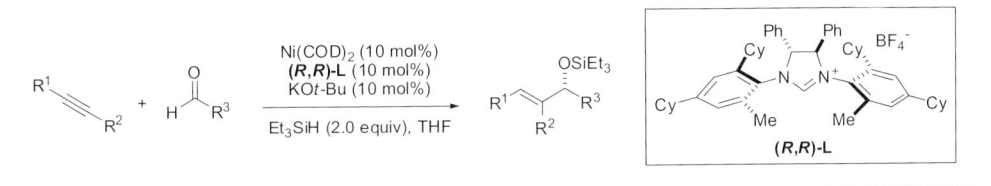

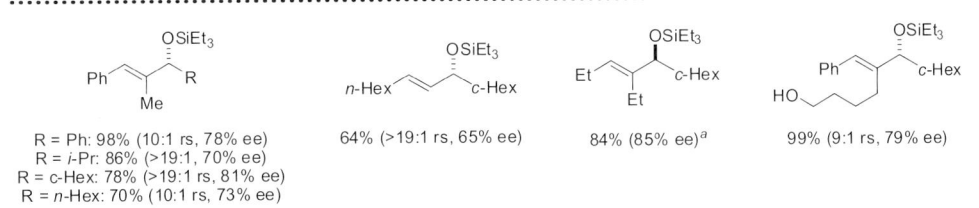

R = Ph: 98% (10:1 rs, 78% ee)
R = i-Pr: 86% (>19:1, 70% ee)
R = c-Hex: 78% (>19:1 rs, 81% ee)
R = n-Hex: 70% (10:1 rs, 73% ee)

64% (>19:1 rs, 65% ee)

84% (85% ee)[a]

99% (9:1 rs, 79% ee)

[a] (S,S)-L was used to obtain this stereoisomer.

Scheme 8.29

8.3.3
Mechanistic Insights

Several mechanistic pathways have been proposed for nickel-catalyzed aldehyde/alkyne reductive couplings, and an overview of the possible mechanisms has been provided elsewhere [3]. Therefore, this description will focus on what is generally believed to be the operative mechanism. The key features of this mechanism are the oxidative cyclization of a zero-valent nickel aldehyde-alkyne complex to form a five-membered oxametallacycle, followed by reductive cleavage of the nickel–carbon bond, and carbon–hydrogen bond formation via reductive elimination (Scheme 8.30).

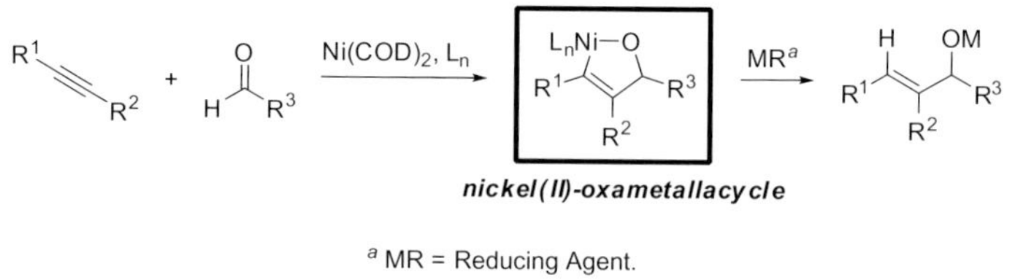

[a] MR = Reducing Agent.

Scheme 8.30

Being able to tune nickel-catalyzed couplings from an alkylative to a reductive pathway has been demonstrated not only in couplings of alkynes and α,β-unsaturated carbonyls but also for alkynes and aldehydes. Using a Ni(COD)$_2$ catalyst system and organozinc reagents, ynals were efficiently cyclized to produce exocyclic allylic alcohols with alkyl transfer at the terminus of the exocyclic olefin (Scheme 8.31) [36]. However, in a Ni(COD)$_2$/PBu$_3$ catalyst system with β-hydrogen-bearing organozincs, a reductive coupling pathway could be achieved with transfer of a hydrogen atom substituent at the distal carbon of the newly formed olefin. This change in reaction mechanism is unique to the intramolecular variant, as identical reaction conditions in the intermolecular variant did not allow crossover from alkylative to reductive coupling. However, the ability to change

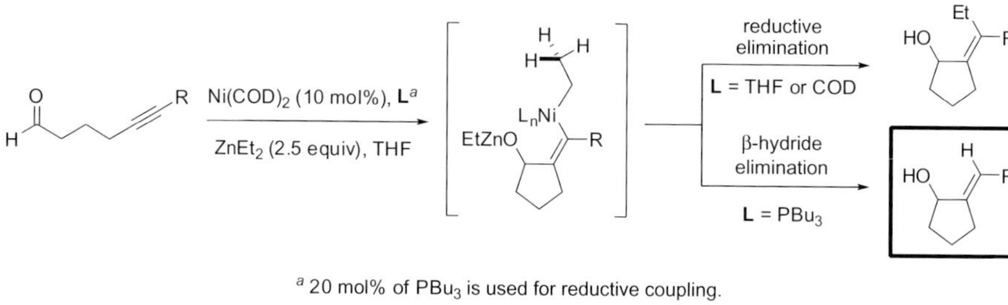

[a] 20 mol% of PBu$_3$ is used for reductive coupling.

Scheme 8.31

mechanism in the intramolecular system from an alkylative to a reductive pathway suggests that the electronic environment on the nickel-center plays an important role in chemical reactivity and reaction outcome. The σ-donation to the metal-center that results from the addition of phosphine ligands can enable the nickel catalyst to undergo β-hydride elimination (as discussed in Section 8.2 for the analogous effect in the couplings of alkynes and α,β-unsaturated carbonyls).

Stereo- and regiocontrol via nickel-catalyzed reductive coupling is illustrated in the construction of the indolidizine skeleton of (+)-allopumiliotoxin 267A via nickel-catalyzed ynal cyclization (Scheme 8.32) [37, 38]. The prerequisite of forming two adjoining rings likely allows for a single, highly ordered nickel-metallacycle 22 to be preferentially formed. σ-Bond metathesis followed by reductive elimination then leads to a single observable diastereomer in 93% yield.

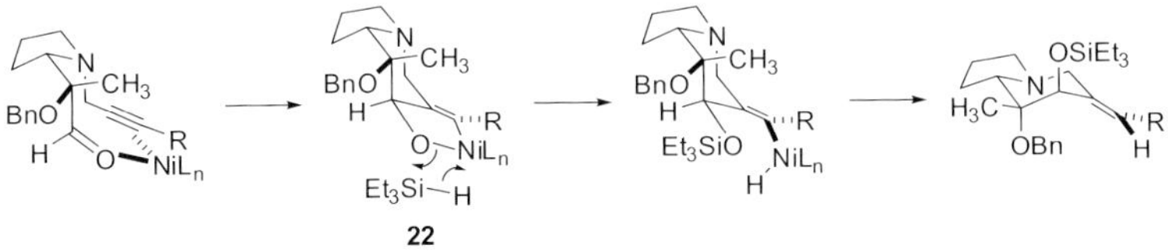

Scheme 8.32

A mechanistic probe was developed involving crossover experiments with Et_3SiD and Pr_3SiH that would allow comparison of reaction mechanisms in different ligand classes (Scheme 8.33) [40]. For the $Ni(COD)_2$/IMes catalyst system, results were similar in both inter- and intra-molecular reductive couplings, with little or no crossover observed. Products 24 and 25 were produced in 55% and 41% yield, respectively, and crossover products 23 and 26 were observed in <2% yield each, consistent with a metallacycle-based mechanism. However, in the $Ni(COD)_2$/PBu_3 catalyst system, significant crossover was observed, with 23, 24, 25, and 26 produced in a 25 : 34 : 23 : 18 ratio. This result implies a change of mechanism, and the origin of this outcome is currently being studied.

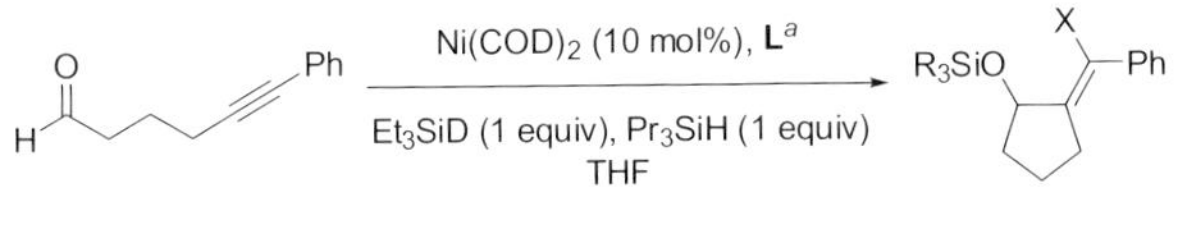

			relative %	
R	X	product	IMes	PBu_3
Et	H	23	<2	25
Et	D	24	55	34
Pr	H	25	41	23
Pr	D	26	<2	18

[a] 10 mol% IMes·HCl with 10 mol% n-BuLi was used. 20 mol% PBu_3 was used.

Scheme 8.33

Evidence for the existence of a nickel-metallacycle intermediate has been demonstrated by the Ogoshi group [53]. Ogoshi and coworkers were able to obtain a crystal structure of oxametallacycle complex 27 (Figure 8.2) generated from an

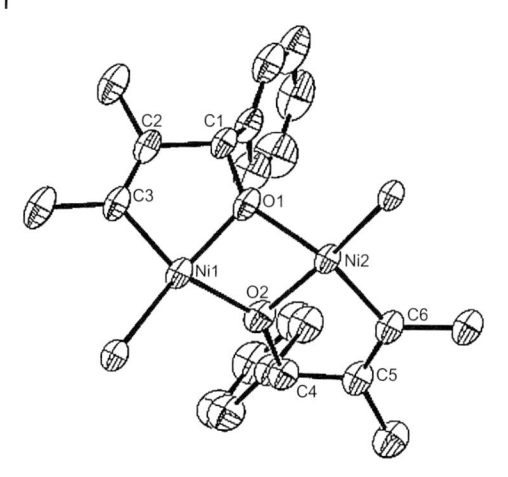

Figure 8.2 X-ray structure of metallacycle **27**.

aldehyde, alkyne, Ni(COD)₂, and PCy₃, which exists as an oxygen-bound dimer (Scheme 8.34).

Scheme 8.34

The formation of an oxametallacycle intermediate has also been proposed in couplings of 1,3- or 1,6-enynes with aldehydes [45–48]. In the use of conjugated enynes, this directed process works for both Ni(COD)₂/NHC/R₃SiH and Ni(COD)₂/PR₃/BEt₃ reaction systems. Precoordination of the enyne **28** to form metallacycle **29** can explain the excellent regioselectivities in this process (Scheme 8.35). 1,6-Enynes have also proven to be interesting both synthetically and mechanistically.

Scheme 8.35

Three reaction pathways were proposed in which regioselectivities are rationalized (Scheme 8.36). In type 1 conditions (no phosphine, **L** = weakly bound ligand), regioselective metallacycle formation is dictated by olefin coordination, leading to regioisomer **30**. In type 2 conditions with a bulky tricyclopentylphosphine (PCyp$_3$) ligand, substitution of the alkyne by the aldehyde occurs, leading to the selective formation of the opposite regioisomer, **31**. However, when tributylphosphine is used in type 3 conditions, the reaction is made unselective, leading to formation of both possible regioisomers since either phosphine in the bis-phosphine adduct can be substituted by the aldehyde.

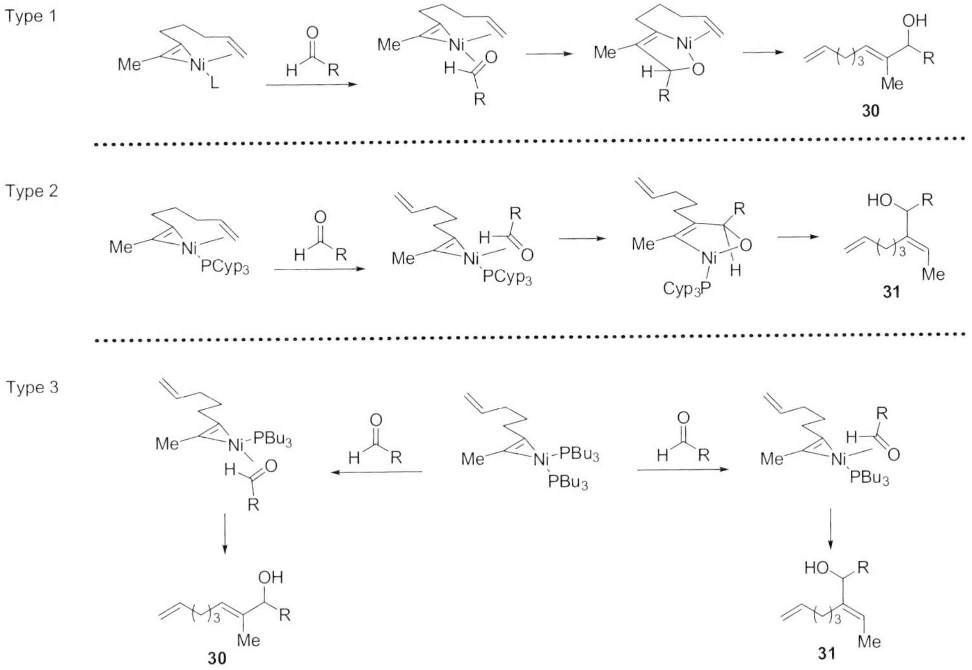

Scheme 8.36

This mechanistic rationale is further supported by transfer of chirality from the 1,6-enyne to product **20**, which was produced in >95:5 regioselectivity and 95:5 diastereoselectivity (discussed previously in Section 8.3.2.3) (Scheme 8.37). When coordination of the olefin was disrupted by addition of PCyp$_3$, regioselectivity was reversed to <5:95 and diastereoselectivity was reduced to 45:55 of **20**. Therefore, it is implicit that olefin coordination leads to metallacycle formation in a regio- and diastereo-selective manner.

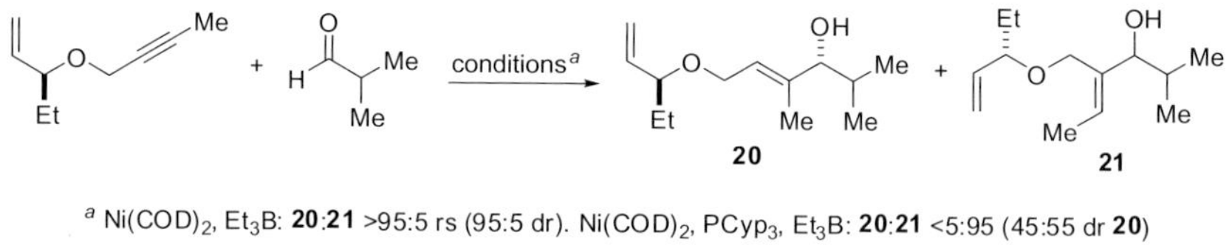

[Scheme 8.37 with reaction scheme showing products 20 and 21]

^{a} Ni(COD)$_2$, Et$_3$B: **20**:**21** >95:5 rs (95:5 dr). Ni(COD)$_2$, PCyp$_3$, Et$_3$B: **20**:**21** <5:95 (45:55 dr **20**)

Scheme 8.37

8.3.4
Cyclocondensations via Hydrogen Gas Extrusion

Based on the previously described crossover studies, it was proposed that the identity of the silane reducing agent would have little effect on the heterocoupling of aldehyde and alkyne π-components using the Ni(COD)$_2$/NHC catalyst system. In an effort to expand the scope and utility of nickel-catalyzed aldehyde–alkyne couplings, dialkylsilane reducing agents were explored. It was shown that, in some instances, the expected silylated allylic alcohol **32** was accompanied by the concurrent formation of silacycle **33** (Scheme 8.38) [54].

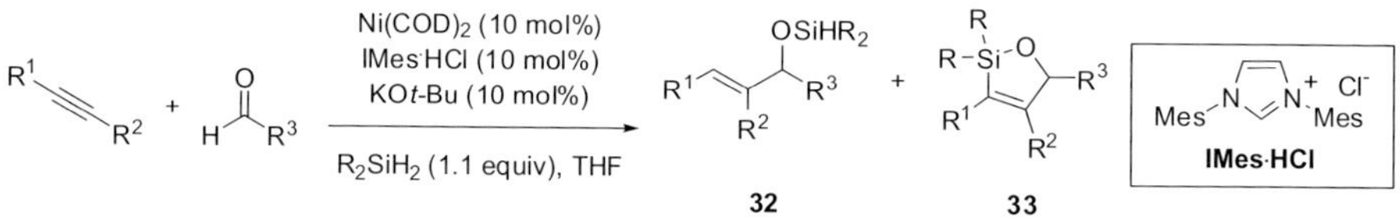

[Scheme 8.38 showing reaction of alkyne and aldehyde with Ni(COD)$_2$ (10 mol%), IMes·HCl (10 mol%), KOt-Bu (10 mol%), R$_2$SiH$_2$ (1.1 equiv), THF to give products **32** and **33**; IMes·HCl structure shown]

Scheme 8.38

The formation of silacycle **33** was further promoted by alkoxy Lewis acids and represents a formal silylene transfer across the two π-components via the extrusion of an equivalent of H$_2$. The scope and generality of the dehydrogenative cyclocondensation of aldehydes, alkynes, and Et$_2$SiH$_2$ is similar to that of the reductive coupling of aldehydes, alkynes, and Et$_3$SiH (Scheme 8.39)[2].

Such similarity can be explained by the presence of vinyl nickel species **34** as a common intermediate in both processes (Scheme 8.40). The direct role of the Lewis acid is unclear, although disruption of a nickel–oxygen interaction in intermediate **34** via Lewis acid coordination to oxygen, as depicted in **35**, may facilitate H$_2$ extrusion.

2) Regioselectivity of internal alkynes was consistent between Et$_2$SiH$_2$ and Et$_3$SiH across various ligand classes.

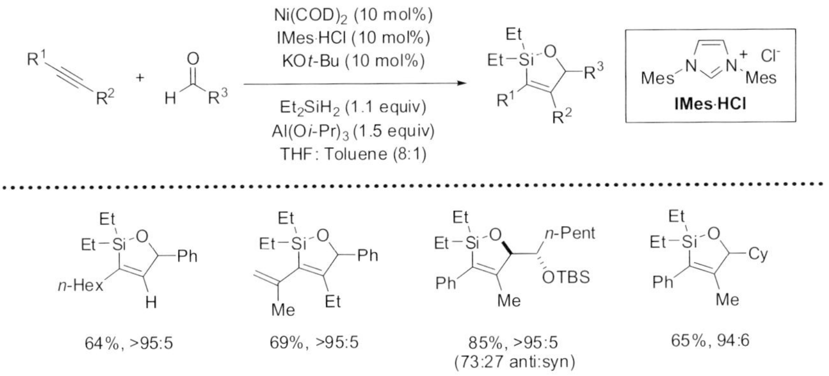

Scheme 8.39

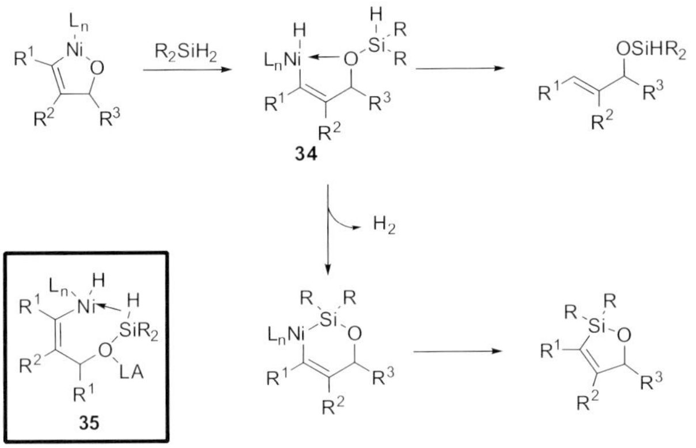

Scheme 8.40

8.3.5
Use in Natural Product Synthesis

A defining feature of any synthetic methodology is its applicability in a complex environment. In this regard, nickel-catalyzed alkylative and reductive couplings of alkynes and aldehydes are especially robust as this methodology has been applied in the total synthesis of several natural products. For each of the following examples the general synthetic plan is briefly described, followed by the key nickel-catalyzed coupling step that is critical for the construction of the overall structural motif of the natural product in question.

The allopumiliotoxin alkaloids are some of most complex members of the pumiliotoxin class of indolizidine alkaloids. Their scarcity, difficulty of isolation, potent cardiotonic and myotonic activity have made them targets of several synthetic efforts, including total syntheses involving nickel-catalyzed reductive coupling as the key step. The first use of silane reducing agents in nickel-catalyzed reductive couplings of aldehydes and alkynes was highlighted in the total synthesis of (+)-allopumiliotoxin 267A, (+)-allopumiliotoxin 339A, and (+)-allopumiliotoxin 339B (Scheme 8.41) [37, 38]. Late-stage nickel-catalyzed cyclization using a Ni(COD)$_2$/PBu$_3$ catalyst system with Et$_3$SiH as a reducing agent proved to be instrumental in arriving at the bicyclic core of the pumiliotoxin class of indolizidine alkaloids in a highly diastereoselective fashion. Ynal **36** was cyclized to **37** in 95% yield as a single observable diastereomer, and subsequent deprotection steps led to the synthesis of (+)-allopumiliotoxin 267A. (+)-Allopumiliotoxin 339A and 339B were synthesized from a common intermediate **38**, which was produced as a single observable diastereomer in 93% yield.

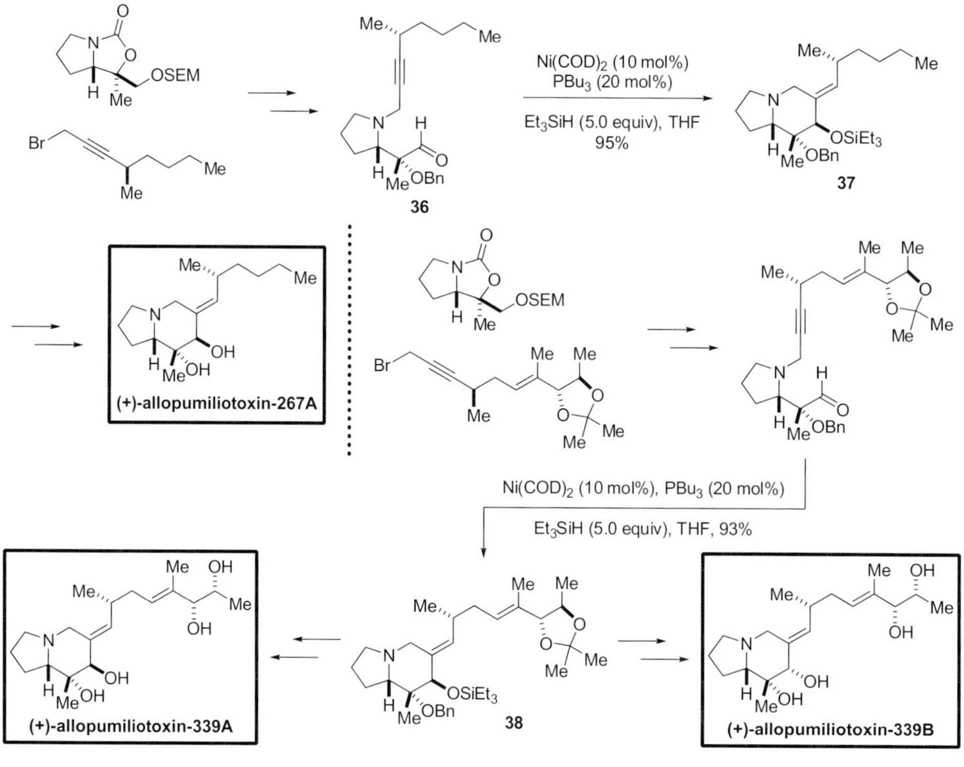

Scheme 8.41

Testudinariol A and B are epimeric triterpene marine natural products possessing a highly functionalized cyclopentanol framework with four contiguous stereocenters appended to a central 3-alkylidene tetrahydrofuran ring. A number of total syntheses of these natural products have been reported, including one that used the C_2 symmetric nature of (+)-testudinariol A to construct the framework in a two-directional synthesis (Scheme 8.42) [55]. Nickel-catalyzed allene/aldehyde alkylative coupling is a closely related reaction to alkyne/aldehyde alkylative coupling and an application of this methodology is illustrated in the total synthesis of (+)-testudinariol A. This strategy proved successful, as **39** underwent alkylative

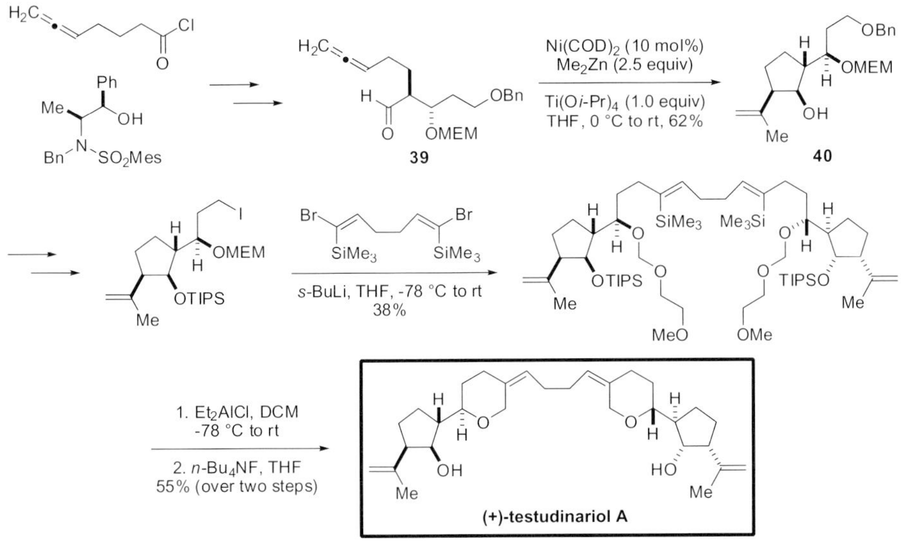

Scheme 8.42

coupling in the presence of a Lewis acid additive to afford **40** in 62% yield as a single observable diastereomer. This was followed by dianion alkylation, a two-directional oxocarbenium ion/vinyl silane cyclization, and silyl deprotection to afford (+)-testudinariol A.

The amphidinolide family of marine natural products has attracted a great deal of interest from the synthetic community due to their structural diversity as well as their potent biological activity. In the synthesis of amphidinolide T1 (Scheme 8.43) and amphidinolide T4 (Scheme 8.44), late-stage nickel-catalyzed

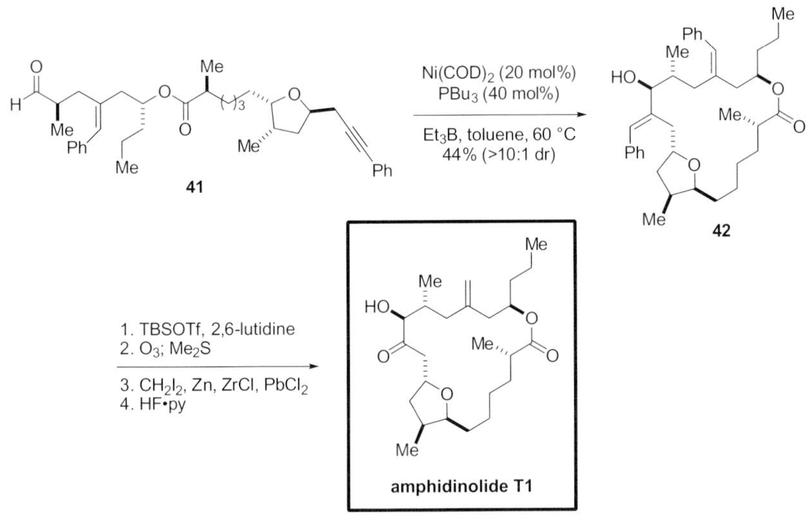

Scheme 8.43

macrocyclization was employed to afford the core natural product structures [41]. Nickel-catalyzed macrocyclization offers a powerful and complementary method to macrolactonization to afford large macrocyclic rings of varying sizes. Several variants exist as the exact reaction conditions are predicated by the molecular scaffold in question (regioselectivity can be attenuated as discussed in Section 8.3.2.1) [44]. As such, this strategy has been applied in the total synthesis of several complex natural products and this section will briefly highlight the synthesis of both amphidinolide T1 and amphidinolide T4. Macrocycle precursor **41** was cyclized to **42** in 44% yield with >10:1 diastereoselectivity. This was followed by protection, ozonolysis, Tebbe methylenation, and deprotection to afford amphidinolide T1. Cyclization product **44** was produced in 31% yield and >10:1 diastereoselectivity over three steps from **43**, followed by a similar endgame sequence to afford amphidinolide T4.

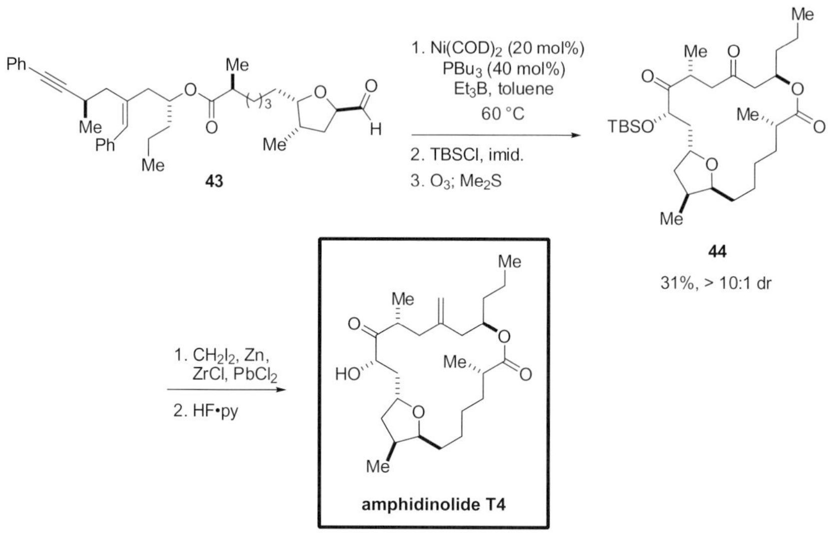

Scheme 8.44

Resorcinylic macrolides are a family of natural products that possess a 14-membered macrolide core with a fused benzenoid subunit. A member of this family that has drawn significant scientific attention due to its antimalarial and cytotoxic activity is aigialomycin D. In one total synthesis of aigialomycin D, late-stage nickel-catalyzed macrocyclization was employed to afford the macrocyclic core [43]. Macrolide precursor **45** was cyclized in 61% yield in a 1:1 diastereoselectivity. Subsequent global deprotection and HPLC purification afforded aigialomycin D and the allylic alcohol epimer in 90% overall yield (Scheme 8.45).

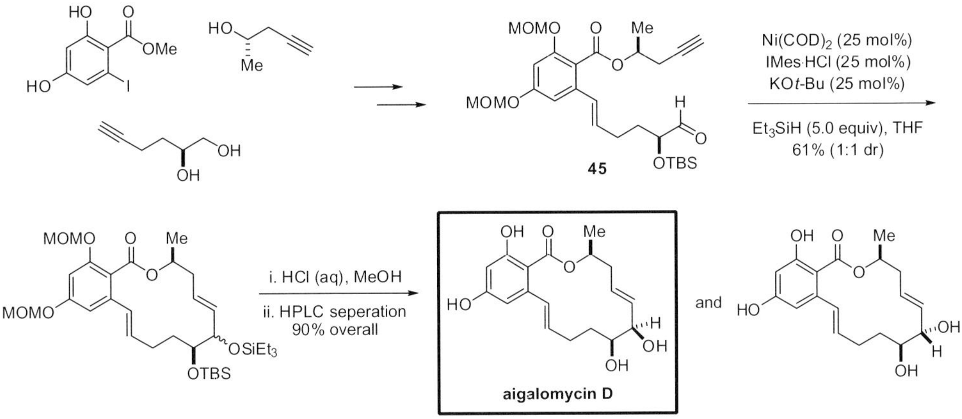

Scheme 8.45

The scientific community has given considerable attention towards the sesterterpene natural product (−)-terpestacin and several related scaffolds. (−)-Terpestacin has attracted synthetic interest due to its unique and challenging structural features along with its ability to inhibit angiogenesis. An enantiospecific total synthesis of both (−)-terpestacin and (+)-11-*epi*-terpestacin was conducted in the context of catalyst-controlled nickel-catalyzed reductive coupling of alkyne **46** and aldehyde **47** (Scheme 8.46) [42]. The product **48** was formed in a 2.6 : 1 mixture of regioisomers, and a 2 : 1 diastereoselectivity in a combined 85% yield. A series of synthetic steps followed to afford both (−)-terpestacin and (+)-11-*epi*-terpestacin, which were differentiated only by their epimeric stereochemistry of the allylic alcohol subunit formed via nickel-catalyzed reductive coupling.

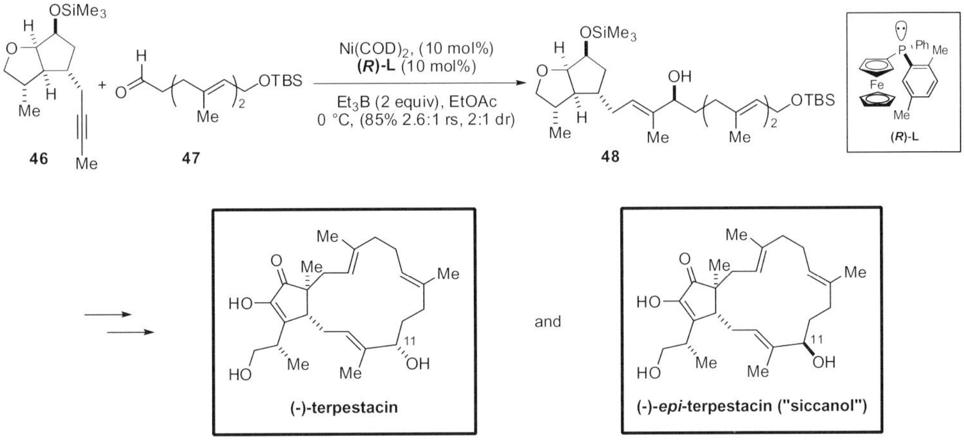

Scheme 8.46

8.4
Conclusions and Outlook

Multi-component coupling of differentiated π-systems in the presence of a zero-valent nickel catalyst can lead to many varied and unique structural motifs. The coupling of alkynes and α,β-unsaturated carbonyls results in the formation of functionalized γ,δ-unsaturated carbonyls containing stereodefined tri- or tetra-substituted alkenes in a highly stereoselective manner (Section 8.2). Allylic alcohols can be accessed by the coupling of alkynes and aldehydes, often with excellent levels of regio-, diastereo-, and enantio-selectivity (Section 8.3). Mechanistic investigation supports the presence of a metallacyclic intermediate being common to many of the nickel-promoted processes (Sections 8.2.3 and 8.3.3) and mechanistic insight has led to the discovery and development of novel nickel-catalyzed processes.

Several issues regarding current methodology still exist, such as control of regioselectivity of internal alkynes, asymmetric induction across a broad range of substrates, and complex mechanistic questions. These concerns are currently being addressed by many research groups and new modifications of these processes and the development of new methodology will continue for some time. There should be a continued drive towards further increasing the simplicity and cost-effectiveness of the nickel pre-catalyst and reducing agents that participate in the types of coupling processes described in this chapter. This trend is evident in the shift from synthetically unfriendly pyrophoric reagents towards bench- and air-stable reagents that are widely commercially available. As these and other issues are addressed, these processes will be utilized with increasing frequency to afford solutions to intricate and interesting chemical challenges.

Acknowledgements

The authors thank the National Institutes of Health (GM-57014) and the National Science Foundation (CHE-0718250) for support of the research described in this review.

References

1 Krische, M.J. (ed.) (2007) *Metal Catalyzed Reductive C-C Bond Formation – A Departure from Preformed Organometallic Reagents*, Springer, Heidelberg.
2 Montgomery, J. (2000) *Acc. Chem. Res.*, **33**, 467–473.
3 Montgomery, J. (2004) *Angew. Chem. Int. Ed.*, **43**, 3890–3908.
4 Montgomery, J., and Sormunen, G.J. (2007) in *Metal Catalyzed Reductive C-C Bond Formation – A Departure from Preformed Organometallic Reagents*, (ed. M.J. Krische), Springer, Heidelberg, pp. 1–23.
5 Moslin, R.M., Miller-Moslin, K., and Jamison, T.F. (2007) *Chem. Commun.*, 4441–4449.

6 Ikeda, S. (2003) *Angew. Chem. Int. Ed.*, **42**, 5120–5122.
7 Jang, H.Y., and Krische, M.J. (2004) *Acc. Chem. Res.*, **37**, 653–661.
8 Ngai, M.Y., Kong, J.R., and Krische, M.J. (2007) *J. Org. Chem.*, **72**, 1063–1072.
9 Sato, F., Urabe, H., and Okamoto, S. (2000) *Chem. Rev.*, **100**, 2835–2886.
10 Lysenko, I.L., Kim, K., Lee, H.G., and Cha, J.K. (2008) *J. Am. Chem. Soc.*, **130**, 15997–16002.
11 Ryan, J., and Micalizio, G.C. (2006) *J. Am. Chem. Soc.*, **128**, 2764–2765.
12 Crowe, W.E., and Rachita, M.J. (1995) *J. Am. Chem. Soc.*, **117**, 6787–6788.
13 Kablaoui, N.M., and Buchwald, S.L. (1996) *J. Am. Chem. Soc.*, **118**, 3182–3191.
14 Kim, I.S., Ngai, M.-Y., and Krische, M.J. (2008) *J. Am. Chem. Soc.*, **130**, 14891–14899.
15 Skucas, E., Ngai, M.-Y., Komanduri, V., and Krische, M.J. (2007) *Acc. Chem. Res.*, **40**, 1394–1401.
16 Sato, Y., Takimoto, M., Hayashi, K., Katsuhara, T., Takagi, K., and Mori, M. (1994) *J. Am. Chem. Soc.*, **116**, 9771–9772.
17 Sawaki, R., Sato, Y., and Mori, M. (2004) *Org. Lett.*, **6**, 1131–1133.
18 Kimura, M., Ezoe, A., Mori, M., Iwata, K., and Tamaru, Y. (2006) *J. Am. Chem. Soc.*, **128**, 8559–8568.
19 Ikeda, S., and Sato, Y. (1994) *J. Am. Chem. Soc.*, **116**, 5975–5976.
20 Ikeda, S., Yamamoto, H., Kondo, K., and Sato, Y. (1995) *Organometallics*, **14**, 5015–5016.
21 Ni, Y., Amarasinghe, K.K.D., and Montgomery, J. (2002) *Org. Lett.*, **4**, 1743–1745.
22 Montgomery, J., and Savchenko, A.V. (1996) *J. Am. Chem. Soc.*, **118**, 2099–2100.
23 Montgomery, J., Oblinger, E., and Savchenko, A.V. (1997) *J. Am. Chem. Soc.*, **119**, 4911–4920.
24 Herath, A., and Montgomery, J. (2006) *J. Am. Chem. Soc.*, **128**, 14030–14031.
25 Herath, A., Thompson, B.B., and Montgomery, J. (2007) *J. Am. Chem. Soc.*, **129**, 8712–8713.
26 Chowdhury, S.K., Amarasinghe, K.K.D., Heeg, M.J., and Montgomery, J. (2000) *J. Am. Chem. Soc.*, **122**, 6775–6776.
27 Herath, A., and Montgomery, J. (2008) *J. Am. Chem. Soc.*, **130**, 8132–8133.
28 Ho, C.-Y., Ohmiya, H., and Jamison, T.F. (2008) *Angew. Chem. Int. Ed.*, **47**, 1893–1895.
29 Trost, B.M., Surivet, J.-P., and Toste, F.D. (2001) *J. Am. Chem. Soc.*, **123**, 2897–2898.
30 Amarasinghe, K.K.D., Chowdhury, S.K., Heeg, M.J., and Montgomery, J. (2001) *Organometallics*, **20**, 370–372.
31 Hratchian, H.P., Chowdhury, S.K., Gutierrez-Garcia, V.M., Amarasinghe, K.K.D., Heeg, M.J., Schlegel, H.B., and Montgomery, J. (2004) *Organometallics*, **23**, 4636–4646.
32 Chevliakov, M.V., and Montgomery, J. (1998) *Angew. Chem. Int. Ed.*, **37**, 3144–3146.
33 Chevliakov, M.V., and Montgomery, J. (1999) *J. Am. Chem. Soc.*, **121**, 11139–11143.
34 Ni, Y., Amarasinghe, K.K.D., Ksebati, B., and Montgomery, J. (2003) *Org. Lett.*, **5**, 3771–3773.
35 Fornicola, R.S., Subburaj, K., and Montgomery, J. (2002) *Org. Lett.*, **4**, 615–617.
36 Oblinger, E., and Montgomery, J. (1997) *J. Am. Chem. Soc.*, **119**, 9065–9066.
37 Tang, X.Q., and Montgomery, J. (2000) *J. Am. Chem. Soc.*, **122**, 6950–6954.
38 Tang, X.-Q., and Montgomery, J. (1999) *J. Am. Chem. Soc.*, **121**, 6098–6099.
39 Huang, W.S., Chan, J., and Jamison, T.F. (2000) *Org. Lett.*, **2**, 4221–4223.
40 Mahandru, G.M., Liu, G., and Montgomery, J. (2004) *J. Am. Chem. Soc.*, **126**, 3698–3699.
41 Colby, E.A., O'Brien, K.C., and Jamison, T.F. (2005) *J. Am. Chem. Soc.*, **127**, 4297–4307.
42 Chan, J., and Jamison, T.F. (2004) *J. Am. Chem. Soc.*, **126**, 10682–10691.
43 Chrovian, C.C., Knapp-Reed, B., and Montgomery, J. (2008) *Org. Lett.*, **10**, 811–814.
44 Knapp-Reed, B., Mahandru, G.M., and Montgomery, J. (2005) *J. Am. Chem. Soc.*, **127**, 13156–13157.
45 Miller, K.M., Luanphaisarnnont, T., Molinaro, C., and Jamison, T.F. (2004) *J. Am. Chem. Soc.*, **126**, 4130–4131.

46 Miller, K.M., and Jamison, T.F. (2004) *J. Am. Chem. Soc.*, **126**, 15342–15343.
47 Moslin, R.M., and Jamison, T.F. (2006) *Org. Lett.*, **8**, 455–458.
48 Moslin, R.M., Miller, K.M., and Jamison, T.F. (2006) *Tetrahedron*, **62**, 7598–7610.
49 Luanphaisarnnont, T., Ndubaku, C.O., and Jamison, T.F. (2005) *Org. Lett.*, **7**, 2937–2940.
50 Sa-ei, K., and Montgomery, J. (2006) *Org. Lett.*, **8**, 4441–4443.
51 Miller, K.M., Huang, W.-S., and Jamison, T.F. (2003) *J. Am. Chem. Soc.*, **125**, 3442–3443.
52 Chaulagain, M.R., Sormunen, G.J., and Montgomery, J. (2007) *J. Am. Chem. Soc.*, **129**, 9568–9569.
53 Ogoshi, S., Arai, T., Ohashi, M., and Kurosawa, H. (2008) *Chem. Commun.*, 1347–1349.
54 Baxter, R.D., and Montgomery, J. (2008) *J. Am. Chem. Soc.*, **130**, 9662–9663.
55 Amarasinghe, K.K.D., and Montgomery, J. (2002) *J. Am. Chem. Soc.*, **124**, 9366–9367.

9
Copper-Catalyzed Ligand Promoted Ullmann-type Coupling Reactions

Yongwen Jiang and Dawei Ma

9.1
Introduction

Ullmann-type coupling reactions between aryl halides and N-containing reactants, phenols and other related nucleophilic agents are useful methods for the preparation of aromatic amines, biaryl ethers and N-aryl heterocycles that are important for the pharmaceutical and material sciences. For a long time, these coupling reactions required high reaction temperatures and stoichiometric amounts of copper. These drawbacks have greatly limited their application in organic synthesis, and have stimulated considerable efforts to develop relatively mild coupling reaction conditions. This campaign has made great progress during the past decade with the utilization of special ligands [1]. Some typical ligands are shown in Figure 9.1, including N,O-bidentate compounds **L1–L7**, S,O-bidentate compounds **L8**, N,N-bidentate compounds **L9–L16** and O,O-bidentate compounds **L17–L22**. In this chapter we detail their application in promoting Ullmann-type coupling reactions, which were classified into C–N, C–O, C–C, C–S and C–P bond formation reactions. It is noteworthy that many reactions discussed here could also be catalyzed with palladium complexes. However, industrial use of Pd-catalyzed reactions is limited in many cases due to their air and moisture sensitivity, as well as the higher costs of Pd and the related ligands.

9.2
C–N Bond Formation

9.2.1
Arylation of Amines

9.2.1.1 Arylation of Aliphatic Primary and Secondary Amines

Copper catalyzed arylation of aliphatic primary amines now has become a test reaction for newly developed ligands. Many ligands were found to be effective for this transformation and some results are summarized in Table 9.1. As a cheap

Catalysis Without Precious Metals. Edited by R. Morris Bullock
© 2010 WILEY-VCH Verlag GmbH & Co. KGaA, Weinheim
ISBN: 978-3-527-32354-8

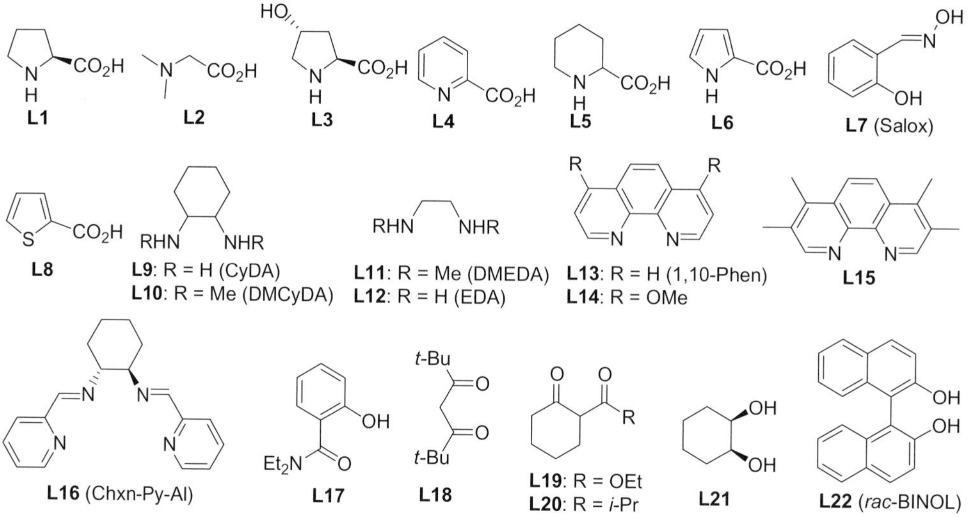

Figure 9.1 Some typical ligands used for Ullmann-type coupling reactions.

Table 9.1 Copper/ligand catalyzed coupling between aryl halides and aliphatic primary amines.

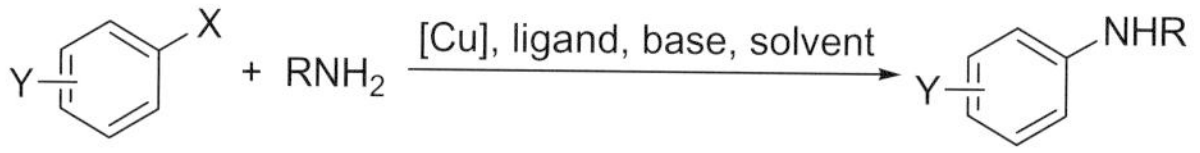

Entry	X	Copper source	Ligand	Base	Reaction conditions	Yield (%)	Ref.
1	I	CuI	L1	K_2CO_3	DMSO, 60 °C	70–89	[2, 3]
2	Br	CuI	L1	K_3PO_4	DMSO, 80–90 °C	81–98	[2, 3]
3	Br	CuI	L17	K_3PO_4	DMF, 90 °C	71–95	[4]
4	I	CuI	L20	Cs_2CO_3	DMF, RT	79–98	[6, 7]
5	Br	CuI	L20	Cs_2CO_3	DMF, 90 °C	87–94	[6]
6	I	CuBr	L22	K_3PO_4	DMF, RT-50 °C	38–94	[8]
7	I	Cu	L22	Cs_2CO_3	DMSO, 90 °C	77–99	[9]

and convenient available ligand, L-proline (**L1**) was shown to be applicable for both aryl iodides and bromides with a working temperature range from 60 to 90 °C (entries 1 and 2) [2, 3]. Buchwald and coworkers discovered that diethylsalicylamide (**L17**) was a suitable ligand for accelerating the reaction of aryl bromides (entry 3) [4], which could overcome the drawback of their ethylene glycol ligand that is only suitable for aryl iodides [5]. Two powerful catalytic systems, CuI/**L20** [6, 7] (entry 4) and CuBr/**L22** [8] (entry 6), which enable the coupling to proceed at room temperature, were disclosed recently. However, only some aryl iodides

Table 9.2 Copper/ligand catalyzed coupling between aryl halides and aliphatic secondary amines.

Entry	X	Copper source	Ligand	Base	Reaction conditions	Yield (%)	Ref.
1	I	CuI	L1	K_2CO_3	DMSO, 65–90 °C	21–100	[3]
2	Br	CuI	L1	K_2CO_3	DMSO, 90 °C	55–99	[3]
3	I	CuBr	L22	K_3PO_4	DMF, RT-40 °C	54–91	[8]
4	I	Cu	L22	Cs_2CO_3	DMSO, 90 °C	75–96	[9]

gave good conversion while most of the aryl halides still required higher reaction temperatures for complete reaction (entry 5). Interestingly, the copper source was found to be important for BINOL ligand **L22**, and heating was required when copper powder was used (entry 7) [9].

Ligand-promoted Ullmann amination was revealed to be very sensitive to steric hindrance of the amines, which contrasts with Pd-catalyzed aryl aminations. For example, when secondary amines were employed, the reaction became sluggish and only less hindered cyclic amines provided satisfactory conversions. Acyclic secondary amines were observed to afford low yields of the coupling products. Some suitable ligands for these couplings are indicated in Table 9.2.

9.2.1.2 Arylation of Aryl Amines

Due to poor reactivity, aryl amines normally require higher reaction temperatures than aliphatic amines to ensure good conversion. In early studies, phenathroline and its Cu(I)-complex were used in the arylation of aryl amine [10, 11], but they were only applicable to the synthesis of triarylamine from secondary aryl amines. L-Proline (**L1**) promoted CuI-catalyzed arylation of primary aryl amines took place at 90 °C (Table 9.1, entry 1) [3]. However, only electron-rich anilines gave complete conversion, while electron-deficient anilines provided low yields. Fu found that this drawback could be overcome by heating at 110 °C and using pipecolinic acid (**L5**) as a ligand (entry 2) [12]. Similar studies were reported by Liu and coworkers in which DMEDA (**L11**) was found to be a better ligand (entry 3) [13]. Recently, Buchwald reported that pyrrole-2-carboxylic acid (**L6**) [14] is an efficient promoter for the synthesis of diarylamines (entry 4) (Table 9.3).

9.2.1.3 Arylation of Ammonia

The first attempt to couple aryl halides with ammonia catalyzed by copper/ligand was reported by a Merck group (Scheme 9.1) [15]. In this case ethylene glycol may serve as both ligand and solvent, and the reaction proceded at 80 °C under elevated pressure. Mild conditions were discovered by Kim and Chang, in which L-proline (**L1**) was employed as the ligand (Scheme 9.2) [16]. Both NH_4Cl and $NH_3 \cdot H_2O$

Table 9.3 Copper/ligand catalyzed coupling between aryl halides and aryl amines.

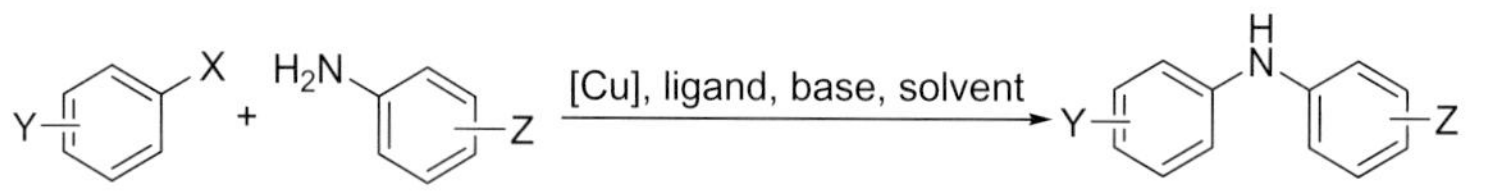

Entry	X	Copper source	Ligand	Base	Reaction conditions	Yield (%)	Ref.
1	I, Br	CuI	L1	K_2CO_3	DMSO, 90 °C	51–97	[3]
2	I, Br	CuI	L5	K_2CO_3	DMF, 110 °C	27–93	[12]
3	I, Br	CuI	L11	K_2CO_3	dioxane, 100 °C	62–84	[13]
4	I, Br	CuI	L6	K_3PO_4	DMSO, 70–100 °C	50–82	[14]

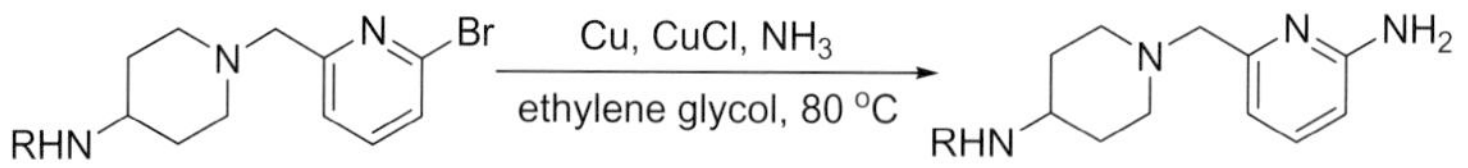

Scheme 9.1

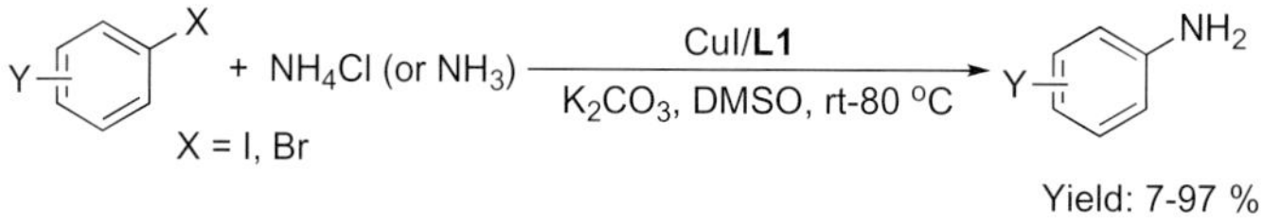

Scheme 9.2

Yield: 7–97 %

could be used as the nitrogen source, and the reaction proceeded smoothly at room temperature for aryl iodides. However, for aryl bromides, only substrates with an electron-withdrawing group gave good yields.

Recently, Fu and coworkers reported an alternative method for assembly of primary aryl amines [17]. In their reaction amidine hydrochlorides were used as the ammonia surrogates (Scheme 9.3). The ligand was also L-proline (**L1**) and heating (110 °C) was required for good yields.

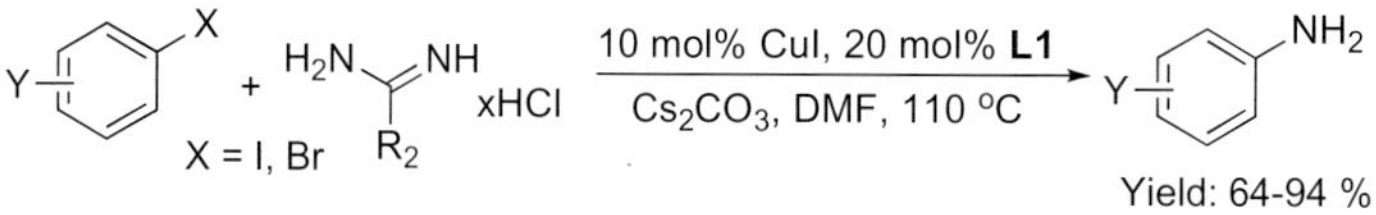

Scheme 9.3

Yield: 64–94 %

9.2.2
Arylation and Vinylation of N-Heterocycles

9.2.2.1 Coupling of Aryl Halides and N-Heterocycles

A stoichiometric quantity of 1,10-phenanthroline (**L13**) was the first ligand used for copper-catalyzed arylation of imidazole with aryl iodides (Table 9.4, entry 1) [18]. As an extension of this methodology, Cu_2O/4,7-dimethoxy-1,10-phenanthroline (**L14**) was found to be more effective, the coupling reaction occurring at 80–110 °C in the presence of Cs_2CO_3 and using n-PrCN as a solvent (entry 2) [19, 20].

Diamine ligands (**L9** and **L10**) were demonstrated to be useful promoters for copper catalyzed coupling of aryl halides with N-heterocycles (entry 3) [21–23]. A wide range of heterocycles, including indoles, pyrroles, pyrazoles, imidazoles and triazoles, were successfully employed in this reaction. Aryl iodides and bromides were also available substrates when the temperature was 110 °C. The Ma group revealed that amino acids were another class of powerful ligands for this transformation. The reaction conditions were slightly different for different substrates (entry 4) [3, 24]. For example, using L-proline (**L1**) as a promoter, the coupling reaction of aryl iodides with indole, pyrrole, carbazole, imidazole or pyrazole could be carried out at 75–90 °C; coupling reactions of electron-deficient aryl bromides with imidazole or pyrazole occur at 60–90 °C. However, for the coupling reaction

Table 9.4 Copper/ligand catalyzed coupling between aryl halides and N-heterocycles amines.

Entry	X	Ligand	N-heterocycle	Reaction conditions	Ref.
1	I	L13	imidazoles	$(CuOTf)_2$, Cs_2CO_3	[18]
2	I Br	L14	imidazoles, benzimidazoles	Cu_2O, Cs_2CO_3, PEG n-PrCN, 80–110 °C	[19] [20]
3	I Br	L9 L10	indoles, pyrroles, pyrazoles, imidazoles, triazoles, indazoles carbazoles	CuI, K_3PO_4 Toluene or dioxane, 110 °C	[21] [22] [23]
4	I Br	L1 L2	indoles, pyrroles, carbazoles, imidazoles, pyrazoles	CuI, K_2CO_3 DMSO, 60–110 °C	[3] [24]
5	I Br	L7 L16	pyrazoles, imidazoles, indoles, pyrroles, triazoles	Cu_2O, CH_3CN, 25–82 °C	[25] [26]
6	I Br	L19	imidazoles, benzimidazoles	CuBr, Cs_2CO_3 DMSO, 45–75 °C	[27]

Table 9.5 Copper/ligand catalyzed coupling between vinyl bromides and N-heterocycles.

$$R\text{-CH=CH-Br} + \text{Het-NH} \xrightarrow{\text{conditions}} R\text{-CH=CH-NHet}$$

Entry	Copper source	Ligand	N-heterocycle	Reaction conditions	Ref.
1	CuI	L1	imidazoles, benzimidazoles	[BMIM]BF$_4$, K$_2$CO$_3$ 110°C	[28]
2	Cu$_2$O	L19	imidazoles, benzimidazoles	Cs$_2$CO$_3$ CH$_3$CN, 80–90°C	[29]
3	CuI	L16	azoles	Cs$_2$CO$_3$, CH$_3$CN, 35–80°C	[30]

of electron-rich aryl bromides with imidazole or pyrazole, N,N-dimethylglycine (**L2**) had to be used and the reaction temperature was raised to 110°C. In the latter case incomplete conversion was observed if **L1** was utilized, which might result from the quick coupling between ligand and aryl bromides at 110°C.

Taillefer and coworkers found that Salox (**L7**) and Chxn-Py-Al (**L16**) were efficient ligands for the Cu$_2$O-catalyzed N-arylation of pyrazoles and imidazoles as well as pyrroles, indoles and triazoles (entry 5) [25, 26]. Both aryl iodides and bromides as substrates gave good results.

In addition, ethyl 2-oxocyclohexanecarboxylate (**L19**) [27] could promote the copper-catalyzed N-arylation of imidazoles and benzimidazoles under relatively mild reaction conditions (45–75°C) (entry 6).

9.2.2.2 Coupling of Vinyl Bromides and N-Heterocycles

Several reports regarding coupling of vinyl bromides with N-heterocycles have appeared. For example, using [BMIM]BF$_4$ as solvent, the recycle of the CuI/L-proline (**L1**) catalytic system became possible for the cross-coupling reaction of vinyl bromides with imidazoles and benzoimidazoles (Table 9.5, entry 1) [28]. Ethyl 2-oxocyclohexanecarboxylate (**L16**) also showed a good accelerating effect for the vinylation of imidazoles and benzimidazoles under the catalysis of Cu$_2$O (entry 2) [29]. Chxn-Py-Al (**L14**) developed by Taillefer promoted the vinylation of azoles at lower reaction temperatures (entry 3) [30].

9.2.3
Aromatic Amidation

The copper-promoted arylation of amides is known as the Goldberg reaction. Similar to the traditional Ullmann reaction, this transformation suffered from high reaction temperatures and strictly limited substrates. The introduction of suitable ligands has changed this situation.

9.2.3.1 Cross-Coupling of aryl Halides with Amides and Carbamates

Buchwald's research indicated that vicinal diamines, such as ethylenediamine (EDA, **L12**), *rac-trans*-cyclohexanediamine (CyDA, **L9**), as well as their N, N'-dimethyl derivatives (DMEDA, **L11** and DMCyDA, **L10**), were excellent ligands for the amidation of aryl halides (Table 9.6, entry 1) [21, 31]. This catalytic system allowed a wide range of amides (including lactams, open chain alkylamides and arylamides) and carbamates to be coupled with aryl halides. The hindered amides also gave good results. The reaction occurred at about 80–110 °C. Remarkably, DMCyDA (**L10**) could even promote the amidation of deactivated aryl chlorides at a little higher reaction temperature (110–130 °C).

Coupling of aryl halides and trifluoroacetamide occurred at 45–75 °C catalyzed by CuI/DMEDA (**L11**), followed by *in situ* hydrolysis to provide the corresponding primary arylamines (entry 2) [32]. The CuI/1,10-phen (**L13**) catalytic system was found to be able to promote the selective coupling of N-Boc hydrazine [33] (entry 3) and the coupling of N- and O- functionalized hydroxylamines with aryl iodides (entry 4) [34]. Hosseinzadeh and coworkers reported the coupling reaction of aryl iodides with amides catalyzed by CuI/1,10-Phen (**L13**) by using KF/Al$_2$O$_3$ as the

Table 9.6 Copper/ligand catalyzed coupling between aryl halides and amides and carbamates.

Entry	X	Ligand	Amides	Reaction conditions	Ref.
1	I, Br, Cl	L9-L12	arylamides, alkylamides, lactams, oxazolidinones carbamates	CuI, K$_3$PO$_4$ (or Cs$_2$CO$_3$) dioxane (or toluene), 80–130 °C	[21] [31]
2	I, Br	L11	trifluoroacetamides	CuI, DMF (or dioxane, K$_3$PO$_4$ (or K$_2$CO$_3$), 45–75 °C	[32]
3	I	L13	N-Boc hydrazine	CuI, Cs$_2$CO$_3$ DMF, 80 °C	[33]
4	I	L13	N- and O- functionalized hydroxylamines	CuI, Cs$_2$CO$_3$ DMF, 80 °C	[34]
5	I	L13	amides	CuI, KF/Al$_2$O$_3$ Toluene, 110 °C	[35]
6	I, Br	L14, L18	2-, or 4-hydroxypyridines	CuI, K$_2$CO$_3$ DMSO, 110–120 °C	[36]
7	I, Br	L19	lactams, amides 2-hydroxypyridine	CuBr, Cs$_2$CO$_3$ DMSO, RT-75 °C	[27]

base (entry 5) [35]. Further investigation by Buchwald revealed 4,7-dimethoxy-1,10-phenanthroline (**L14**) and 2,2,6,6-tetramethylheptane-3,5-dione (TMHD, **L18**) with good acceleration effects for the selective coupling of 2- and 4-hydroxypyridines (entry 6) [36].

Recently, ethyl 2-oxocyclohexanecarboxylate (**L19**) [27] also showed good activity for the arylation of amides (entry 7). Interestingly, the reaction even worked at room temperature for some aryl iodides.

9.2.3.2 Cross-Coupling of Vinyl Halides with Amides or Carbamates

By using the Liebeskind catalyst copper(I) thiophene-2-carboxylate (copper salt of **L8**), the coupling reaction of vinyl iodides and amides proceeded smoothly to give the corresponding enamides (Table 9.7, entry 1) [37]. CuI/DMEDA (**L11**) was another efficient catalytic system for the coupling of vinyl halides with amides (entry 2) [38].

Although the CuI/L-proline (**L1**) catalytic system showed no obvious acceleration effect for the coupling of aryl halide and amides, the Ma group found that N,N-dimethylglycine (**L2**) was a suitable promoter for the coupling reaction of vinyl halides with amides, and, in many cases, the reaction worked at room temperature (entry 3) [39]. Another combination (Cu(CH$_3$CN)$_4$PF$_6$/**L15**/Rb$_2$CO$_3$) was recently found effective for this transformation (entry 4) [40].

9.2.3.3 Cross-Coupling of Alkynl Halides with Amides or Carbamates

Under the catalysis of CuCN/DMEDA (**L11**), ynamides could be conveniently synthesized via N-alkynylation of amides, including oxazolidinones, lactams and

Table 9.7 Copper/ligand catalyzed coupling between vinyl halides and amides and carbamates.

Entry	X	Ligand	N-heterocycle	Reaction conditions	Ref.
1	I	L8	arylamide, alkylamides	CuTC, Cs$_2$CO$_3$ NMP, 90°C	[37]
2	I, Br	L11	arylamide, alkylamides lactam, oxazolidinones	CuI, Cs$_2$CO$_3$ (or K$_2$CO$_3$), THF (or toluene), RT-110°C	[38]
3	I, Br	L2	arylamide, alkylamides lactam, oxazolidinones	CuI, Cs$_2$CO$_3$, dioxane, RT-80°C	[39]
4	I	L15	arylamide, alkylamides lactam, oxazolidinones	Cu(CH$_3$CN)$_4$PF$_6$, Rb$_2$CO$_3$, DMA, 45–60°C	[40]

9.2 C–N Bond Formation

Table 9.8 Copper/ligand catalyzed coupling between alkynl halides and amides and or carbamates.

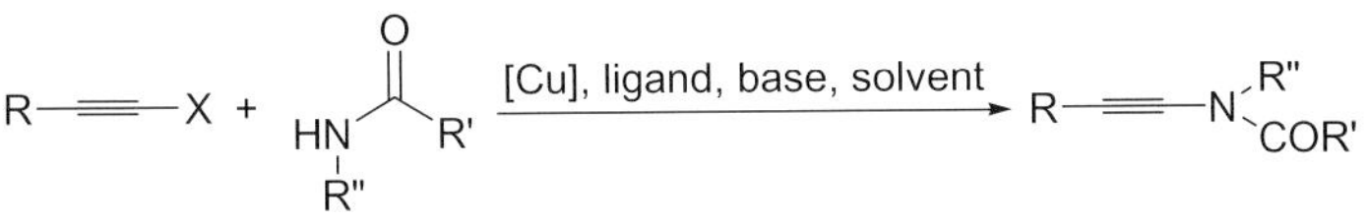

$$R\!\!=\!\!\!=\!\!X + \underset{R''}{\underset{|}{HN}}\!\!-\!\!\overset{O}{\underset{}{C}}\!\!-\!\!R' \xrightarrow{\text{[Cu], ligand, base, solvent}} R\!\!=\!\!\!=\!\!N\!\!\underset{COR'}{\overset{R''}{\diagup}}$$

Entry	X	Ligand	N-heterocycle	Reaction conditions	Ref.
1	I, Br	L11	oxazolidinones, lactams, carbamates	CuCN, K$_3$PO$_4$ toluene, 110–150 °C	[41]
2	Br	L13	oxazolidinones, sulfonamides, carbamates	CuSO$_4$·5H$_2$O, K$_3$PO$_4$ toluene, 60–95 °C	[42]

carbamates (Table 9.8, entry 1) [41]. The reaction worked well at 110 °C, but for hindered amides, higher reaction temperatures were required. 1,10-Phenanthroline (**L13**) was found to be a more efficient ligand for this reaction (entry 2) [42], enabling the reaction to proceed at 50–95 °C. In addition, the use of cheap and stable CuSO$_4$·5H$_2$O makes the catalytic system simpler and more practical. Based on this investigation, functionalized oxazolones were synthesized [43].

9.2.4
Azidation

With the assistance of L-proline (**L1**), the coupling of aryl halides with sodium azide proceeded smoothly to give aryl azides (Scheme 9.4) [44]. Both aryl iodides and bromides were suitable substrates, but they required different reaction conditions. For iodides, using DMSO as solvent gave better results, while EtOH/water was suitable for bromides.

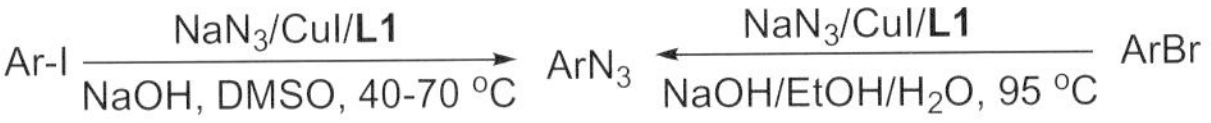

$$\text{Ar-I} \xrightarrow[\text{NaOH, DMSO, 40-70 °C}]{\text{NaN}_3/\text{CuI/L1}} \text{ArN}_3 \xleftarrow[\text{NaOH/EtOH/H}_2\text{O, 95 °C}]{\text{NaN}_3/\text{CuI/L1}} \text{ArBr}$$

Scheme 9.4

DMCyDA (**L10**) could also promote the synthesis of aryl azides [45]. Sodium ascorbate was found to have a positive effect on stabilization of the catalytic system. The aryl azides produced *in situ* could react with terminal alkynes to provide 1,2,3-triazoles [46, 47].

9.3
C–O Bond Formation

9.3.1
Synthesis of Diaryl Ethers

Diaryl ethers are an important class of compounds in pharmaceuticals and agricultural chemicals. The Ullmann ether formation reaction provides direct access to diaryl ethers. However, harsh reaction conditions, such as high temperatures (125–220 °C), stoichiometric quantities of the copper catalyst, and low to moderate yields, have greatly limited the utility of the reaction. The use of soluble copper salts and some effective ligands made it possible to carry out these reactions under mild reaction conditions.

Some ligands like 2,2,6,6-tetramethylheptane-3,5-dione (TMHD, **L18**) (Table 9.9, entry 1) [48], N,N-dimethylglycine hydrochloride (**L2**) (entry 2) [49], Salox (**L7**) (entry 3) and Chxn-Py-Al (**L16**) (entries 4, 5) [50, 51] were found effective for promoting this coupling reaction, leading to complete reaction at relatively low reaction temperatures. Usually, aryl halides with electron-withdawing groups are more active for this reaction. In contrast, phenols with electron-donating groups gave better results. It is noteworthy that the CuI/**L2** catalytic system developed by Ma and coworkers was mild enough to afford non-racemizing diaryl ether bearing amino ester moieties, which has been used successfully for assembling (S,S)-isodityrosine and K-13 [52]. Using Chxn-Py-Al (**L16**) as a leading ligand, Taillefer and coworkers further investigated the structure–activity relationship of a series of ligands for the copper-catalyzed arylation of phenols [53]. This led them to find

Table 9.9 Copper/ligand catalyzed diaryl ether formation from aryl halides and phenols.

Entry	X	[Cu]	Ligand	Base	Reaction conditions	Yield (%)	Ref.
1	I, Br	CuCl	L18	Cs_2CO_3	NMP, 120 °C	51–85	[48]
2	I, Br	CuI	L2	Cs_2CO_3	dioxane, 90 °C	33–97	[49, 52]
3	I	Cu_2O	L7	Cs_2CO_3	CH_3CN, 82 °C	45–100	[50]
4	I Br	Cu_2O	L16	Cs_2CO_3	CH_3CN, 82 °C or DMF, 110 °C	40–100	[50]
5	I, Br	CuI	L16	K_3PO_4	CH_3CN, 60–80 °C	55–97	[51]
6	Cl	CuBr	L18	Cs_2CO_3	DMF, 135 °C	40–99	[54]
7	I, Br	CuBr	L19	Cs_2CO_3	DMSO, 60–80 °C	72–97	[27]

a more powerful ligand that contains an iminopyridine moiety. Recently, **L18** was shown to be able to make inactive aryl chlorides work for the arylation of phenols (entry 6) [54], and β-keto ester (**L19**) [27] was revealed as an excellent promoter for diaryl ether formation (entry 7).

Room temperature diaryl ether formation for some special substrates has been observed by Ma and coworkers (Scheme 9.5). This may result from the combination of an ortho-substituent (directed by an amido group) and ligand effects [55]. These mild reaction conditions showed excellent prospects in the total synthesis of some important natural products.

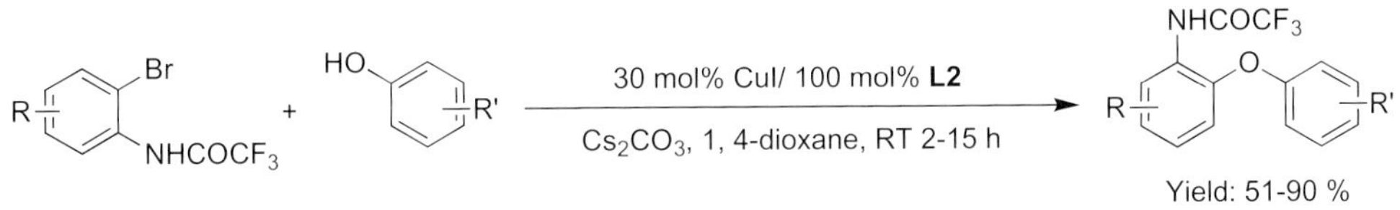

Scheme 9.5

9.3.2
Aryloxylation of Vinyl Halides

Aryloxylation of vinyl halides is a convenient route to vinyl aryl ether. Subsequently, CuI/N,N-dimethylglycine (**L2**) [56] and CuI/Chxn-Py-Al (**L16**) [30] were revealed to have similar abilities (Table 9.10).

9.3.3
Cross-Coupling of Aryl Halides with Aliphatic Alcohols

Aliphatic alcohols were found to be less reactive toward ligand promoted copper-catalyzed arylation. But some successful examples have been reported by using special ligands. Buchwald and coworkers discovered that if 1,10-phenanthroline (**L13**) was used as a ligand, alkoxylation of aryl iodides with a wide range of

Table 9.10 Copper/ligand catalyzed coupling of vinyl halides and phenols.

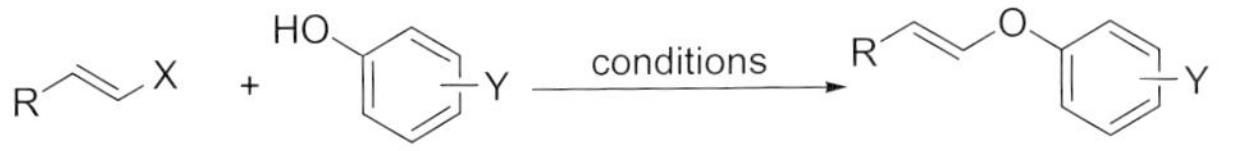

Entry	X	[Cu]	Ligand	Base	Reaction conditions	Yield (%)	Ref.
1	I, Br	CuI	L2	Cs_2CO_3	Dioxane, 60–90 °C	65–90	[56]
2	Br	CuI	L16	Cs_2CO_3	CH_3CN, 50 °C	74–90	[30]

Table 9.11 Copper/ligand catalyzed coupling of aryl halides and aliphatic alcohols.

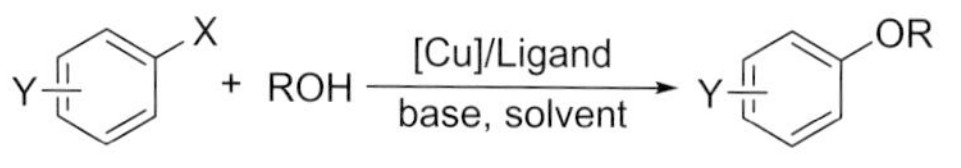

Entry	X	[Cu]	Ligand	Base	Reaction conditions	Yield (%)	Ref.
1	I	CuI	L13	Cs_2CO_3	alcohol, 110 °C	40–97	[57]
2	I	CuI	L13	KF/Al_2O_3	alcohol, 110 °C	25–98	[58]
3	I, Br	CuI	L15	Cs_2CO_3	toluene, 80–130 °C	59–99	[59]
4	I	CuI	L2	Cs_2CO_3	alcohol, 110 °C	25–98	[60]

aliphatic alcohols (normal and branched, saturated and unsaturated) worked well to give the corresponding products in good yields (Table 9.11, entry 1) [57]. Switching the base to KF/Al_2O_3 gave similar results (entry 2) [58].

Use of 3,4,7,8-tetramethyl-1,10-phenanthroline (**L15**) afforded milder reaction conditions for the alkoxylation of aryl halides (entry 3) [59]. Aryl bromides also worked although only a limited number of cases were described. More importantly, the amount of alcohol could be decreased in these cases.

CuI/N,N-dimethylglycine (**L2**) was found effective for the coupling reaction of aryl iodides with aliphatic alcohols (entry 4) [60], although excess alcohol was required to ensure complete consumption of the coupling partners.

Recently, the Buchwald group revealed that selective O/N-arylation of amino alcohols could be realized by modulating different ligand or reaction conditions (Scheme 9.6) [7, 61], which may prompt further ligand design.

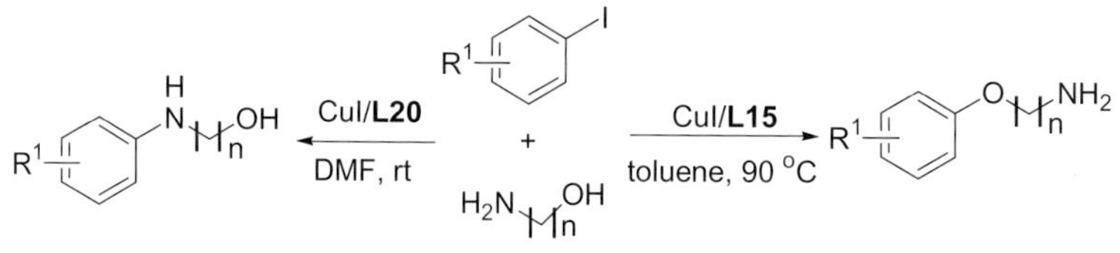

Scheme 9.6

9.4
C–C Bond Formation

9.4.1
Cross-Coupling with Terminal Acetylene

Cross-coupling of aryl and vinyl halides with terminal acetylenes (the Sonogashira reaction) is among the fundamental methods of Pd-catalyzed C–C bond forming reactions. Cu–Pd co-catalysis is considered a standard requirement for this method.

Table 9.12 Copper/ligand catalyzed coupling of aryl halides with terminal acetylenes.

Entry	X	[Cu]	Ligand	base	Reaction conditions	Yield (%)	Ref.
1	I	Cu(phen)(PPh$_3$)Br	L13	K$_2$CO$_3$	toluene, 110 °C	70–97	[11]
2	I, Br	CuI	L2	K$_2$CO$_3$	DMF, 100–120 °C	60–98	[62]
3	I, Br	CuBr	L22	Cs$_2$CO$_3$	DMF, 130 °C	23–87	[63]

Table 9.13 Copper/ligand catalyzed coupling of vinyl iodides with terminal acetylenes.

Entry	X	[Cu]	Ligand	Base	Reaction conditions	Yield (%)	Ref.
1	I	CuI	L2	Cs$_2$CO$_3$	dioxane, 80 °C	60–91	[64]
2	I	Cu(phen)(PPh$_3$)NO$_2$	L13	Cs$_2$CO$_3$	toluene, 110 °C	78–99	[65]

By using suitable ligands, some palladium-free copper-catalyzed procedures for this transformation have been developed, which provide more practical conditions for Sonogashira coupling reactions. As shown in Table 9.12, the soluble complex Cu(phen)(PPh$_3$)Br (**L13**) enabled the coupling reaction of aryl iodides with terminal acetylenes to take place at 110 °C (Table 9.12, entry 1) [11], CuI/N,N-dimethylglycine (**L2**) enabled either aryl iodides or aryl bromides to react at 100 °C (entry 2) [62], while rac-BINOL (**L22**) [63] was found to require a higher reaction temperature to promote this reaction (entry 3).

Vinyl halides were found to be more reactive toward this coupling, as evident from the fact that the CuI/N,N-dimethylglycine (**L2**) catalyzed reaction of vinyl iodides and terminal acetylenes proceeded smoothly at 80 °C (Table 9.13, entry 1) [64]. When Cu(phen)(PPh$_3$)$_2$NO$_3$ was used (entry 2) [65], a higher reaction temperature was still needed.

9.4.2
The Arylation of Activated Methylene Compounds

The arylation of activated methylene compounds mediated by copper salts is a well-established process, dating back to the Hurtly reaction in 1929. In the following reports of this process, high yields were only obtained with aryl halides bearing electron-withdrawing groups or ortho-substituents that could be

Table 9.14 Copper/ligand catalyzed coupling between aryl halides and activated methylene compounds.

$$\text{Y}\underset{}{\overset{}{\bigcirc}}-X + \underset{O}{\overset{O}{\underset{\|}{R'}}}R \xrightarrow[\text{base, solvent}]{\text{[Cu]/Ligand}} \text{Y}\underset{}{\overset{}{\bigcirc}}\underset{O}{\overset{O}{\underset{\|}{R'}}}R$$

Entry	X	Ligand	Activated methylene compounds	Reaction conditions	Ref.
1	I, Br	L1	diethyl malonate, ethyl acetoacetate, ethyl benzoyl acetate	CuI, Cs_2CO_3 DMSO, 40–50 °C	[67]
2	I	L1	acetylacetone, ethyl cyanoacetate, malononitrile	CuI, K_2CO_3 DMSO, 90 °C	[68]
3	I, Br	L4	diethyl malonate	CuI, Cs_2CO_3, dioxane, RT–70 °C	[69]
4	I	L16	diethyl malonate, ethyl cyanoacetate, malononitrile	CuI, Cs_2CO_3, CH_3CN, 50–70 °C	[26]

coordinated with copper. In addition, the high reaction temperatures also limited its application.

The first milder method for this transformation, developed by Buchwald and Hennessy in 2002 [66], employed CuI/2-phenylphenol as a catalytic system to perform the arylation of diethyl malonates. The reaction worked well at 70 °C, but for some substrates such as isopropylidene malonate, 1,3-cyclohexanedione, and 1,3-cyclopentanedione, no desired products were observed. In addition, only aryl iodides were compatible with these reaction conditions.

CuI/L-proline (L1) was proved to be a quite efficient catalytic system for this coupling. The reaction proceeded smoothly in DMSO at 40–50 °C in the presence of Cs_2CO_3 (Table 9.14, entry 1) [67]. Both aryl iodides and bromides were suitable substrates. A wide range of activated methylene compounds, including ethyl acetoacetate, ethyl benzoyl acetate, and diethyl malonate were compatible with these reaction conditions. This catalytic system could also be used for the arylation of aryl iodides with acetylacetone and ethyl cyanoacetate (entry 2) [68]. Recently, it was found that 2-picolinic acid (L4) could lead to the arylation of aryl iodides and diethyl malonate worked at room temperature (entry 3) [69]. For hindered substrates or aryl bromides, higher reaction temperatures were still required. Additionally, Chxn-Py-Al (L16) was revealed to accelerate the CuI-catalyzed coupling of iodobenzene with diethyl malonate, ethyl cyanoacetate and malononitrile (entry 4) [26].

The combination of ortho-substitution, ligand and solvent effects enabled the reaction of 2-iodotrifluoroacetanilides with 2-methylacetoacetates to proceed at

–45 °C (Scheme 9.7) [70]. This represents the lowest reaction temperature for an Ullmann-type reaction to date. The corresponding coupling products were obtained in good yields and excellent enantioselectivity using CuI/*trans*-4-OH-L-proline (**L3**).

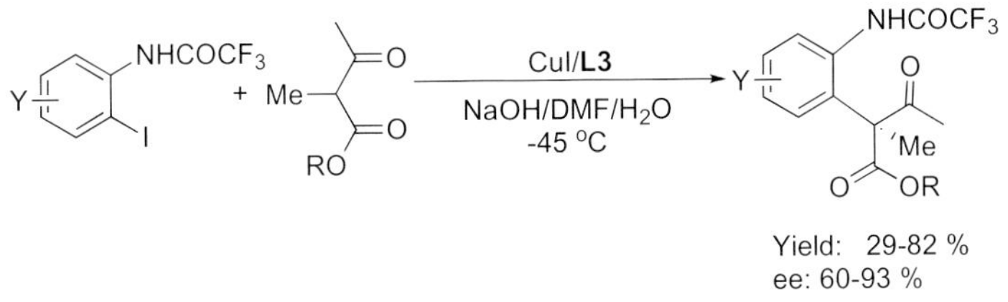
NHCOCF3
Y
I
+ Me
O
O
RO
CuI/L3
NaOH/DMF/H2O
-45 °C
NHCOCF3
O
Me
O
OR
Yield: 29-82 %
ee: 60-93 %

Scheme 9.7

9.4.3
Cyanation

The exchange of a halogen for a cyano group by copper (I) cyanide, the Rosenmund-von Braun reaction, is a classical method in organic synthesis. The reaction is performed by prolonged heating with a stoichiometric amount or excess of CuCN in polar solvents at high temperatures. By using suitable ligands, some milder protocols have been developed. The first such method was reported by Buchwald's group (Scheme 9.8) [71]. The reaction involved exchange of bromine by iodine and used DMEDA (**L11**) as a promoter. These conditions showed a high degree of tolerance to functional groups, but stoichiometric amounts of the ligand were needed.

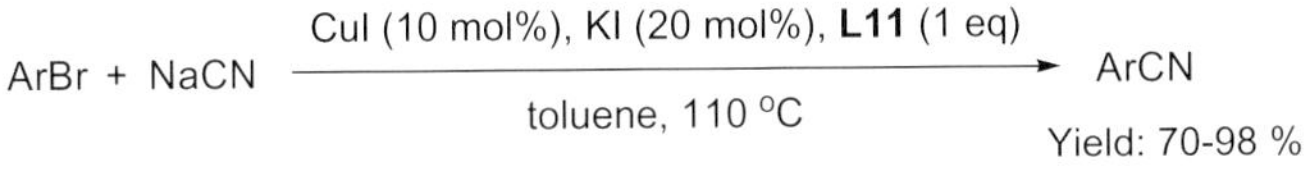
ArBr + NaCN
CuI (10 mol%), KI (20 mol%), L11 (1 eq)
toluene, 110 °C
ArCN
Yield: 70-98 %

Scheme 9.8

A catalytic version was developed by Cristau's group (Scheme 9.9) [72]. Aryl nitriles were obtained from the copper-catalyzed reaction of aryl bromides and acetonecynanohydrin in the presence of a catalytic amount of 1,10-phenanthroline (**L13**). Similar to Buchwald's report, this reaction also involved a halogen-exchanging process. In addition, the use of cyanohydrin was recognized to avoid the deactivation of the catalyst by cynanide ions.

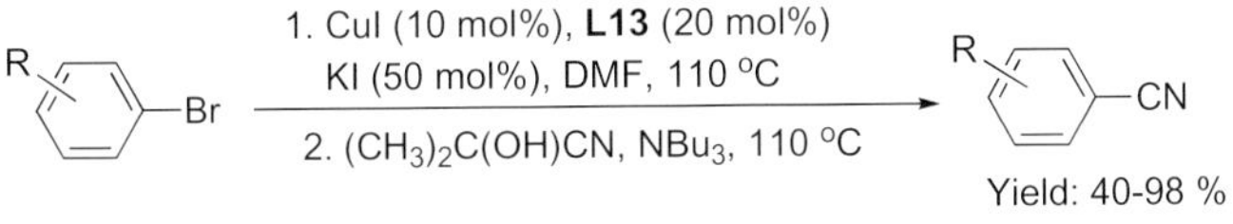
R
Br
1. CuI (10 mol%), L13 (20 mol%)
KI (50 mol%), DMF, 110 °C
2. (CH3)2C(OH)CN, NBu3, 110 °C
R
CN
Yield: 40-98 %

Scheme 9.9

L-Proline (**L1**) was also identified as an effective additive to promote the reaction of aryl halides and CuCN at relatively low temperatures (80–120 °C), affording the corresponding arylnitriles in good yields (Scheme 9.10) [73]. In this case, aryl bromides could be converted directly into arylnitriles, without the requirement of the extra halogen-exchanging process.

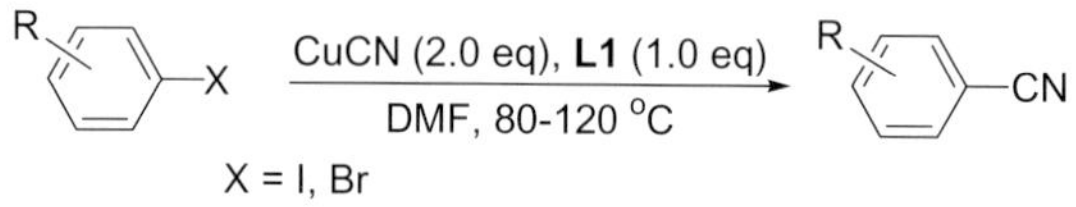

$$R\text{-Ar-}X \xrightarrow[\text{DMF, 80-120 °C}]{\text{CuCN (2.0 eq), } \mathbf{L1}\text{ (1.0 eq)}} R\text{-Ar-CN}$$

X = I, Br

Scheme 9.10

9.5
C–S Bond Formation

9.5.1
The Formation of Bisaryl- and Arylalkyl-Thioethers

Although thiols are generally stronger nucleophiles than alcohols and amines, the sensitivity of sulfides towards oxidation as well as the formation of disulfide made the cross-coupling between aryl halides and thiols difficult. From all previously reported results, it was found that only aryl iodides were suitable for this copper-catalyzed coupling process.

In 2002, Venkataraman published the synthesis of bisaryl- and arylalkyl-thioethers catalyzed by the CuI/neocuproine system [74]. At the same time, using excess ethylene glycol as a ligand, CuI-catalyzed coupling of aryl iodides with thiols was found to take place at 80 °C [75]. Soon after, N,N-dimethylglycine (**L2**) (Table 9.15, entry 1) [76] and L-proline (**L1**) (entry 2) [77] were revealed to be effective for the same transformation. Recently, CuBr/β-keto ester (**L19**) was reported to be able to promote this cross-coupling reaction to afford the corresponding diaryl thioethers at 60–75 °C (entry 3) [27].

Table 9.15 Copper/ligand catalyzed coupling between aryl halides and thiols.

$$Y\text{-Ar-}X + RSH \xrightarrow[\text{base, solvent}]{\text{[Cu]/ligand}} Y\text{-Ar-SR}$$

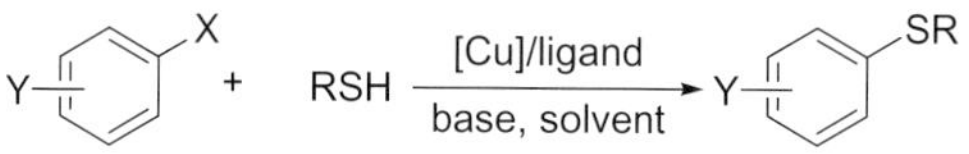

Entry	X	Ligand	RSH	Reaction conditions	Ref.
1	Br	L2	thiols, thiophenols	CuI, K_3PO_4, DMF, 120 °C	[76]
2	I	L1	thiols, thiophenols	CuI, K_2CO_3, DME, 80 °C	[77]
3	I	L19	thiophenols	CuBr, Cs_2CO_3, DMSO, 60–75 °C	[27]

9.5.2
The Synthesis of Alkenylsulfides

As shown in Table 9.16, assembly of alkenylsulfides could be achieved through coupling of vinyl halides with thiols. The suitable catalytic systems are [Cu(phen)(PPh$_3$)$_2$]NO$_3$ (**L13**) (Table 9.16, entry 1) [78], CuX/amino acid (**L1** and **L2**) (entries 2 and 3) [79, 80], and CuI/cis-1,2-cyclohexanediol (**L21**) that was found recently to be a quite efficient system (entry 4) [81]. Among them, CuI/**L21** led the reaction to occur at 30–60 °C, providing corresponding vinyl sulfides in excellent yields, even from hindered and functional thiols.

9.5.3
Assembly of aryl Sulfones

The Ma group revealed that CuI-catalyzed coupling of aryl halides and sulfinic acid salts could be promoted by L-proline (**L1**), leading to reaction at 80–95 °C to afford aryl sulfones in good yields (Table 9.17, entry 1) [82]. Both aryl/heteroaryl iodides and bromides were found to be suitable substrates. Prior to that (CuOTf)$_2$·PhH/DMEDA (**L11**) was demonstrated to effect this reaction at 110 °C (entry 2) [83].

Table 9.16 Copper/ligand catalyzed coupling between vinyl halides and thiols.

Entry	X	[Cu]	Ligand	Base	Reaction conditions	Yield (%)	Ref.
1	I	[Cu(phen)(PPh$_3$)$_2$]NO$_3$	L13	K$_3$PO$_4$	toluene, 110 °C	80–99	[78]
2	Br	CuBr	L1	K$_2$CO$_3$	[BMIM]BF$_4$, 110–120 °C	75–96	[79]
3	Br	CuI	L2	Cs$_2$CO$_3$	[BMIM]L2, 90 °C	76–93	[80]
4	I	CuI	L21	K$_3$PO$_4$	DMF, 30–50 °C	84–98	[81]

Table 9.17 Copper/ligand catalyzed coupling between aryl halides and sulfinic acid salts.

Entry	X	[Cu]	Ligand	Reaction conditions	Yield (%)	Ref.
1	I, Br	CuI	L1 (as Na$^+$ salt)	DMSO, 80–95 °C	46–93	[82]
2	I, Br	(CuOTf)$_2$·PhH	L11	DMSO, 110 °C	24–96	[83]

9.6
C–P Bond Formation

Venkataraman and coworkers reported that the cross-coupling of aryl iodides with Ph_2PH could be effected using CuI as a catalyst, providing the corresponding triarylphophines in good yields (Scheme 9.11) [84]. As phosphines themselves were good ligands, no extra ligand was required in this case.

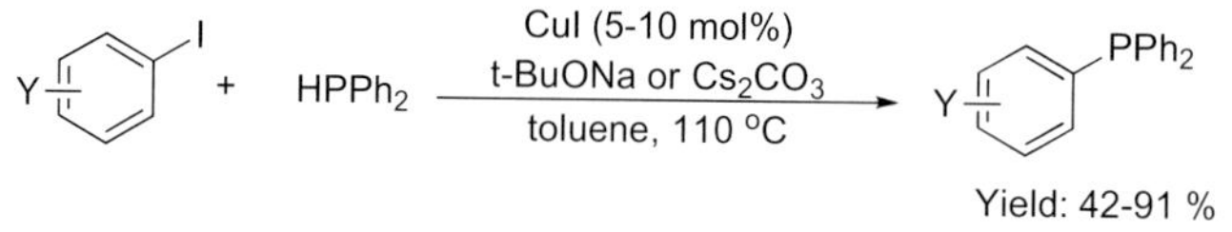

$$Y\text{-}C_6H_4\text{-}I + HPPh_2 \xrightarrow[\text{toluene, 110 °C}]{\text{CuI (5-10 mol\%)}, \text{ t-BuONa or } Cs_2CO_3} Y\text{-}C_6H_4\text{-}PPh_2$$

Yield: 42-91 %

Scheme 9.11

Two class of ligands, DMEDA (**L11**) (Scheme 9.12) [85], and amino acids (**L1** and **L5**) (Scheme 9.13) have been tested for copper-catalyzed C–P bond formation [86]. The former was effective for cross-coupling of aryl iodides with R_2PH as well as dibutyl phosphite, while the latter could promote the formation of arylphosphonates, arylphosphinates and arylphosphine oxides.

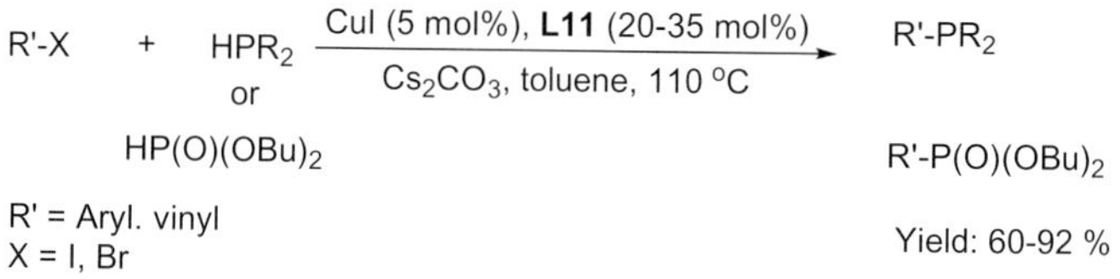

$$R'\text{-}X + HPR_2 \text{ or } HP(O)(OBu)_2 \xrightarrow[\text{Cs}_2\text{CO}_3\text{, toluene, 110 °C}]{\text{CuI (5 mol\%), L11 (20-35 mol\%)}} R'\text{-}PR_2 \text{ or } R'\text{-}P(O)(OBu)_2$$

R' = Aryl, vinyl
X = I, Br

Yield: 60-92 %

Scheme 9.12

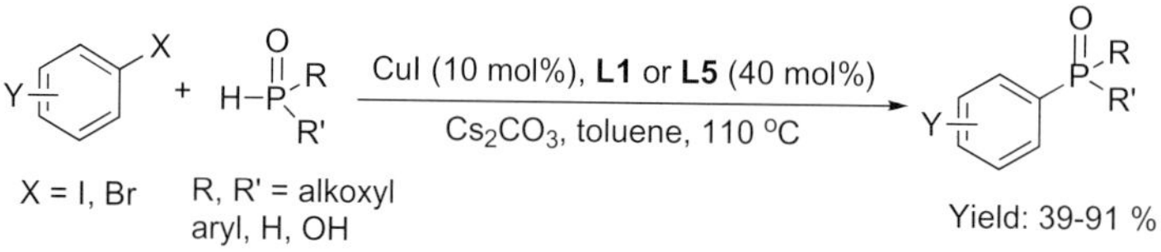

$$Y\text{-}C_6H_4\text{-}X + H\text{-}P(O)(R)(R') \xrightarrow[\text{Cs}_2\text{CO}_3\text{, toluene, 110 °C}]{\text{CuI (10 mol\%), L1 or L5 (40 mol\%)}} Y\text{-}C_6H_4\text{-}P(O)(R)(R')$$

X = I, Br
R, R' = alkoxyl aryl, H, OH

Yield: 39-91 %

Scheme 9.13

9.7
Conclusion

Several classes of ligands, such as amino acids (**L1–3, L5**), diamines (**L9–12**), phenanthrolines (**L13–15**), and diketone (**L18, L20**) (or β-keto ester (**L19**)), have shown remarkable acceleration effects for almost all copper-catalyzed Ullmann-type coupling reactions. These newly developed conditions have found extensive applications in the synthesis of natural products [1b], designed bioactive compounds and molecules for materials applications. Based on these reactions, several cascade reaction processes for the assembly of heterocycles have been discovered [87]. This opens a new avenue for the assessment of these pharmaceutically important compounds. These studies will further prompt the discovery of new ligands for these copper-catalyzed coupling reactions.

References

1 For reviews, see: (a) Monnier, F., Taillefer, M. (2008) *Angew. Chem. Int. Ed.*, **47**, 3096; (b) Evano, G., Blanchard, N., and Toumi, M. (2008) *Chem. Rev.*, **108**, 3054; (c) Finet, J.-P., Fedorov, A.Y., Combes, S., and Boyer, G. (2002) *Current Org. Chem.*, **6**, 597; (d) Ley, S.V., and Thomas, A.W. (2003) *Angew. Chem. Int. Ed.*, **42**, 5400; (e) Kunz, K., Scholz, U., and Ganzer, D. (2003) *Synlett*, 2428; (f) Beletskaya, I.P., and Cheprakov, A.V. (2004) *Coord. Chem. Rev.*, **248**, 2337.
2 Ma, D., Cai, Q., and Zhang, H. (2003) *Org. Lett.*, **5**, 2453.
3 Zhang, H., Cai, Q., and Ma, D. (2005) *J. Org. Chem.*, **70**, 5164.
4 Kwong, F.Y., and Buchwald, S.L. (2003) *Org. Lett.*, **5**, 793.
5 Kwong, F.Y., Klapars, A., and Buchwald, S.L. (2002) *Org. Lett.*, **4**, 581.
6 Shafir, A., and Buchwald, S.L. (2006) *J. Am. Chem. Soc.*, **128**, 8742.
7 Shafir, A., Lichtor, P.A., and Buchwald, S.L. (2007) *J. Am. Chem. Soc.*, **129**, 3490.
8 Jiang, D., Fu, H., Jiang, Y. and Zhao, Y. (2007) *J. Org. Chem.*, **72**, 672.
9 Zhu, D., Wang, R., Mao, J., Xu, L., Wu, F., and Wan, B. (2006) *J. Mol. Catal. A: Chem*, **256**, 256.
10 Goodbrand, H.B., and Hu, N.-X. (1999) *J. Org. Chem.*, **64**, 670.
11 Gujadhur, R.K., Bates, C.G., and Venkataraman, D. (2001) *Org. Lett.*, **3**, 4315.
12 Guo, X., Rao, H., Fu, H., Jiang, Y., and Zhao, Y. (2006) *Adv. Synth. Catal.*, **348**, 2197.
13 Liu, Y., Bai, Y., Zhang, J., Li, Y., Jiao, J., and Qi, X. (2007) *Eur. J. Org. Chem.*, 6084.
14 Altman, R.A., Anderson, K.W., and Buchwald, S.L. (2008) *J. Org. Chem.*, **73**, 5167.
15 Lang, F., Zewge, D., Houpis, I.N., and Volante, R.P. (2001) *Tetrahedron Lett.*, **42**, 3251.
16 Kim, J., and Chang, S. (2008) *Chem. Commun.*, 3052.
17 Gao, X., Fu, H., Qiao, R., Jiang, Y., and Zhao, Y. (2008) *J. Org. Chem.*, **73**, 6864.
18 Kiyomori, A., Marcoux, J.-F., and Buchwald, S.L. (1999) *Tetrahedron Lett.*, **40**, 2657.
19 Altman, R.A., and Buchwald, S.L. (2006) *Org. Lett.*, **8**, 2779.
20 Altman, R.A., Koval, E.D., and Buchwald, S.L. (2007) *J. Org. Chem.*, **72**, 6190.
21 Klapars, A., Antilla, J.C., Huang, X., and Buchwald, S.L. (2001) *J. Am. Chem. Soc.*, **123**, 7727.
22 Antilla, J.C., Klapars, A., and Buchwald, S.L. (2002) *J. Am. Chem. Soc.*, **124**, 11684.
23 Antilla, J.C., Baskin, J.M., Barder, T.E., and Buchwald, S.L. (2004) *J. Org. Chem.*, **69**, 5578.
24 Ma, D., and Cai, Q. (2004) *Synlett*, 128.
25 Cristau, H.-J., Cellier, P.P., Hamada, S., Spindler, J.-F., and Taillefer, M. (2004) *J. Org. Chem.*, 695.
26 Cristau, H.-J., Cellier, P.P., Hamada, S., Spindler, J.-F., and Taillefer, M. (2004) *Chem. Eur. J.*, **10**, 5607.
27 Lv, X., and Bao, W. (2007) *J. Org. Chem.*, **72**, 3863.
28 Wang, Z., Bao, W., and Jiang, Y. (2005) *Chem. Commun.*, 2849.
29 Shen, G., Lv, X., Qian, W., and Bao, W. (2008) *Tetrahedron Lett.*, **49**, 4556.

30 Taillefer, M., Quali, A., Renard, B., and Spindler, J.-F. (2006) *Chem. Eur. J.*, **12**, 5301.
31 Klapars, A., Huang, X., and Buchwald, S.L. (2002) *J. Am. Chem. Soc.*, **124**, 7421.
32 Tao, C.-Z., Li, J., Fu, Y., Liu, L., and Guo, Q.-X. (2008) *Tetrahedron Lett.*, **49**, 70.
33 Wolter, M., Klapars, A., and Buchwald, S.L. (2001) *Org. Lett.*, **3**, 3803.
34 Jones, K.L., Porzelle, A., Hall, A., Woodrow, M.D., and Tomkinson, N.C.O. (2008) *Org. Lett.*, **10**, 797.
35 Hosseinzadeh, R., Tajbakhsh, M., Mohadjerani, M., and Mehdinejad, H. (2004) *Synlett*, **9**, 1517.
36 Altman, R.A., and Buchwald, S.L. (2007) *Org. Lett.*, **9**, 643.
37 Shen, R., and Porco, J.A. (2000) *Org. Lett.*, **2**, 1333.
38 Jiang, L., Job, G.E., Klapars, A., and Buchwald, S.L. (2003) *Org. Lett.*, **5**, 3667.
39 Pan, X., Cai, Q., and Ma, D. (2004) *Org. Lett.*, **6**, 1809.
40 Han, C., Shen, R., Su, S., and Porco, J.A. (2004) *Org. Lett.*, **6**, 27.
41 Frederick, M.O., Mulder, J.A., Tracey, M.R., Hsung, R.P., Huang, J., Kurtz, K.C.M., Shen, L., and Douglas, C.J. (2003) *J. Am. Chem. Soc.*, **125**, 2368.
42 Zhang, Y., Hsung, R.P., Tracey, M.R., Kurtz, K.C.M., and Vera, E.L. (2004) *Org. Lett.*, **6**, 1151.
43 Istrate, F.M., Buzas, A.K., Jurberg, I.D., Odabachian, Y., and Gagosz, F. (2008) *Org. Lett.*, **10**, 925.
44 Zhu, W., and Ma, D. (2004) *Chem. Commun.*, 888.
45 Anderson, J., Madsen, U., Bjorkling, F., and Liang, X. (2005) *Synlett*, **14**, 2209.
46 Anderson, J., Bolvig, S., and Liang, X. (2005) *Synlett*, **19**, 2941.
47 Ackermann, L., Potukuchi, H.K., Landsberg, D., and Vicente, R. (2008) *Org. Lett.*, **10**, 3081.
48 Buck, E., Song, Z.J., Tshaen, D., Dormer, P.G., Volane, R.P., and Reider, P.J. (2002) *Org. Lett.*, **4**, 1623.
49 Ma, D., and Cai, Q. (2003) *Org. Lett.*, **5**, 3799.
50 Cristau, H.-J., Cellier, P.P., Hamada, S., Spindler, J.-F., and Taillefer, M. (2004) *Org. Lett.*, **6**, 913.
51 Ouali, A., Spindler, J.-F., Cristau, H.-J., and Taillefer, M. (2006) *Adv. Synth. Catal.*, **348**, 499.
52 Cai, Q., He, G., and Ma, D. (2006) *J. Org. Chem.*, **71**, 5268.
53 Quali, A., Splindler, J.-F., Jutand, A., and Taillefer, M. (2007) *Adv. Synth. Catal.*, **349**, 1906.
54 Xia, N., and Taillefer, M. (2008) *Chem. Eur. J.*, **14**, 6037.
55 Cai, Q., Zou, B., and Ma, D. (2006) *Angew. Chem. Int. Ed.*, **45**, 1276.
56 Ma, D., Cai, Q., and Xie, X. (2005) *Synlett*, **11**, 1767.
57 Wolter, M., Nordmann, G., Job, G.E., and Buchwald, S.L. (2002) *Org. Lett.*, **4**, 973.
58 Hosseinzadeh, R., Tajbakhsh, M., Mohadjerani, M., and Alikarami, M. (2005) *Synlett*, **7**, 1101.
59 Altman, R.A., Shafir, A., Choi, A., Lichtor, P.A., and Buchwald, S.L. (2008) *J. Org. Chem.*, **73**, 284.
60 Zhang, H., Ma, D., and Cao, W. (2007) *Synlett*, 243.
61 Job, G.E., and Buchwald, S.L. (2002) *Org. Lett.*, **4**, 3703.
62 Ma, D., and Liu, F. (2004) *Chem. Commun.*, 1934.
63 Mao, J., Guo, J., and Ji, S. (2008) *J. Mol. Catal. A: Chem.*, **284**, 85.
64 Liu, F., and Ma, D. (2007) *J. Org. Chem.*, **72**, 4844.
65 Bates, C.G., Saejueng, P., and Venkataraman, D. (2004) *Org. Lett.*, **6**, 1441.
66 Hennessy, E.J., and Buchwald, S.L. (2002) *Org. Lett.*, **4**, 269.
67 Xie, X., Cai, G., and Ma, D. (2005) *Org. Lett.*, **7**, 4693.
68 Jiang, Y., Wu, N., Wu, H., and He, M. (2005) *Synlett*, **18**, 2731.
69 Yip, S.F., Cheung, H.Y., Zhou, Z., and Kwong, F.Y. (2007) *Org. Lett.*, **9**, 3469.
70 Xie, X., Chen, Y., and Ma, D. (2006) *J. Am. Chem. Soc.*, **128**, 16050.
71 Zanon, J., Klapars, A., and Buchwald, S.L. (2003) *J. Am. Chem. Soc.*, **125**, 2890.
72 Cristau, H.-J., Ouali, A., Spindler, J.-F., and Taillefer, M. (2005) *Chem. Eur. J.*, **11**, 2483.
73 Wang, D., Kuang, L., Li, Z., and Ding, K. (2008) *Synlett*, 69.

74 Bates, C.G., Gujadhur, R.K., and Venkataraman, D. (2002) *Org. Lett.*, **4**, 2803.
75 Kwong, F.Y., and Buchwald, S.L. (2002) *Org. Lett.*, **4**, 3517.
76 Deng, W., Zou, Y., Wang, Y.-F., Liu, L., and Guo, Q.-X. (2004) *Synlett*, **7**, 1254.
77 Zhang, H., Cao, W., and Ma, D. (2007) *Syn. Commun.*, **37**, 25.
78 Bates, C.G., Saejueng, P., Doherty, M.Q., and Venkataraman, D. (2004) *Org. Lett.*, **6**, 5005.
79 Zheng, Y., Du, X., and Bao, W. (2006) *Tetrahedron Lett.*, **47**, 1217.
80 Wang, Z., Mo, H., and Bao, W. (2007) *Synlett*, 91.
81 Kabir, M.S., Van Linn, M.L., Monte, A., and Cook, J.M. (2008) *Org. Lett.*, **10**, 3363.
82 Zhu, W., and Ma, D. (2005) *J. Org. Chem.*, **70**, 2696.
83 Baskin, J.M., and Wang, Y. (2002) *Org. Lett.*, **4**, 4423.
84 Allen, D.V., and Venkataraman, D. (2003) *J. Org. Chem.*, **68**, 4590.
85 Gelman, D., Jiang, L., and Buchwald, S.L. (2003) *Org. Lett.*, **5**, 2315.
86 Huang, C., Tang, X., Fu, H., Jiang, Y., and Zhao, Y. (2006) *J. Org. Chem.*, **71**, 5020.
87 For selected examples, see: (a) Yuan, Q., Ma, D. (2008) *J. Org. Chem.*, **73**, 5159;
(b) Chen, Y., Wang, Y., Sun, Z., and Ma, D. (2008) *Org. Lett.*, **10**, 625;
(c) Wang, B., Lu, B., Jiang, Y., Zhang, Y., and Ma, D. (2008) *Org. Lett.*, **10**, 2761;
(d) Minatti, A., and Buchwald, S.L. (2008) *Org. Lett.*, **10**, 2721;
(e) Viirre, R.D., Evindar, G., and Batey, R.A. (2008) *J. Org. Chem.*, **73**, 3452;
(f) Yang, D., Fu, H., Hu, L., Jiang, Y., and Zhao, Y. (2008) *J. Org. Chem.*, **73**, 7841;
(g) Bao, W., Liu, Y., Lv, X., and Qian, W. (2008) *Org. Lett.*, **10**, 3899;
(h) Lu, B., Wang, B., Zhang, Y., and Ma, D. (2007) *J. Org. Chem.*, **72**, 5337;
(i) Zou, B., Yuan, Q., and Ma, D. (2007) *Angew. Chem. Int. Ed.*, **46**, 2598;
(j) Zou, B., Yuan, Q., and Ma, D. (2007) *Org. Lett.*, **9**, 4291;
(k) Chen, Y., Xie, X., and Ma, D. (2007) *J. Org. Chem.*, **72**, 9329;
(l) Zheng, N., and Buchwald, S.L. (2007) *Org. Lett.*, **9**, 4749;
(m) Rivero, M.R., and Buchwald, S.L. (2007) *Org. Lett.*, **9**, 973;
(n) Martín, R., Cuenca, A., and Buchwald, S.L. (2007) *Org. Lett.*, **9**, 5521;
(o) Jones, C.P., Anderson, K.W., and Buchwald, S.L. (2007) *J. Org. Chem.*, **72**, 7968;
(p) Yuan, X., Xu, X., Zhou, X., Yuan, J., Mai, L., and Li, Y. (2007) *J. Org. Chem.*, **72**, 1510–1513;
(q) Martin, R., Rodríguez, R., and Buchwald, S.L. (2006) *Angew. Chem. Int. Ed.*, **45**, 7079;
(r) Klapars, A., Parris, S., Andersom, K.W., and Buchwald, S.L. (2004) *J. Am. Chem. Soc.*, **126**, 3529;
(s) Bates, C.G., Saejueng, P., Murphy, J.M., and Venkataraman, D. (2002) *Org. Lett.*, **4**, 4727.

10
Copper-Catalyzed Azide-Alkyne Cycloaddition (CuAAC)
M.G. Finn and Valery V. Fokin

10.1
Introduction

1,3-Dipolar cycloaddition reactions provide direct access to a wide range of useful heterocyclic systems. Many such heterocycles are found in natural products, man-made drugs, agrochemicals, and materials. As "fusion" processes, dipolar cycloadditions are atom-economical, and the variety of readily accessible dipoles and dipolarophiles make these transformations particularly suitable for the synthesis of structurally and functionally diverse collections of compounds. This large family of reactions has been a subject of intensive research, notably by Rolf Huisgen and coworkers, whose work led to the formulation of the general concept of 1,3-dipolar cycloadditions in 1958 [1]. Since then, dipolar cycloaddition chemistry has found extensive applications in organic synthesis and has been the subject of several reviews [2, 3].

Organic azides represent a unique component of cycloaddition processes because $R-N_3$ is the only 1,3-dipolar reagent that can be made, handled, and stored as a stable entity. Until recently, the cycloaddition reactions of azides with olefins received greater attention among organic chemists than the corresponding reactions with alkynes. Yet, the azide–alkyne combination is uniquely useful for three main reasons. First, the product 1,2,3-triazole is a remarkably stable aromatic structure that can serve, among other roles, as a structural analogue of a peptide linkage. Second, the azide–alkyne reaction is remarkably slow for a process that is thermodynamically favorable by more than 50kcal mol^{-1}. Finally, azides and alkynes are relatively nonpolar, neither acidic nor basic, and are devoid of substantial hydrogen bonding ability, properties that make them quite "invisible" to and unreactive with most other chemical functionalities in nature and the laboratory. Azides and alkynes can, thereby, be installed on structures that one wishes to link together, and kept in place through many operations of synthesis and elaboration. They are ideal connectors for such a modular approach to synthesis, *if* effective methods to catalyze their cycloaddition are available. This chapter describes the catalytic role of copper complexes that has allowed this general scheme to be put in operation for a wide array of applications and settings.

Catalysis Without Precious Metals. Edited by R. Morris Bullock
© 2010 WILEY-VCH Verlag GmbH & Co. KGaA, Weinheim
ISBN: 978-3-527-32354-8

The Cu-catalyzed azide–alkyne cycloaddition (CuAAC) reaction has emerged as the premier example of click chemistry, a term coined in 2001 by Sharpless and colleagues to describe a set of "near-perfect" bond-forming reactions useful for rapid assembly of molecules with desired function [4]. Click transformations are easy to perform, give rise to their intended products in very high yields with little or no byproducts, work well under many conditions (usually especially well in water), and are unaffected by the nature of the groups being connected to each other. The "click" moniker is meant to signify that, with the use of these methods, joining molecular fragments is as easy as "clicking" together the two pieces of a buckle. The buckle works no matter what is attached to it, as long as its two pieces can reach each other, and the components of the buckle can make a connection only with each other. The potential of organic azides as highly energetic, yet very selective functional groups in organic synthesis was highlighted, and their dipolar cycloadditions with olefins and alkynes were placed among the reactions fulfilling the click criteria. However, the inherently low reaction rates of the azide–alkyne cycloaddition did not make it very useful in the click context, and its potential was not revealed until its catalysis by copper under broadly useful conditions was discovered.

The copper-catalyzed reaction was reported simultaneously and independently by the groups of Meldal in Denmark [5] and Fokin and Sharpless in the US [6]. It transforms organic azides and terminal alkynes exclusively to the corresponding 1,4-triazoles, in contrast to the uncatalyzed reaction which requires much higher temperatures and provides mixtures of 1,4- and 1,5-triazole isomers. Meldal and coworkers used the transformation for the synthesis of peptidotriazoles in organic solvents starting from alkynylated amino acids attached to solid supports, whereas the Scripps group immediately turned to aqueous systems and devised a straightforward and practical procedure for the covalent "stitching" of virtually any fragments containing an azide and an alkyne functionality, noting the broad utility and versatility of the novel process "for those organic synthesis endeavors which depend on the creation of covalent links between diverse building blocks" [6]. Within months, applications of this reaction began to appear across the chemical disciplines, rapidly establishing the CuAAC process as a leading method for the covalent assembly of complex molecules. It has since been widely employed in synthesis, medicinal chemistry, molecular biology, and materials science. Unsurprisingly, it is often equated with click chemistry, but we stress that the CuAAC reaction is but a representative of this type of transformation. Its success in enabling the creation of new and useful functions serves for us, and we hope for others, as an inspiration for the development of new click reactions.

Numerous applications of the CuAAC reaction reported during the last several years have been regularly reviewed [7–13], and are continually enriched by investigators in many fields [14]. We focus here on the fundamental aspects of the CuAAC process and on its mechanism, with an emphasis on the qualities of copper that enable this unique mode of reactivity.

10.2 Azide–Alkyne Cycloaddition: Basics

The thermal reaction of terminal or internal alkynes with organic azides (Scheme 10.1A) has been known for more than a century, the first 1,2,3-triazole being synthesized by A. Michael from phenyl azide and diethyl acetylenedicarboxylate in 1893 [15]. The reaction has been most thoroughly investigated by Huisgen and coworkers in the 1950s–70s in the course of their studies of the larger family of 1,3-dipolar cycloaddition reactions [16–22]. Although the reaction is highly exothermic (ΔG^0 between −50 and −65 kcal mol^{-1}), its high activation barrier (approximately 25 kcal mol^{-1} for methyl azide and propyne [23]) results in exceedingly low reaction rates for unactivated reactants, even at elevated temperature. Furthermore, since the differences in HOMO–LUMO energy levels for both azides and alkynes are of similar magnitude, both dipole-HOMO- and dipole-LUMO-controlled pathways operate in these cycloadditions. As a result, a mixture of regioisomeric 1,2,3-triazole products is usually formed when an alkyne is unsymmetrically substituted. These mechanistic features can be altered, and the reaction dramatically accelerated, by making the alkyne electron-deficient, such as in propiolate or acetylene dicarboxylate esters, at the cost of opening up competitive side

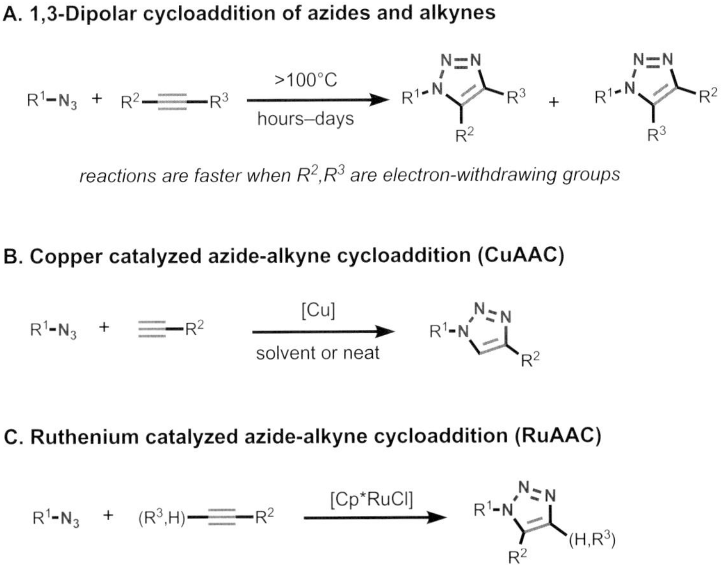

A. 1,3-Dipolar cycloaddition of azides and alkynes

R^1-N_3 + $R^2-\equiv-R^3$ →(>100°C, hours–days) 1,4- and 1,5-regioisomeric triazoles

reactions are faster when R^2, R^3 are electron-withdrawing groups

B. Copper catalyzed azide-alkyne cycloaddition (CuAAC)

R^1-N_3 + $\equiv-R^2$ →([Cu], solvent or neat) 1,4-disubstituted triazole

C. Ruthenium catalyzed azide-alkyne cycloaddition (RuAAC)

R^1-N_3 + $(R^3,H)-\equiv-R^2$ →([Cp*RuCl]) 1,5-disubstituted triazole

Scheme 10.1 Thermal cycloaddition of azides and alkynes usually requires prolonged heating and results in mixtures of both 1,4- and 1,5-regioisomers (A), whereas CuAAC produces only 1,4-disubstituted-1,2,3-triazoles at room temperature in excellent yields (B). The RuAAC reaction proceeds with both terminal and internal alkynes and gives 1,5-disubstituted and fully, 1,4,5-trisubstituted-1,2,3-triazoles.

reactions such as conjugate addition. Since both the lack of side reactions and the ability of catalysts to turn on AAC activity are central to many applications, we and others have focused on unactivated alkynes.

Copper catalysts (Scheme 10.1B) accelerate the cycloaddition reactions of azides with such terminal alkynes by at least a factor of 10^6 relative to the uncatalyzed version [23], making it conveniently fast at and below room temperature. The reaction is not significantly affected by the steric and electronic properties of the groups attached to the azide and alkyne centers. For example, primary, secondary, and even tertiary, electron-deficient and electron-rich, aliphatic, aromatic, and heteroaromatic azides usually react well with variously substituted terminal alkynes. The reaction proceeds well in most protic and aprotic solvents, including water, and is unaffected by most organic and inorganic functional groups, therefore all but eliminating the need for protecting group chemistry [24]. The 1,2,3-triazole heterocycle has the advantageous properties of high chemical stability (generally inert to severe hydrolytic, oxidizing, and reducing conditions, even at high temperature), strong dipole moment (4.8–5.6 D), aromatic character, and hydrogen bond accepting ability [25, 26]. Thus it can interact productively in several ways with biological molecules and can serve as a hydrolytically-stable replacement for the amide bond [27–29]. Compatibility of the CuAAC reaction with a broad range of functional groups and reaction conditions [30–34] has made it broadly useful across the chemical disciplines, and its applications include the synthesis of biologically active compounds, the preparation of conjugates to proteins and polynucleotides, the synthesis of dyes, the elaboration of known polymers and the synthesis of new ones, the creation of responsive materials, and the covalent attachment of desired structures to surfaces [7, 35–39].

10.3
Copper-Catalyzed Cycloadditions

10.3.1
Catalysts and Ligands

A wide range of experimental conditions for the CuAAC have been employed since its discovery, underscoring the robustness of the process and its compatibility with most functional groups, solvents, and additives, regardless of the source of the catalyst. The most commonly used protocols and their advantages and limitations are discussed below, and representative experimental procedures are described at the end of this chapter.

Different copper(I) sources can be utilized in the reaction, as recently summarized in necessarily partial fashion by Meldal and Tornøe [12]. Copper(I) salts (iodide, bromide, chloride, acetate) and coordination complexes such as $[Cu(CH_3CN)_4]PF_6$ and $[Cu(CH_3CN)_4]OTf$ [40, 41] have been commonly employed [6]. In general, however, we recommend against the use of cuprous iodide because

10.3 Copper-Catalyzed Cycloadditions

of the ability of iodide to act as a ligand for the metal. As we shall discuss below, control of Cu(I) coordination chemistry is paramount in the optimization of CuAAC catalysts, and iodide tends to interfere. Similarly, high concentrations of chloride ion in water (0.5 M or above) can be deleterious. For aqueous-rich reactions, cuprous bromide and acetate are favored, as is the sulfate from *in situ* reduction of $CuSO_4$; for organic reactions, the acetate salt is generally a good choice.

Copper(II) salts and coordination complexes are not competent catalysts, and reports describing Cu(II)-catalyzed cycloadditions [42–44] are not accurate. Copper(II) is a well-known oxidizing agent for organic compounds [45]. Alcohols, amines, aldehydes, thiols, phenols, and carboxylic acids may be oxidized by the cupric ion, reducing it to the catalytically active copper(I) species in the process. Especially relevant is the family of oxidative acetylenic couplings catalyzed by the cupric species [46], with the venerable Glaser coupling [47, 48] being the most studied example. Since terminal acetylenes are necessarily present in the CuAAC reaction, their oxidation is an inevitable side reaction which would, in turn, produce the needed catalytically active copper(I) species.

It is also true that Cu(I) can be readily oxidized to catalytically inactive Cu(II) species, or can disproportionate to a mixture of Cu(II) and Cu(0). The standard potential of the Cu^{2+}/Cu^+ couple is 159 mV, but can vary widely depending on the solvent and the ligands coordinated to the metal, and is especially complex in water [49]. When present in significant amounts, the ability of Cu(II) to mediate the aforementioned Glaser-type alkyne coupling processes can result in the formation of undesired byproducts while impairing triazole formation. When the cycloaddition is performed in organic solvents using copper(I) halides as catalysts, the conditions originally reported by the group of Meldal [5], the reaction is plagued by the formation of oxidative coupling byproducts **4a-d** unless the alkyne is bound to a solid support (**5**, Scheme 10.2A and B). If the alkyne is present in solution and the azide is immobilized on the resin, only traces of the desired triazole product are formed. The above side reactions are a consequence of the thermodynamic instability of the Cu(I) oxidation state. Therefore, when copper(I) catalyst is used directly, whether by itself or in conjunction with amine ligands, exclusion of oxygen may be required.

The use of a reducing agent, most commonly sodium ascorbate, introduced by Fokin and coworkers [6], is a convenient and practical alternative to oxygen-free conditions. Its combination with a copper(II) salt, such as the readily available and stable copper(II) sulfate pentahydrate or copper(II) acetate, has become the method of choice for preparative synthesis of 1,2,3-triazoles. Water appears to be an ideal solvent, capable of supporting copper(I) acetylides in their reactive state, especially when they are formed *in situ*. The "aqueous ascorbate" procedure often furnishes triazole products in nearly quantitative yield and greater than 90% purity, without the need for ligands or protection of the reaction mixture from oxygen (Scheme 10.2C). Of course, copper(I) salts can also be used in combination with ascorbate, wherein it converts any oxidized copper(II) species back to the catalytically active +1 oxidation state.

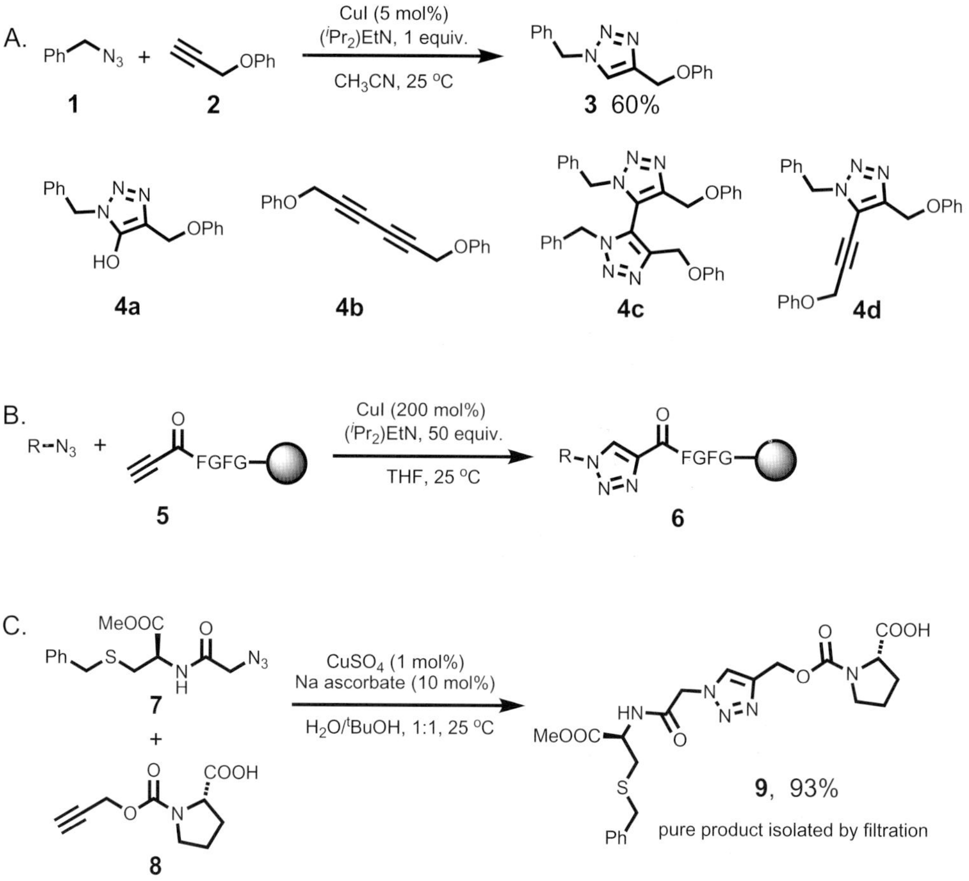

Scheme 10.2 **A**: Oxidative coupling byproducts in the CuAAC reactions catalyzed by copper(I) salts; **B**: CuAAC with immobilized alkyne avoids the formation of the oxidative byproducts but requires a large excess of the catalyst; reactions with immobilized azide fail; **C**: solution-phase CuAAC in the presence of sodium ascorbate.

The reaction can also be catalyzed by Cu(I) ions supplied by elemental copper, thus further simplifying the experimental procedure – a small piece of copper metal (wire or turning) is all that is added to the reaction mixture, followed by shaking or stirring for 12–48 h [6, 23, 50]. Aqueous alcohols (methanol, ethanol, *tert*-butanol), tetrahydrofuran, and dimethylsulfoxide can be used as solvents in this procedure. Cu(II) sulfate may be added to accelerate the reaction; however, this is not necessary in most cases, as copper oxides and carbonates, the patina on the metal surface, are sufficient to initiate the catalytic cycle. Although the procedure based on copper metal requires longer reaction times when performed at ambient temperature, it usually provides access to very pure triazole products with low levels of copper contamination. Alternatively, the reaction can be performed under microwave irradiation at elevated temperature, reducing the reaction time to 10–30 min [50–56].

The copper metal procedure is also very simple experimentally and is particularly convenient for high-throughput synthesis of compound libraries for biological screening. The reaction is very selective, and triazole products are generally isolated in >85–90% yields, and can often be submitted for screening directly. When required, trace quantities of copper remaining in the reaction mixture can be removed with an ion-exchange resin or using solid-phase extraction techniques. Other heterogeneous copper(0) and copper(I) catalysts, such as copper nanoclusters [57], copper/cuprous oxide nanoparticles [58], and copper nanoparticles adsorbed onto charcoal [59] have also shown good catalytic activity.

Many other copper complexes involving ligands or solid supports have been reported as catalysts or mediators of the CuAAC reaction. Providing quantitative comparisons between them is difficult since they are employed under widely differing conditions [12]. Instead, it is perhaps useful to organize them into "soft" and "hard" classes by virtue of the properties of their donor centers [60]. Cu(I) can be regarded as borderline "soft" in its acceptor character [61, 62], making for a wide variety of potentially effective ligands. A small but representative sample of reported systems is as follows.

The "soft" ligand class is exemplified by phosphine-containing CuAAC-active species such as the simple coordination complexes $Cu(P(OMe)_3)_3Br$ [63] and $Cu(PPh_3)_3Br$ [64, 65]. These species are favored for reactions in organic solvents, in which cuprous salts have limited solubility. A very recent report describes the bis(phosphine) complex $Cu(PPh_3)_2OAc$ as an excellent catalyst for the CuAAC reaction in toluene and dichloromethane [66]. Monodentate phosphoramidite and related donors have also been evaluated [67]. Chelating complexes involving phosphines have not found favor, except for bidentate combinations of phosphine with the relatively weakly-binding triazole unit [68]. Related to the phosphines are the N-heterocyclic carbenes, several Cu(I) complexes of which have been described as CuAAC catalysts at elevated temperature in organic solvents [69–71]. Thiols are a potent poison of the CuAAC reaction in water, but thioethers are an underexplored member of the "soft" ligand class [72].

The category of "hard" donor ligands of CuAAC-active systems is dominated by amines (see reference [12] and the Supporting Information of reference [73]). Indeed, in many cases, amines are labeled as "additives" rather than "ligands," a distinction made with the assumption that they aid in the deprotonation of the terminal alkyne rather than coordinate to the metal center. However, this may not be accurate, as the formation of copper(I) acetylides is very facile even in highly acidic media, such as 25% H_2SO_4 [74]. As we shall discuss below, as good ligands for copper(I), amines may facilitate ligand exchange and prevent formation of unreactive polymeric copper(I) acetylides. In several cases of amine-based chelates, it is probable that metal binding is at least part of the productive role of the additive. A notable example is the use of a hydrophobic tren ligand for CuAAC catalysis in organic solvent at elevated temperature [75].

As befits the borderline nature of the Cu(I) ion, by far the largest and most successful class of ligands is those of intermediate character between "hard" and "soft," particularly those containing heterocyclic donors. With rare exceptions [76], these also contain a central tertiary amine center, which can serve as a coordinating donor or base. The need for these ligands was particularly evident for reactions involving biological molecules that are handled in water in low concentrations and are not stable to heating. The appeal of azide–alkyne cycloaddition for "bioorthogonal" connectivity in aqueous media, amply precedented by the utility of the Staudinger ligation involving organic azides and phosphine-esters [77], propelled us to investigate CuAAC reactivity in water early on. Chemical transformations used in the synthesis of bioconjugates impose additional demands on the efficiency and selectivity. They must be exquisitely chemoselective, biocompatible, and fast. Despite the experimental simplicity and efficiency of the "ascorbate" procedure, the CuAAC reactions in the forms described above are simply not fast enough when the concentrations of the reactants are low, particularly in aqueous media.

The first general solution to the bioconjugation problem was provided by tris(benzyltriazolyl)methyl amine ligand **10** (TBTA, Scheme 10.3), prepared using the CuAAC reaction and introduced soon after its discovery [78]. This ligand was shown to significantly accelerate the reaction and stabilize the Cu(I) oxidation state in water-containing mixtures. After its utility in bioconjugation was demonstrated by the efficient attachment of 60 alkyne-containing fluorescent dye molecules to the azide-labeled cowpea mosaic virus [31], it was widely adopted for use with such biological entities as nucleic acids [79, 80], proteins [30, 81], *E. coli* bacteria [32, 82, 83], and mammalian cells [84, 85]. A resin-immobilized version of TBTA [86] has also been shown to be very useful in combinatorial and related experiments. The tris(*tert*-butyl) analog, TTTA, **10b**, shows superior activity to TBTA in organic solvents.

The poor solubility of TBTA in water prompted the development of more polar analogues such as **11a-c** (Scheme 10.3) [78, 87, 88]. At the same time, a combinatorial search for alternatives led to the identification of the commercially-available sulfonated bathophenanthroline **12** as the ligand component of the fastest water-soluble CuAAC catalyst under dilute aqueous conditions [89]. The unmatched activity of this system made it very useful for demanding bioconjugation tasks [41, 90–95]. However, Cu•**12** complexes are strongly electron-rich and are therefore highly susceptible to oxidation in air. Ascorbate can be used to keep the metal in the +1 oxidation state, but when exposed to air, reduction of O_2 is very fast and can easily use up all of the available reducing agent. Thus, **12** is usually used under inert atmosphere, which can be inconvenient, particularly when small amounts of biomolecule samples are used. A procedural solution was found in the use of an electrochemical cell to provide the reducing equivalents to scrub O_2 out of such reactions and maintain Cu•**12** in the cuprous oxidation state, but this was again less than optimal due to the need for extra equipment and electrolyte salts [87].

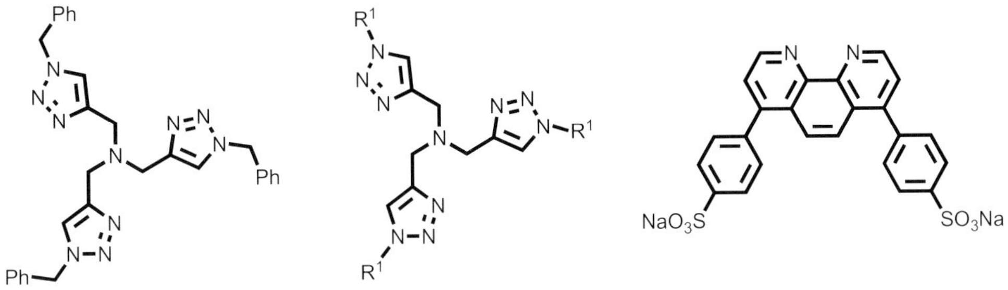

10 (**TBTA**)

11a-c
a) R = $CH_2CH_2CH_2OH$ (**THPTA**)
b) R = $CH_2CH_2CO_2H$
c) R = $CH_2C_6H_4CO_2H$

12

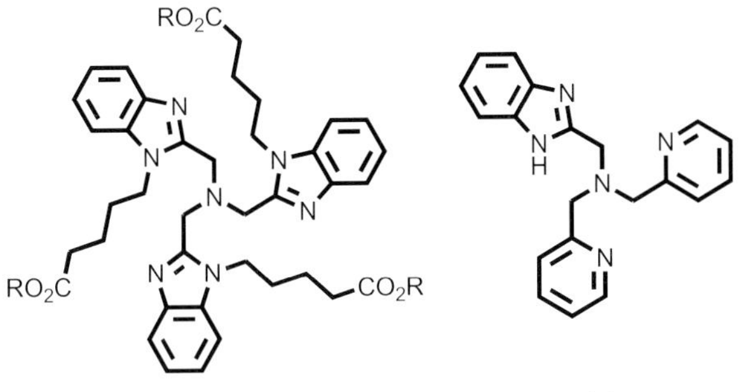

13a,b
a) R = H
b) R = Et

14

Scheme 10.3 CuAAC-accelerating ligands of choice: tris(1,2,3-triazolyl)methyl amine (TBTA), water-soluble analogues **11**, sulfonated bathophenanthroline **12**, tris(benzimidazole)methyl amine (TBIA) **13**, and "hybrid" ligand **14**.

The polydentate trimethylamine theme has been extended to benzimidazole, benzothiazole, oxazoline, and pyridine substituents [73, 96]. Several have provided significantly faster catalysis when quantitative rates are measured, particularly the pendant ester and water-soluble acid derivatives of the tris(benzimidazole) motif, **13a,b**. Recent studies have also led to the development of effective "hybrid" heterocyclic chelates such as **14**. This work has also provided significant mechanistic understanding (see below), which has led to the ability to tailor the ligand to the application. Thus, ligands **11a** and **11b** are recommended for bioconjugation reactions in media containing low concentrations of competitive Cu(I)-binding donors, whereas ligand **14** is superior when competing ligands are present [88, 97].

10.3.2
CuAAC with In Situ Generated Azides

Although organic azides are generally stable and safe compounds, those of low molecular weight can spontaneously decompose and, therefore, could be difficult or dangerous to handle. This is especially true for small molecules with several azide functionalities that would be of much interest for the generation of polyfunctionalized structures. Indeed, small-molecule azides should never be isolated away from solvent, for example by distillation, precipitation, or recrystallization. Fortunately, the CuAAC reaction is highly tolerant of all manner of additives and spectator compounds, including inorganic azide even in large excess. The process can therefore be performed in a one-pot two-step sequence, whereby an *in situ* generated organic azide is immediately consumed in a reaction with a copper acetylide (Scheme 10.4). This type of process has been implemented many times in our laboratories and others from alkyl halides or arylsulfonates by S_N2 reaction with sodium azide (Scheme 10.4A). In a recent example, Pfizer chemists developed a continuous flow process wherein a library of 1,4-disubstituted 1,2,3-triazoles was synthesized from alkyl halides, sodium azide, and terminal acetylenes, with the copper catalyst required for cycloaddition being supplied from the walls of the heated copper tubing through which the reaction solution was passed (Scheme 10.4B) [98].

Aryl and vinyl azides can also be accessed in one step from the corresponding halides or triflates via a copper-catalyzed reaction with sodium azide in the presence of catalytic amount of L-proline (Scheme 10.4C) [99]. In this fashion, a range of 1,4-disubstituted 1,2,3-triazoles can be prepared in excellent yields [100–102]. This reaction sequence can be performed at elevated temperature under microwave irradiation, reducing the reaction time to 10–30 min [50]. Anilines can also be converted to aryl azides by the reaction with *tert*-butyl nitrite and azidotrimethylsilane [103]. The resulting azides can be submitted to the CuAAC conditions without isolation, furnishing triazole products in excellent yields. Microwave heating further improves the reaction, significantly reducing reaction time [56].

10.3.3
Mechanistic Aspects of the CuAAC

We are often asked if copper is the only element that can mediate the AAC ligation. In 2005, ruthenium cyclopentadienyl complexes were found to catalyze the formation of the complementary 1,5-disubstituted triazole from azides and terminal alkynes, and also to engage internal alkynes in the cycloaddition [104]. As one would imagine from these differences, this sister process, designated RuAAC (ruthenium-catalyzed azide–alkyne cycloaddition), is mechanistically quite distinct from its cuprous cousin. Although the scope and functional group compatibility of RuAAC are excellent [105–107], the reaction is more sensitive to the solvents and the steric demands of the azide substituents than CuAAC. Applications of

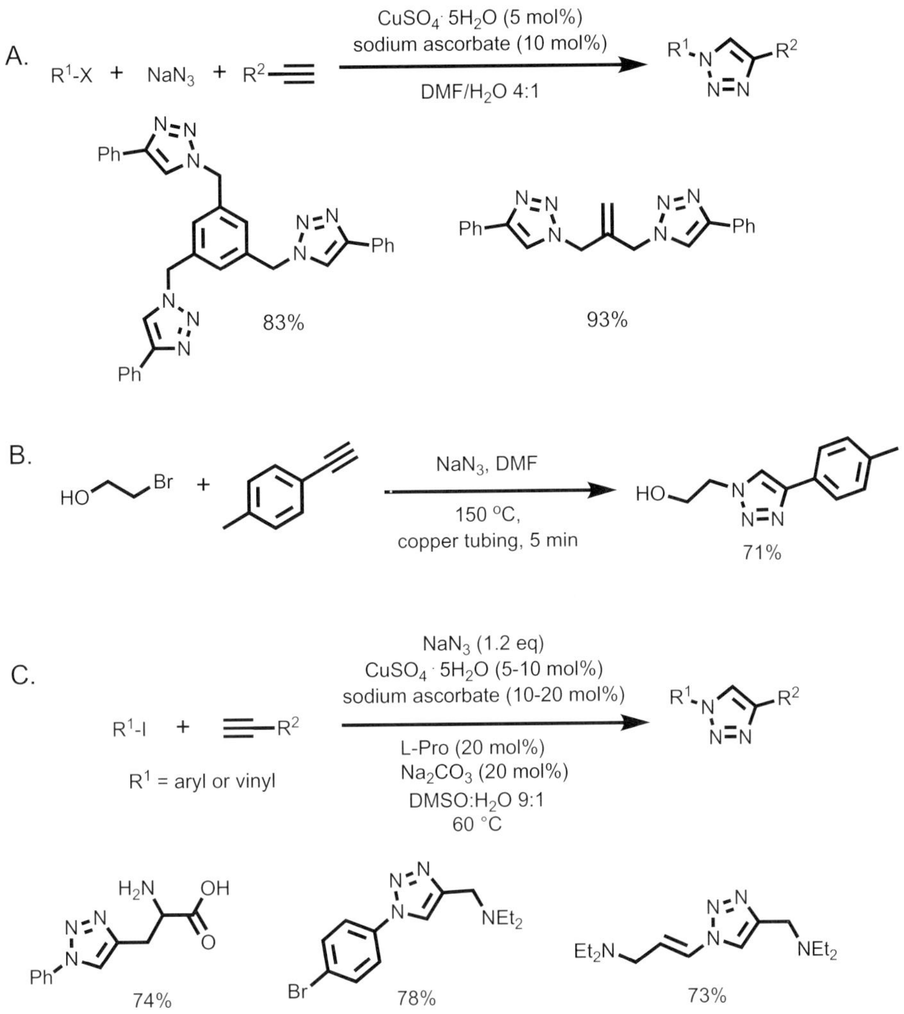

Scheme 10.4 One-pot syntheses of triazoles from halides at (A, B) sp³- and (C) sp²- carbon centers. Reaction B was performed in a flow reactor in 0.75 mm diameter Cu tubing with no added CuAAC catalyst.

RuAAC are only in their infancy [108–110]. It is not discussed further here since the focus of this book is on non-precious metals.

For the conversion of azides and terminal alkynes to 1,4-triazoles, Cu(I) remains the only active species identified so far. Surveys in our laboratories of complexes of all of the first-row transition elements and such ions as Ag(I), Pd(0/II), Pt(II), Au(I/III), and Hg(II), among others, have all failed to produce triazoles, although interesting and complex reactivity has occasionally been seen. We attribute the unique AAC-catalyzing function of Cu(I) to the fortuitous combination of an ability to engage terminal alkynes in both σ- and π-interactions (a property shared by many transition elements, mostly of the second and third row) and the ability to rapidly exchange these and other ligands in its coordination sphere (a characteristic of first-row transition metals).

The involvement of copper(I) acetylides in CuAAC was postulated early on, based on the lack of reactivity of internal alkynes under copper catalysis conditions. While the history of copper(I) acetylides dates back as far as Glaser's discovery in 1869 of oxidative dimerization of Cu-phenylacetylide [47], the precise nature of the reactive alkynyl copper species in CuAAC is not well understood. The chief complications are the tendency of copper species to form polynuclear clusters [74, 111, 112] and the great facility of the ligand exchange at the copper center. As a result, mixtures of Cu(I), terminal alkynes, and other ligands usually contain multiple organocopper species in rapid equilibrium with each other. While this may make life difficult for the mechanistic investigator, it is undoubtedly a major contributor to the remarkable adaptability of the reaction to widely different conditions.

Whatever the details of the interactions of Cu with alkyne during the CuAAC reaction, it is clear that Cu-acetylide species are easily formed and are productive components of the reaction mechanism. Early indications that azide activation was rate-determining came from the CuAAC reaction of diazide **15**, shown in Scheme 10.5, which afforded ditriazole **17** as the predominant product, even when **15** was used in excess [113]. The same phenomenon was observed for 1,1-, and 1,2-diazides, but not for 1,4-, 1,5-, and conformationally flexible 1,3-diazide analogues. The dialkyne **18**, in contrast to its diazide analogue **15**, gave statistical mixtures of mono- and di-triazoles **19** and **20** under similar conditions. Independent kinetics measurements showed that the CuAAC reaction of **16** was slightly slower than that of **15**, ruling out the intermediacy of **16** in the efficient production of **17**. The Cu-triazolyl precursor **21** is, therefore, likely to be converted to **17** very rapidly,

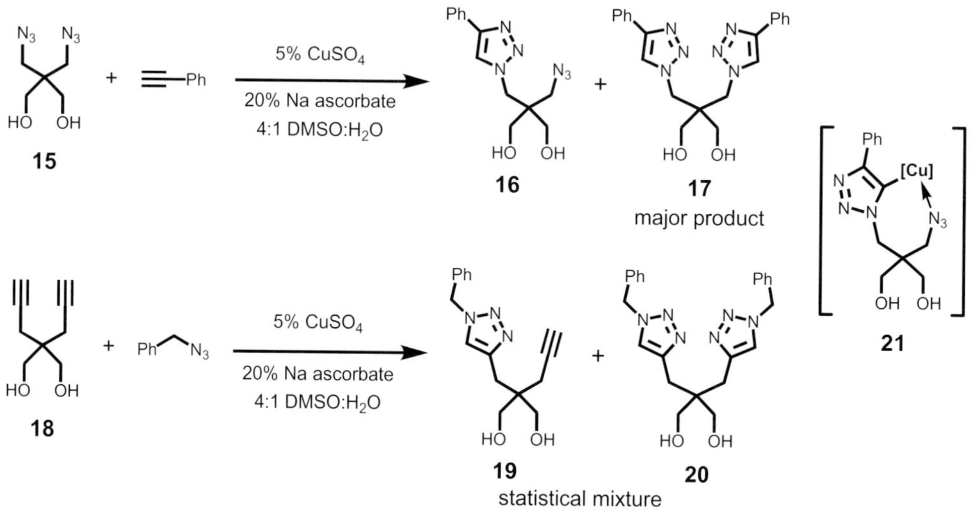

Scheme 10.5 Selective formation of ditriazole from a 1,3-diazide.

showing that azide activation, presumably by intramolecular interaction with a Cu(I) center which may already bear another alkyne as a ligand, thus dramatically accelerates the reaction.

The initial computational treatment of the CuAAC focused on the possible reaction pathways available to mononuclear copper(I) acetylides and organic azides; propyne and methyl azide were chosen for simplicity [23]. The key bond-making steps are shown in Scheme 10.6. Formation of Cu-acetylide **16** (step A) was calcu-

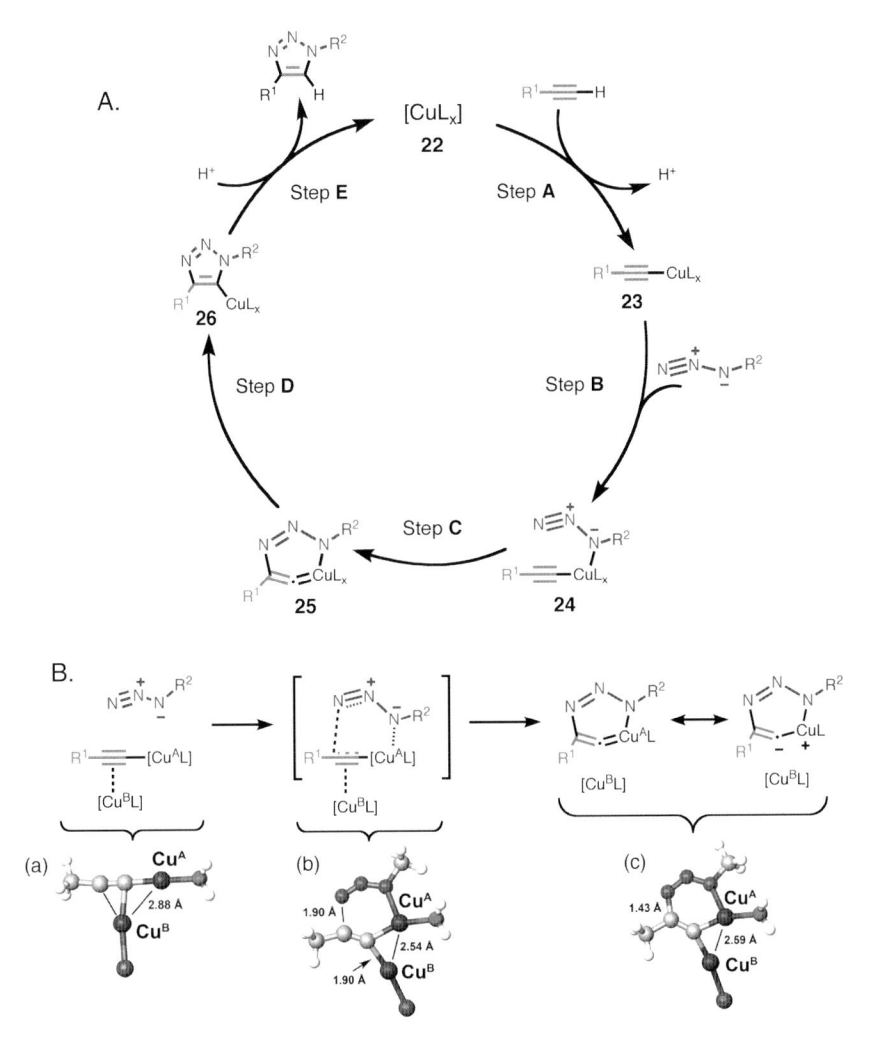

Scheme 10.6 (A) Early proposed catalytic cycle for the CuAAC reaction based on DFT calculations. (B) Introduction of a second copper(I) atom favorably influences the energetic profile of the reaction (L–H$_2$O in DFT calculations). At the bottom are shown the optimized structures for dinuclear Cu forms of (a) the starting acetylide (corresponding to **16**), (b) the transition state for the key C–N bond-forming step, and (c) the metallacycle **18**. The calculated structures are essentially identical when acetylide instead of chloride is used as the ancillary ligand on the second copper center (Cu^B).

lated to be exothermic by 11.7 kcal mol^{-1}, consistent with the well-known facility of this step which probably occurs through a π-alkyne copper complex intermediate. π-Coordination of alkyne to copper is calculated to bring the pK_a of the terminal proton down by about 10 units, bringing it into the proper range to be deprotonated in an aqueous medium. A concerted 1,3-dipolar cycloaddition of azide to the Cu-acetylide has a high calculated barrier (23.7 kcal mol^{-1}), thus the metal must play an additional role. In the proposed sequence, the azide is activated by coordination to copper (step B), forming intermediate **17**. This ligand exchange step is nearly thermoneutral computationally (2.0 kcal mol^{-1} uphill when L is water). The key bond-forming event takes place in the next step (C), when **17** is converted to the unusual 6-member copper metallacycle **18**. This step is endothermic by 12.6 kcal mol^{-1} with a calculated barrier of 18.7 kcal mol^{-1}, which corresponds roughly to the observed rate increase and is considerably lower than the barrier for the uncatalyzed reaction (approximately 26.0 kcal mol^{-1}), thus accounting for the enormous rate acceleration accomplished by Cu(I). The CuAAC reaction is, therefore, not a true concerted cycloaddition, and its regiospecificity is explained by binding of both azide and alkyne to copper prior to C–C bond formation. From **18**, the barrier for ring contraction (step D), which forms the triazolyl-copper derivative **19**, is quite low, 3.2 kcal mol^{-1}. Proteolysis (step E) of **19** releases the triazole product, thereby completing the catalytic cycle. When protected by steric bulk, Cu-triazolyls can be isolated from CuAAC reactions [114], and in rare cases of low catalyst loading and high catalytic rate, step E can be turnover-limiting [96].

The DFT investigation described above was soon followed by a study of the kinetics of the copper-mediated reaction between benzyl azide and phenylacetylene in DMSO. Under conditions of low catalyst loading and in the presence of the triazole product, the initial rate law of the reaction was second order in catalyst [113]. Second-order dependence on catalyst concentration was also observed for CuAAC reactions accelerated by *tris*(benzimidazolyl)methyl amine ligands such as **13** [96, 114]. When the CuAAC pathway was investigated by density functional theory (B3LYP) taking into account the possibility of the involvement of dinuclear copper(I) acetylides, a substantial further drop in the activation barrier (by approximately 3–6 kcal mol^{-1}) was revealed (Scheme 10.6B) [115, 116]. In this study, two transition states containing a second copper(I) center were located, differing only in the spectator ligand (acetylide or chloride) on the second copper atom (Scheme 10.6C). In both structures the second copper atom, Cu^B, engages in a strong interaction with the reacting proximal acetylide carbon (C^1), indicated by the short Cu–C distances of 1.93 and 1.90 Å [117].

It is, therefore, possible that the first job of a CuAAC-supporting ligand is to channel the coordination chemistry of Cu(I) to allow the formation of a σ,π-dinuclear acetylide complex with an open coordination site to accommodate the relatively weak azide donor. Given the dynamic and complex nature of Cu(I) ligand exchange processes, this is a delicate balancing act that gives rise to discontinuous kinetic behavior, wherein the nature of the catalyst, its concentration, and, consequently, the observed rate law, are changing as the reaction progresses [118].

General conclusions that have emerged are as follows. First, successful ligands need to balance the competing requirements of binding Cu(I) tightly in a way that allows a binuclear complex to assemble and allowing azide to access the coordination sphere of the Cu center. TBTA and related families of ligands successfully meet these challenges by bringing to bear a *high local concentration of a weak donor ligand* on the third "arm" of the structure. Too potent a chelator would tie up the necessary Cu coordination sites and too weak a ligand would not break up non-productive coordination complexes. The minimal motif that we believe provides for CuAAC acceleration, and which is accessible using tripodal molecules such as **11** and **14**, is depicted in Scheme 10.7. These ligands also accomplish the delicate task of providing enough electron density to the metal to promote the stepwise cycloaddition process while not destabilizing the Cu(I) oxidation state.

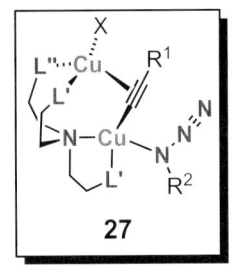

Scheme 10.7 Two general types of CuAAC conditions defined by the presence of competing donors for the metal centers, and recommended CuAAC-accelerating ligands. At the right is shown a generalized proposed structure of the active binuclear complex promoting C–N bond formation.

Second, the strength of the donor ligands must be assessed relative to competing donors in solution. Among the heterocycle components used in ligands, triazole is a weaker donor than benzimidazole or 2-pyridyl, as judged by their relative protic basicities [96]. DMSO, DMF, and NMP are examples of water-compatible solvents that compete well for Cu coordination sites. In solvent mixtures rich in these components ("strong-donor" conditions), ligands containing stronger donor arms are best, with the benzimidazle-pyridine compound **14** recently identified as a superior example [97]. Contrariwise, in solvent mixtures dominated by water ("weak-ligand" conditions), the tris(triazolylmethyl) amines such as water-soluble **11a** (tris(hydroxypropyltriaolylmethyl)amine, THPTA) work well. In addition to donor solvent moieties, strong-donor conditions can be established by the presence of metal-binding groups in the substrate, as sometimes occurs with proteins, or by the use of high concentrations of alkyne (usually >20 mM, which is rare in bioconjugation applications). These two general classes of reaction conditions are summarized in Scheme 10.7.

An additional advantage occasions the use of relatively weak chelates such as **11a** in the context of bioconjugation. While ascorbate is a surpassingly effective and convenient reducing agent for Cu(II) in water, a possible side product of the reaction is hydrogen peroxide. Its formation has been observed in acetate and phosphate buffer solutions (but not in the presence of chloride). These oxidizing byproducts can be harmful to the integrity of proteins and other biomolecules [88]. When ligands such as **11a** are present, they act as sacrificial reductants, intercepting the reactive oxygen species in the coordination sphere of the metal and protecting the other sensitive molecules in solution. An excess of ligand is therefore necessary, and, fortunately, such excesses do not inhibit the CuAAC reaction rate very much. With this and a few other modifications to the procedure, a robust and highly effective bioconjugative CuAAC protocol has been reported [88].

10.3.4
Reactions of Sulfonyl Azides

Sulfonyl azides participate in unique CuAAC reactions with terminal alkynes. Depending on the conditions and reagents, products other than the expected triazole **28** [119] can be obtained, as shown in Scheme 10.8. For example, N-sulfonyl azides are converted to N-sulfonyl amidines **29** when the reaction is conducted in the presence of amines [120]. In the aqueous conditions, N-acyl sulfonamides **30** are the major products [121, 122].

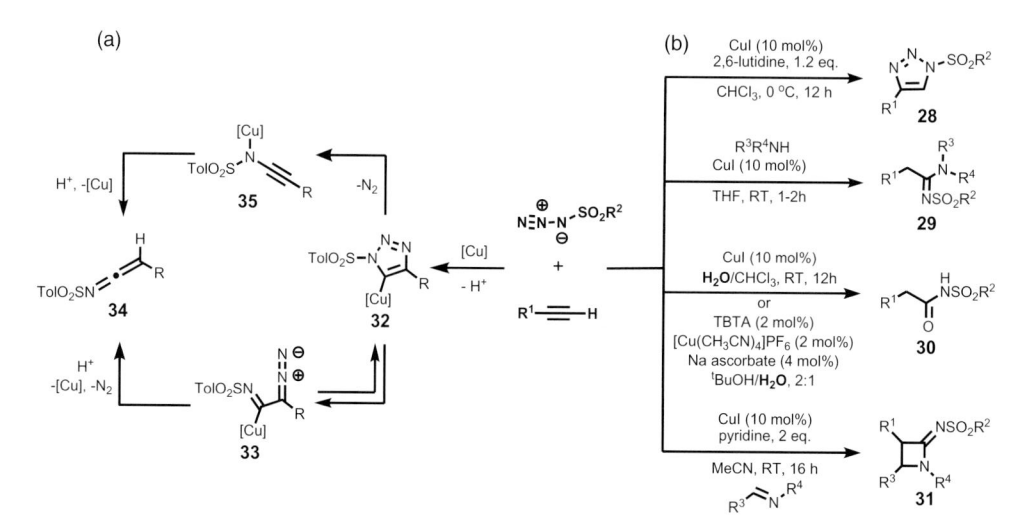

Scheme 10.8 (a) Products of CuAAC reactions with sulfonyl azides. (b) Possible pathways leading to ketenimine intermediates.

These products are thought to derive from the cuprated triazole intermediate **32**, which is destabilized by the strong electron-withdrawing character of the *N*-sulfonyl substituent. Ring-chain isomerization can occur to form the cuprated diazoimine **33** which, upon the loss of a molecule of dinitrogen, furnishes the *N*-sulfonyl ketenimine **34** [123]. Alternatively, copper(I) alkynamide **35** can be generated with a concomitant elimination of N_2 and, after protonation, would again generate the reactive ketenimine species **34**. In addition to amines and water, the latter can be trapped with imines, furnishing an *N*-sulfonyl azetidinimines **31** (Scheme 10.8) [123].

10.3.5
Copper-Catalyzed Reactions with Other Dipolar Species

In the mechanism of the CuAAC reaction described above, the metal catalyst activates terminal alkyne for reaction with a Cu-coordinated azide. This mode of reactivity operates with other dipolar reagents as well. In fact, the first example of a copper-catalzyed 1,3-dipolar cycloaddition reaction of alkynes was reported for nitriones by Kinugasa in 1972 [124]. An asymmetric version of the Kinugasa reaction was developed by Fu *et al.* in 2002 [125, 126].

Another early example that followed the discovery of CuAAC, the copper-catalyzed reaction of nitrile oxides, is shown in Scheme 10.9. Similarly to azides, the uncatalyzed 1,3-dipolar cycloaddition of nitrile oxides and acetylenes has long been known, but its applications to the synthesis of the corresponding heterocycle (isoxazoles) are scarce. Yields of isoxazole products are often quite low, side reactions are common, and both regioisomers may be formed (although the selectivity of nitrile oxide cycloadditions is usually higher than in reactions of azides, favoring the 3,5-isomer) [127]. Furthermore, nitrile oxides are not very stable and readily dimerize.

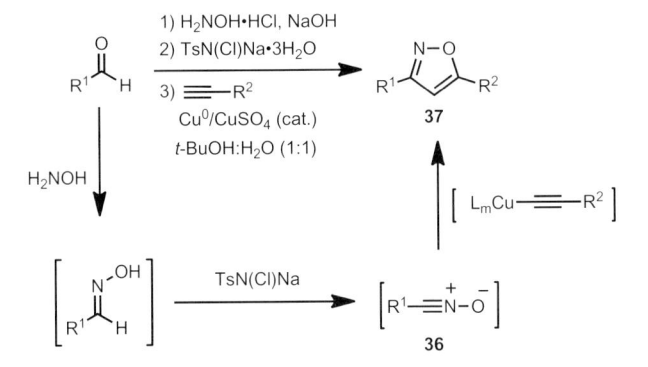

Scheme 10.9 Synthesis of 3,5-disubstituted isoxazoles from aldehydes, hydroxylamine, and terminal alkynes.

In contrast, Cu(I) catalysis makes possible the efficient synthesis of 3,5-disubstituted isoxazoles **57** from aromatic or aliphatic aldehydes and alkynes. Stable nitrile oxides can be isolated and subsequently submitted to the reaction [128] in isolated form and submitted to the reaction in a one-pot, three-step process [129]. Here, nitrile oxide intermediates (**56**) are generated *in situ* via the corresponding aldoxime and halogenation/deprotonation by Chloramine-T [130]. Capture of the intermediate nitrile oxide by copper(I) acetylides occurs, presumably, before dimerization. In this case, the Cu catalyst was obtained from copper metal and copper(II) sulfate, and the products were isolated by simple filtration or aqueous work-up. Trace amounts of toluenesulfonamide and unreacted acetylene are easily removed by recrystallization or by passing the product through a short plug of silica gel.

Reactions of azomethine imines with alkynes catalyzed by copper yielding fused nitrogen heterocycles have also been reported [131].

10.3.6
Examples of Application of the CuAAC Reaction

The CuAAC reaction has been applied to a remarkable array of problems in synthetic chemistry, chemical biology, materials science, and other fields. A comprehensive or even representative list is beyond the scope of this chapter. Instead, we will highlight two examples involving metallic copper as the source of CuAAC catalyst. While decidedly not typical in the body of CuAAC applications in the literature, we believe that this most convenient form of this inexpensive metal should receive greater attention in azide–alkyne ligation reactions. We also hope that these examples will give the reader some indication of the facility with which the CuAAC process can be applied.

10.3.6.1 Synthesis of Compound Libraries for Biological Screening

The CuAAC process performs well in most common laboratory solvents and usually does not require protection from oxygen and water. Indeed, as noted above, aqueous solvents are commonly used and, in many cases, result in cleaner isolated products. The reaction is therefore an ideal tool for the synthesis of libraries for initial screening as well as focused sets of compounds for structure–activity profiling. The lack of byproducts and high conversions often permit testing of CuAAC reaction mixtures without further purification. When necessary, traces of copper can be removed by solid phase extraction utilizing a metal-scavenging resin or simple filtration through a plug of silica gel.

In a study aimed at the discovery of inhibitors of human fucosyltransferase, CuAAC was used to link 85 azides to the GDP-derived alkyne **37** with excellent yields (Scheme 10.10) [132]. The library was screened directly, and a nanomolar inhibitor **38** was identified. Testing the purified hit compound **38** against several glycosyltransferases and nucleotide binding enzymes revealed that it was the most potent and selective inhibitor of human α-1,3-fucosyltransferase VI at that time.

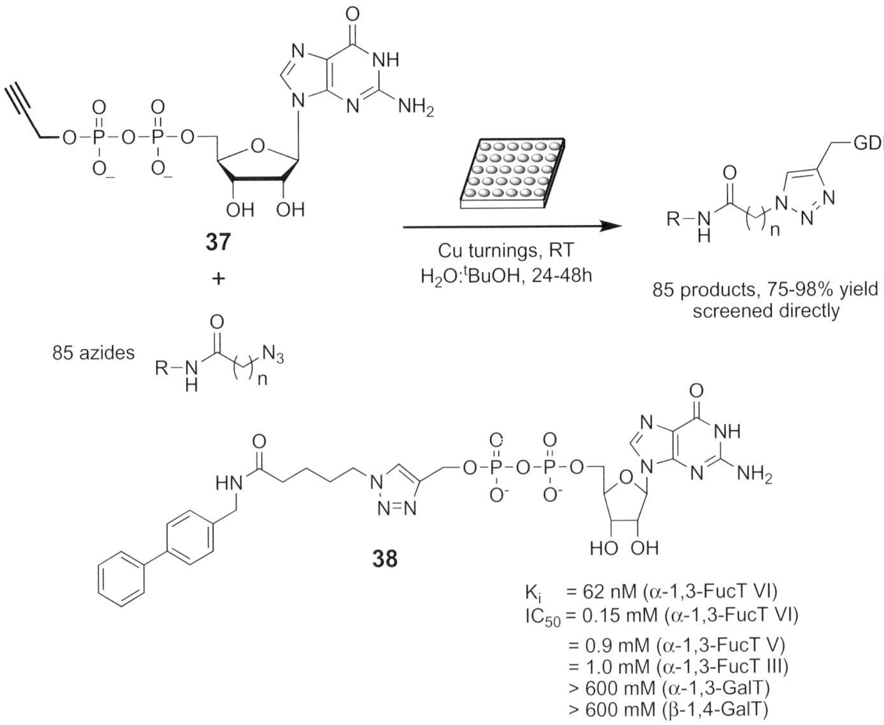

Scheme 10.10 Cu-mediated synthesis and direct screening of fucosyl transferase inhibitors.

Scheme 10.11 shows another example, in which a novel family of potent HIV-1 protease inhibitors was discovered using the CuAAC reaction [133]. A focused library of azide-containing fragments was united with a diverse array of functionalized alkyne-containing building blocks. After the direct screening of the crude reaction products, a lead structure **39** with K_i = 98 nM was identified. Optimization of both azide and alkyne fragments was equally facile (compound **40**, K_i = 23 nM). Further functionalization of the triazole at C-5 gave a series of compounds with increased activity, exhibiting K_i values as low as 8 nM (compound **41**).

10.3.6.2 Copper-Binding Adhesives

Since Cu metal provides useful amounts of Cu(I) catalyst, and triazoles are major components of metal adhesives and coatings, we investigated the formation of crosslinked polymeric materials by the deposition of multivalent azides and alkynes onto Cu-containing metallic surfaces [134–136]. The general design is shown in Scheme 10.12. Strong adhesives are formed over curing times ranging from many hours in the absence of added accelerating ligand to minutes in the

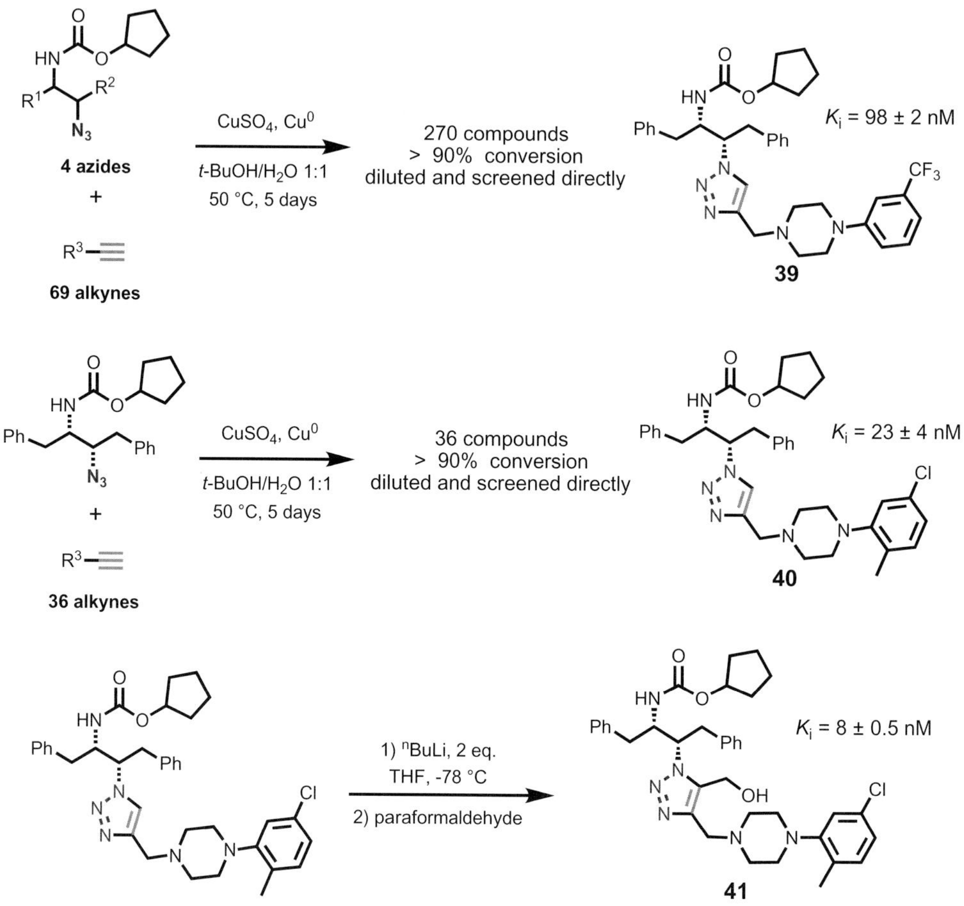

Scheme 10.11 Cu-mediated synthesis and direct screening of HIV protease inhibitors.

presence of added Cu, ligands, and at elevated temperatures. Copper ions are extracted from the metal surface and distribute throughout the polymer matrix, presumably forming active catalytic sites as triazoles are created, which function as accelerating ligands as well as stable linkages and adhesive units to the metal surface. This results in very high levels of crosslinking, giving rise to glass transition temperatures that are significantly greater than the curing temperature at which the adhesives are formed [136]. In the context of macroscopic materials, therefore, as well as among competing molecules in solution, a kinetically labile metal catalyst can have significant advantages.

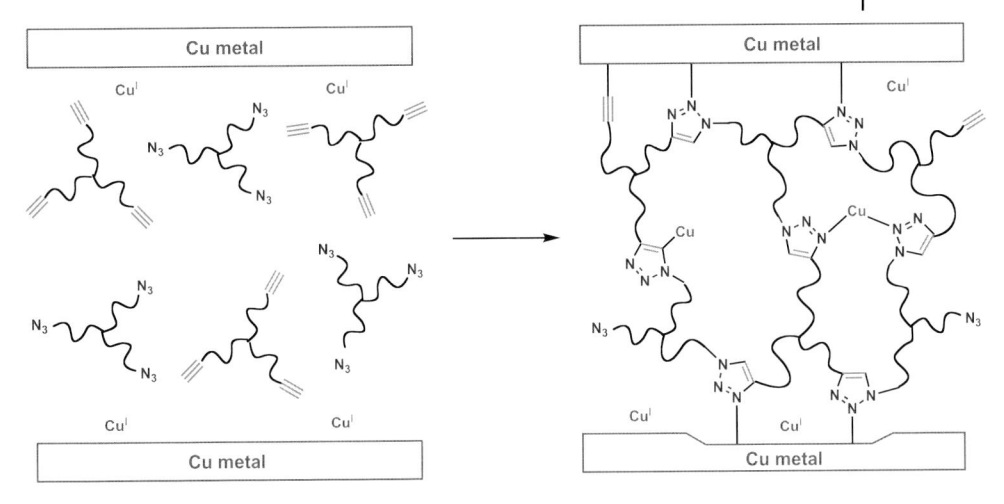

Scheme 10.12 Spontaneous formation of Cu-binding adhesives from polyvalent azides and alkynes placed between two metal surfaces.

10.3.7
Representative Experimental Procedures

General procedure 1, *"aqueous ascorbate" method for the synthesis of 1,2,3-triazoles* [6, 23]. 17-Ethynyl estradiol (888 mg, 3 mmol) and (S)-3-azidopropane-1,2-diol (352 mg, 3 mmol) were suspended in 12 mL of 1:1 water/*tert*-butanol mixture. Sodium ascorbate (0.3 mmol, 300 μL of freshly prepared 1 M solution in water) was added, followed by copper(II) sulfate pentahydrate (7.5 mg, 0.03 mmol, in 100 μL of water). The heterogeneous mixture was stirred vigorously overnight, at which point it cleared and TLC analysis indicated complete consumption of the reactants. The reaction mixture was diluted with 50 mL of water, cooled in ice, and the white precipitate was collected by filtration. After washing with cold water (2 × 25 mL), the precipitate was dried under vacuum to afford 1.17 g (94%) of pure product as off-white powder.

General procedure 2, *copper metal-catalyzed synthesis of 1,2,3-triazoles* [6, 23]. Phenylacetylene (2.04 g, 20 mmol) and 2,2-bis(azidomethyl)propane-1,3-diol (1.86 g, 10 mmol) were dissolved in 1:2 *tert*-butanol/water mixture (50 mL). About 1 g of copper metal turnings was added, and the reaction mixture was stirred for 24 h, after which time TLC analysis indicated complete consumption of starting materials. Copper was removed, and the white product was filtered off, washed with water, and dried to yield 3.85 g (98%) of pure bis-triazole product.

General procedure 3, *general bioconjugation procedure in aqueous buffer* [88]. The example here is generalized to a "biomolecule-alkyne" and a "cargo-azide," where the biomolecule can be peptide, protein, or polynucleotide, and the "cargo" can be any small- or large-molecule moiety that the user wishes to couple to the biomolecule. To 25 μL of a stock solution of a biomolecule-alkyne in a 2 mL Eppendorf tube was added (in the following order) 407.5 μL of 100 mM phosphate buffer (pH 7), 10 μL of the "cargo"-azide, 7.5 μL of a premixed solution of $CuSO_4$ (2.5 μL of 20 mM aqueous stock) and ligand **11a** (5.0 μL of 50 mM aqueous stock), 25 μL of aminoguanidine stock solution (100 mM in water), and 25 μL of sodium ascorbate (100 mM freshly-made solution in water). The final concentrations of the biomolecule and cargo can be approximately 2 μM and higher with this procedure; the final Cu concentration here is 0.1 mM but can be adjusted between 0.05 and 0.25 mM; and the final concentration of ligand **11a** is five times that of Cu. Workup depends on the application, but often the product is purified directly in such a way as to leave small molecules behind. Copper ions can be removed by washing or dialysis with aqueous EDTA.

General Procedure 4, copper-catalyzed synthesis of isoxazoles [137]. *trans*-Cinnamaldehyde (2.6 g, 20 mmol) was added to a solution of hydroxylamine hydrochloride (1.46 g, 21 mmol) in 80 mL of 1:1 *t*-BuOH:H_2O. To this was added NaOH (0.84 g, 21 mmol), and after stirring for 30 min at ambient temperature, TLC analysis indicated that oxime formation was complete. Chloramine-T trihydrate (5.9 g, 21 mmol) was added in small portions over 5 min, followed by $CuSO_4 \cdot 5H_2O$ (0.15 g, 0.6 mmol) and copper turnings (*ca.* 50 mg). 1-Ethynylcyclohexene (2.23 g, 21 mmol) was added, pH was adjusted to about 6 by addition of a few drops of 1 M NaOH, and stirring was continued for another 6 h. The reaction mixture was poured into ice/water (150 mL), and 10 mL of dilute NH_4OH was added to remove all copper salts. The product was collected by filtration, redissolved, and passed through a short plug of silica gel (ethyl acetate:hexanes 1:6, $R_F = 0.6$) affording 3.6 g (72%) of 5-cyclohex-1-enyl-3-styrylisoxazole as an off-white solid, m.p. 138–139 °C.

Acknowledgements

We are very grateful to the many coworkers and colleagues who have contributed to the work cited here (notably Maarten Ahlquist, Michael Cassidy, Timothy Chan, Sukbok Chang, Christoph Fahrni, Alina Feldman, Luke Green, Jason Hein, Fahmi Himo, Vu Hong, Warren Lewis, John Muldoon, Louis Noodleman, Stanislav Presolski, Jessica Raushel, Valentin Rodionov, Vsevolod Rostovtsev, Matthew Whiting, and Peng Wu), and to the many colleagues around the world who have given us feedback about the performance of the CuAAC process in their hands. Most important of all has been Professor K. Barry Sharpless, whose inspiration and insights on matters large and small are responsible for the development of click chemistry in general and of this reaction in particular.

References

1 Huisgen, R. (1963) *Angew. Chem. Int. Ed.*, **2**, 565–598.
2 Padwa, A. (ed.) (1984) *1,3-Dipolar Cycloaddition Chemistry*, John Wiley & Sons, Inc., New York.
3 Padwa, A., and Pearson, W.H. (eds) (2002) *Synthetic Applications of 1,3 Dipolar Cycloaddition Chemistry Toward Heterocycles and Natural Products*, (eds), John Wiley & Sons, Inc., New York.
4 Kolb, H.C., Finn, M.G., and Sharpless, K.B. (2001) *Angew. Chem. Int. Ed. Engl.*, **40**, 2004–2021.
5 Tornøe, C.W., Christensen, C., and Meldal, M. (2002) *J. Org. Chem.*, **67**, 3057–3062.
6 Rostovtsev, V.V., Green, L.G., Fokin, V.V., and Sharpless, K.B. (2002) *Angew. Chem. Int. Ed. Engl.*, **41**, 2596–2599.
7 Bock, V.D., Hiemstra, H., and van Maarseveen, J.H. (2006) *Eur. J. Org. Chem.*, 51–68.
8 Fokin, V.V. (2007) *ACS Chem. Biol.*, **2**, 775–778.
9 Moses, J.E., and Moorhouse, A.D. (2007) *Chem. Soc. Rev.*, **36**, 1249–1262.
10 Johnson, J.A., Koberstein, J.T., Finn, M.G., and Turro, N.J. (2008) *Macromol. Rapid Commun.*, **29**, 1052–1072.
11 Lutz, J.-F., and Zarafshani, Z. (2008) *Adv. Drug Deliv. Rev.*, **60**, 958–970.
12 Meldal, M., and Tornøe, C.W. (2008) *Chem. Rev.*, **108**, 2952–3015.
13 Tron, G.C., Pirali, T., Billington, R.A., Canonico, P.L., Sorba, G., and Genazzani, A.A. (2008) *Med. Res. Rev.*, **28**, 278–308.
14 For a continuously updated list of applications of the CuAAC and a compehensive list of reaction conditions, the reader is referred to http://www.scripps.edu/chem/fokin/cuaac.
15 Michael, A. (1893) *J. Prakt. Chem./Chem. Ztg.*, **48**, 94.
16 Huisgen, R., and Blaschke, H. (1965) *Chem. Ber.*, **98**, 2985–2997.
17 Huisgen, R., Knorr, R., Moebius, L., and Szeimies, G. (1965) *Chem. Ber.*, **98**, 4014–4021.
18 Huisgen, R., Moebius, L., and Szeimies, G. (1965) *Chem. Ber.*, **98**, 1138–1152.
19 Huisgen, R., Szeimies, G., and Moebius, L. (1966) *Chem. Ber.*, **99**, 475–490.
20 Huisgen, R., Szeimies, G., and Moebius, L. (1967) *Chem. Ber.*, **100**, 2494–2507.
21 Huisgen, R. (1984) *1,3-Dipolar Cycloaddition Chemistry*, vol. 1 (ed. A. Padwa), John Wiley & Sons, Inc., New York, pp. 1–176.
22 Huisgen, R. (1989) *Pure Appl. Chem.*, **61**, 613–628.
23 Himo, F., Lovell, T., Hilgraf, R., Rostovtsev, V.V., Noodleman, L., Sharpless, K.B., and Fokin, V.V. (2005) *J. Am. Chem. Soc.*, **127**, 210–216.
24 Wu, P., and Fokin, V.V. (2007) *Aldrichim. Acta*, **40**, 7–17.
25 Tomé, A.C. (2004) Five-membered hetarenes with three or more heteroatoms, in *Science of Synthesis: Houben-Weyl Methods of Molecular Transformations*, vol. 13 (eds R.C. Storr and T.L. Gilchrist), Thieme, Stuttgart, pp. 415–601.
26 Krivopalov, V.P., and Shkurko, O.P. (2005) *Russ. Chem. Rev.*, **74**, 339–379.
27 Brik, A., Muldoon, J., Lin, Y.-c., Elder, J.H., Goodsell, D.S., Olson, A.J., Fokin, V.V., Sharpless, K.B., and Wong, C.-h. (2003) *Chembiochem*, **4**, 1246–1248.
28 Brik, A., Alexandratos, J., Lin, Y.-C., Elder, J.H., Olson, A.J., Wlodawer, A., Goodsell, D.S., and Wong, C.-H. (2005) *Chembiochem*, **6**, 1167–1169.
29 Tam, A., Arnold, U., Soellner, M.B., and Raines, R.T. (2007) *J. Am. Chem. Soc.*, **129**, 12670–12671.
30 Speers, A.E., Adam, G.C., and Cravatt, B.F. (2003) *J. Am. Chem. Soc.*, **125**, 4686–4687.
31 Wang, Q., Chan, T.R., Hilgraf, R., Fokin, V.V., Sharpless, K.B., and Finn, M.G. (2003) *J. Am. Chem. Soc.*, **125**, 3192–3193.
32 Link, A.J., Vink, M.K.S., and Tirrell, D.A. (2004) *J. Am. Chem. Soc.*, **126**, 10598–10602.
33 Binder, W.H., and Kluger, C. (2006) *Curr. Org. Chem.*, **10**, 1791–1815.

34 Nebhani, L., and Barner-Kowollik, C. (2009) *Adv. Mater.*, **21**, 1–27.
35 Kolb, H.C., and Sharpless, K.B. (2003) *Drug Disc. Today*, **8**, 1128–1137.
36 Gierlich, J., Burley, G.A., Gramlich, P.M.E., Hammond, D.M., and Carell, T. (2006) *Org. Lett.*, **8**, 3639–3642.
37 Goodall, G.W., and Hayes, W. (2006) *Chem. Soc. Rev.*, **35**, 280–312.
38 Nandivada, H., Chen, H.-Y., Bondarenko, L., and Lahann, J. (2006) *Angew. Chem. Int. Ed.*, **45**, 3360–3363.
39 Spruell, J.M., Sheriff, B.A., Rozkiewicz, D.I., Dichtel, W.R., Rohde, R.D., Reinhoudt, D.N., Stoddart, J.F., and Heath, J.R. (2008) *Angew. Chem. Int. Ed.*, **47**, 9927–9932.
40 Kubas, G.J. (1979) *Inorg. Synth.*, **19**, 90–92.
41 Sen Gupta, S., Kuzelka, J., Singh, P., Lewis, W.G., Manchester, M., and Finn, M.G. (2005) *Bioconjug. Chem.*, **16**, 1572–1579.
42 Reddy, K.R., Rajgopal, K., and Kantam, M.L. (2006) *Synlett*, 957–959.
43 Fukuzawa, S., Shimizu, E., and Kikuchi, S. (2007) *Synlett*, 2436–2438.
44 Reddy, K.R., Rajgopal, K., and Kantam, M.L. (2007) *Catal. Lett.*, **114**, 36–40.
45 Nigh, W.G. (1973) *Oxidation in Organic Chemistry*, vol. Part B (ed. W.S. Trahanovsky), Academic Press, New York, pp. 1–95.
46 Siemsen, P., Livingston, R.C., and Diederich, F. (2000) *Angew. Chem. Int. Ed.*, **39**, 2632–2657.
47 Glaser, C. (1869) *Chem. Ber.*, **2**, 422.
48 Glaser, C. (1870) *Ann. Chem. Pharm.*, **154**, 159.
49 Fahrni, C.J. (2007) *Curr. Opin. Chem. Biol.*, **11**, 121–127.
50 Appukkuttan, P., Dehaen, W., Fokin, V.V., and Van der Eycken, E. (2004) *Org. Lett.*, **6**, 4223–4225.
51 Ermolat'ev, D., Dehaen, W., and Van der Eycken, E. (2004) *QSAR Comb. Sci.*, **23**, 915–918.
52 Khanetskyy, B., Dallinger, D., and Kappe, C.O. (2004) *J. Comb. Chem.*, **6**, 884–892.
53 Bouillon, C., Meyer, A., Vidal, S., Jochum, A., Chevolot, Y., Cloarec, J.-P., Praly, J.-P., Vasseur, J.-J., and Morvan, F. (2006) *J. Org. Chem.*, **71**, 4700–4702.
54 Appukkuttan, P., and Van der Eyeken, E. (2008) *Eur. J. Org. Chem.*, 1133–1155.
55 Lucas, R., Neto, V., Bouazza, A.H., Zerrouki, R., Granet, R., Krausz, P., and Champavier, Y. (2008) *Tetrahedron Lett.*, **49**, 1004–1007.
56 Moorhouse, A.D., and Moses, J.E. (2008) *Synlett*, 2089–2092.
57 Pachon, L.D., van Maarseveen, J.H., and Rothenberg, G. (2005) *Adv. Synth. Catal.*, **347**, 811–815.
58 Molteni, G., Bianchi, C.L., Marinoni, G., Santo, N., and Ponti, A. (2006) *New J. Chem.*, **30**, 1137–1139.
59 Lipshutz, B.H., and Taft, B.R. (2006) *Angew. Chem. Int. Ed.*, **45**, 8235–8238.
60 Pearson, R.G. (1988) *J. Am. Chem. Soc.*, **110**, 7684–7690.
61 Hathaway, B.J. (1987) *Comprehensive Coordination Chemistry*, vol. **5** (ed. G. Wilkinson), Pergamon Press, Oxford, pp. 533–757.
62 Sivasankar, C., Sadhukhan, N., Bera, J.K., and Samuelson, A.G. (2007) *New J. Chem.*, **31**, 385–393.
63 Perez-Balderas, F., Ortega-Munoz, M., Morales-Sanfrutos, J., Hernandez-Mateo, F., Calvo-Flores, F.G., Calvo-Asin, J.A., Isac-Garcia, J., and Santoyo-Gonzalez, F. (2003) *Org. Lett.*, **5**, 1951–1954.
64 Wu, P., Feldman, A.K., Nugent, A.K., Hawker, C.J., Scheel, A., Voit, B., Pyun, J., Frechet, J.M.J., Sharpless, K.B., and Fokin, V.V. (2004) *Angew. Chem. Int. Ed.*, **43**, 3928–3932.
65 Malkoch, M., Schleicher, K., Drockenmuller, E., Hawker, C.J., Russell, T.P., Wu, P., and Fokin, V.V. (2005) *Macromolecules*, **38**, 3663–3678.
66 Gonda, Z., and Novák, Z. (2010) *Dalton Trans.*, **39**, 726–729.
67 Campbell-Verduyn, L.S., Mirfeizi, L., Dierckx, R.A., Elsinga, P.H., and Feringa, B.L. (2009) *Chem. Commun.*, 2139–2141.
68 Detz, R.J., Heras, S.A., De Gelder, R., van Leeuwen, P.W.N.M., Hiemstra, H., Reek, J.N.H., and Van Maarseveen, J.H. (2006) *Org. Lett.*, **8**, 3227–3230.
69 Díez-González, S., Correa, A., Cavallo, L., and Nolan, S.P. (2006) *Chem. Eur. J.*, **12**, 7558–7564.

70 Díez-González, S., and Nolan, S.P. (2008) *Angew. Chem. Int. Ed.*, **46**, 9013–9016.
71 Li, P., Wang, L., and Zhang, Y. (2008) *Tetrahedron*, **64**, 10825–10830.
72 Bai, S.-Q., Koh, L.L., and Hor, T.S.A. (2009) *Inorg. Chem.*, **48**, 1207–1213.
73 Rodionov, V.O., Presolski, S.I., Gardinier, S., Lim, Y.-H., and Finn, M.G. (2007) *J. Am. Chem. Soc.*, **129**, 12696–12704.
74 Mykhalichko, B.M., Temkin, O.N., and Mys'kiv, M.G. (2001) *Russ. Chem. Rev.*, **69**, 957–984.
75 Candelon, N., Lastécoueres, D., Diallo, A.K., Aranzaes, J.R., Astruc, D., and Vincent, J.-M. (2008) *Chem. Commun.*, 741–743.
76 Özcubukcu, S., Ozkal, E., Jimeno, C., and Pericás, M.A. (2009) *Org. Lett.*, **11**, 4680–4683.
77 Saxon, E., and Bertozzi, C.R. (2000) *Science*, **287**, 2007–2010.
78 Chan, T.R., Hilgraf, R., Sharpless, K.B., and Fokin, V.V. (2004) *Org. Lett.*, **6**, 2853–2855.
79 Weller, R.L., and Rajski, S.R. (2005) *Org. Lett.*, **7**, 2141–2144.
80 Burley, G.A., Gierlich, J., Mofid, M.R., Nir, H., Tal, S., Eichen, Y., and Carell, T. (2006) *J. Am. Chem. Soc.*, **128**, 1398–1399.
81 Speers, A.E., and Cravatt, B.F. (2004) *Chem. Biol.*, **11**, 535–546.
82 Link, A.J., and Tirrell, D.A. (2003) *J. Am. Chem. Soc.*, **125**, 11164–11165.
83 Beatty, K.E., Xie, F., Wang, Q., and Tirrell, D.A. (2005) *J. Am. Chem. Soc.*, **127**, 14150–14151.
84 Dieterich, D.C., Link, A.J., Graumann, J., Tirrell, D.A., and Schuman, E.M. (2006) *Proc. Natl. Acad. Sci. USA*, **103**, 9482–9487.
85 Sawa, M., Hsu, T.-L., Itoh, T., Sugiyama, M., Hanson Sarah, R., Vogt Peter, K., and Wong, C.-H. (2006) *Proc. Natl. Acad. Sci. USA*, **103**, 12371–12376.
86 Chan, T.R., and Fokin, V.V. (2007) *QSAR Comb. Sci.*, **26**, 1274–1279.
87 Hong, V., Udit, A.K., Evans, R.A., and Finn, M.G. (2008) *Chembiochem*, **9**, 1481–1486.
88 Hong, V., Presolski, S.I., Ma, C., and Finn, M.G. (2009) *Angew. Chem. Int. Ed.*, **48**, 9879–9883.
89 Lewis, W.G., Magallon, F.G., Fokin, V.V., and Finn, M.G. (2004) *J. Am. Chem. Soc.*, **126**, 9152–9153.
90 Prasuhn, D.E., Jr., Yeh, R.M., Obenaus, A., Manchester, M., and Finn, M.G. (2007) *Chem. Commun.*, 1269–1271.
91 Zeng, Q., Li, T., Cash, B., Li, S., Xie, F., and Wang, Q. (2007) *Chem. Commun.*, 1453–1455.
92 Megiatto, J.D., Jr., and Schuster, D.I. (2008) *J. Am. Chem. Soc.*, **130**, 12872–12873.
93 Prasuhn, D.E., Jr., Singh, P., Strable, E., Brown, S., Manchester, M., and Finn, M.G. (2008) *J. Am. Chem. Soc.*, **130**, 1328–1334.
94 Schoffelen, S., Lambermon, M.H.L., Van Eldijk, M.B., and Van Hest, J.C.M. (2008) *Bioconjug. Chem.*, **19**, 1127–1131.
95 Strable, E., Prasuhn, D.E., Jr, Udit, A.K., Brown, S., Link, A.J., Ngo, J.T., Lander, G., Quispe, J., Potter, C.S., Carragher, B., Tirrell, D.A., and Finn, M.G. (2008) *Bioconjug. Chem.*, **19**, 866–875.
96 Rodionov, V.O., Presolski, S.I., Diaz, D.D., Fokin, V.V., and Finn, M.G. (2007) *J. Am. Chem. Soc.*, **129**, 12705–12712.
97 Presolski, S., Hong, V., Cho, S.-H., and Finn, M.G., unpublished results.
98 Bogdan, A.R., and Sach, N.W. (2009) *Adv. Synth. Catal.*, **351**, 849–854.
99 Zhu, W., and Ma, D. (2004) *Chem. Commun.*, 888–889.
100 Feldman, A.K., Colasson, B., and Fokin, V.V. (2004) *Org. Lett.*, **6**, 3897–3899.
101 Chittaboina, S., Xie, F., and Wang, Q. (2005) *Tetrahedron Lett.*, **46**, 2331–2336.
102 Kacprzak, K. (2005) *Synlett*, 943–946.
103 Barral, K., Moorhouse, A.D., and Moses, J.E. (2007) *Org. Lett.*, **9**, 1809–1811.
104 Zhang, L., Chen, X., Xue, P., Sun, H.H.Y., Williams, I.D., Sharpless, K.B., Fokin, V.V., and Jia, G. (2005) *J. Am. Chem. Soc.*, **127**, 15998–15999.
105 Majireck, M.M., and Weinreb, S.M. (2006) *J. Org. Chem.*, **71**, 8680–8683.

106 Rasmussen, L.K., Boren, B.C., and Fokin, V.V. (2007) *Org. Lett.*, **9**, 5337–5339.

107 Boren, B.C., Narayan, S., Rasmussen, L.K., Zhang, L., Zhao, H., Lin, Z., Jia, G., and Fokin, V.V. (2008) *J. Am. Chem. Soc.*, **130**, 8923–8930.

108 Oppilliart, S., Mousseau, G., Zhang, L., Jia, G., Thuéry, P., Rousseau, B., and Cintrat, J.-C. (2007) *Tetrahedron*, **63**, 8094–8098.

109 Tam, A., Arnold, U., Soeliiner, M.B., and Raines, R.T. (2007) *J. Am. Chem. Soc.*, **129**, 12670–12671.

110 Horne, W.S., Olsen, C.A., Beierle, J.M., Montero, A., and Ghadiri, M.R. (2009) *Angew. Chem. Int. Ed.*, **48**, 4718–4724, S4718/4711–S4718/4711.

111 Vrieze, K., and van Koten, G. (1987) *Comprehensive Coordination Chemistry*, vol. **2**, Pergamon, Oxford, pp. 189–245.

112 Abu-Salah, O.M. (1998) *J. Organomet. Chem.*, **565**, 211–216.

113 Rodionov, V.O., Fokin, V.V., and Finn, M.G. (2005) *Angew. Chem. Int. Ed.*, **44**, 2210–2215.

114 Nolte, C., Mayer, P., and Straub, B.F. (2007) *Angew. Chem. Int. Ed.*, **46**, 2101–2103.

115 Ahlquist, M., and Fokin, V.V. (2007) *Organometallics*, **26**, 4389–4391.

116 Straub, B.F. (2007) *Chem. Commun.*, 3868–3870.

117 Chui, S.S.Y., Ng, M.F.Y., and Che, C.-M. (2005) *Chem. Eur. J.*, **11**, 1739–1749.

118 Hein, J.E., and Fokin, V.V. (2010) *Chem. Soc. Rev.*, **39**, 1302–1315.

119 Yoo, E.J., Ahlquist, M., Kim, S.H., Bae, I., Fokin, V.V., Sharpless, K.B., and Chang, S. (2007) *Angew. Chem. Int. Ed.*, **46**, 1730–1733.

120 Bae, I., Han, H., and Chang, S. (2005) *J. Am. Chem. Soc.*, **127**, 2038–2039.

121 Cho, S.H., Yoo, E.J., Bae, I., and Chang, S. (2005) *J. Am. Chem. Soc.*, **127**, 16046–16047.

122 Cassidy, M.P., Raushel, J., and Fokin, V.V. (2006) *Angew. Chem. Int. Ed. Engl.*, **45**, 3154–3157.

123 Whiting, M., and Fokin, V.V. (2006) *Angew. Chem. Int. Ed.*, **45**, 3157–3161.

124 Kinugasa, M., and Shizunobu, H. (1972) *J. Chem. Soc., Chem. Commun.*, 466–467.

125 Lo, M.M.C., and Fu, G.C. (2002) *J. Am. Chem. Soc.*, **124**, 4572–4573.

126 Shintani, R., and Fu, G.C. (2003) *Angew. Chem. Int. Ed.*, **42**, 4082–4085.

127 Lang, S.A., and Lin, Y.I. (1984) *Comprehensive Heterocyclic Chemistry*, vol. VI/4B (eds A.R. Katritzky and C.W. Rees), Pergamon, Oxford, pp. 1–130.

128 Himo, F., Lovell, T., Hilgraf, R., Rostovtsev, V.V., Noodleman, L., Sharpless, K.B., and Fokin, V.V. (2005) *J. Am. Chem. Soc.*, **127**, 210–216.

129 Hansen, T.V., Wu, P., and Fokin, V.V. (2005) *J. Org. Chem.*, **70**, 7761–7764.

130 Hassner, A., and Rai, K.M. (1989) *Synthesis*, **1**, 57–59.

131 Shintani, R., and Fu, G.C. (2003) *J. Am. Chem. Soc.*, **125**, 10778–10779.

132 Lee, L.V., Mitchell, M.L., Huang, S.-J., Fokin, V.V., Sharpless, K.B., and Wong, C.-H. (2003) *J. Am. Chem. Soc.*, **125**, 9588–9589.

133 Whiting, M., Tripp, J.C., Lin, Y.C., Lindstrom, W., Olson, A.J., Elder, J.H., Sharpless, K.B., and Fokin, V.V. (2006) *J. Med. Chem.*, **49**, 7697–7710.

134 Díaz, D.D., Punna, S., Holzer, P., McPherson, A.K., Sharpless, K.B., Fokin, V.V., and Finn, M.G. (2004) *J. Polym. Sci. A Polym. Chem.*, **42**, 4392–4403.

135 Liu, Y., Díaz, D.D., Accurso, A., Sharpless, K.B., Fokin, V.V., and Finn, M.G. (2006) *J. Polym. Sci. A Polym. Chem.*, **45**, 5182–5189.

136 Le Baut, N., Díaz, D.D., Punna, S., Finn, M.G., and Brown, H. (2007) *Polymer*, **48**, 239–244.

137 Hansen, T.V., Wu, P., and Fokin, V.V. (2005) *J. Org. Chem.*, **70**, 7761–7764.

11
"Frustrated Lewis Pairs": A Metal-Free Strategy for Hydrogenation Catalysis

Douglas W. Stephan

One strategy to avoid or replace costly precious metal catalysts is the development of new metal-free processes. While such developments would also eliminate the negative environmental impact of trace heavy metal pollutants, they are limited by the inability of non-transition metal species to activate small molecules for common catalytic transformations. In particular, the process of hydrogenation of organic substrates has been the purview of transition metal catalysts for almost 50 years. Ru-based catalysts, as well as a variety of well known Rh-based systems, are commonly employed in hydrogenation catalysis in both academic and industrial settings. Nonetheless, the notion of employing non-metal catalysts in hydrogenation is not a new concept. Indeed, use of non-metal systems to effect hydrogenation date back to the early days of transition-metal-based hydrogenations. For example, in the 1960s, the first metal-free hydrogenation of alkenes was achieved with borane and H_2 [1]. In these cases, the reaction proceeds via hydroboration of the olefin followed by hydrogenolysis of the B–C bond to regenerate the borane and give the alkane product (Scheme 11.1). The latter step proved to be slow and thus comparatively high temperatures (225–235 °C) and pressures of H_2 (2500 psi) were required [1]. Nonetheless, this approach was utilized to hydrogenate a series of olefins and polymers containing olefinic units [1–4]. This approach has also drawn recent attention. Haenel *et al.* have employed I_2, BI_3 or combinations of I_2 and $NaBH_4$ to effect the metal-free hydrogenation of 1,2-dinapthylethane and pyrene (Scheme 11.1) at 350 °C and 15 MPa of H_2 to give a mixture of the possible partial hydrogenation products. Such transformations have also been applied to the liquefaction of coal [2].

An alternative approach, also initially discovered in 1961 and detailed in 1964, was reported by Walling and Bollyky [5, 6]. In this method, base was used to effect the metal-free hydrogenation of benzophenone (Scheme 11.2) [7], although forcing conditions (20 mol% KO*t*-Bu, 200 °C and >100 bar H_2) were required. In a report some 40 years later, Berkessel *et al.* [7] examined the mechanism and the rate dependence on the alkali ion. These authors noted the analogy to the heterolytic H_2 mechanism mediated by Ru-catalyzed asymmetric ketone hydrogenation catalysts developed by Noyori [8, 9].

Catalysis Without Precious Metals. Edited by R. Morris Bullock
© 2010 WILEY-VCH Verlag GmbH & Co. KGaA, Weinheim
ISBN: 978-3-527-32354-8

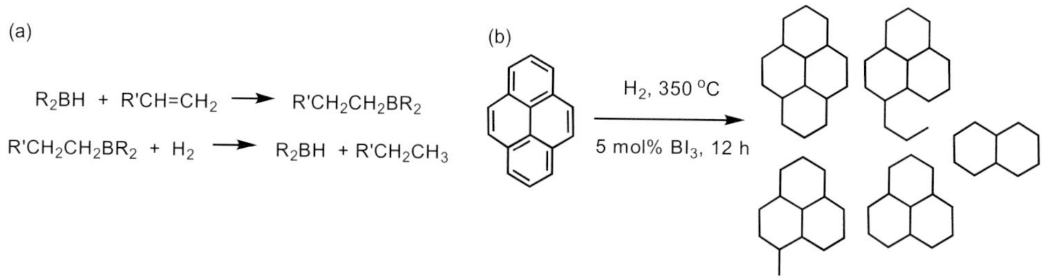

Scheme 11.1 (a) Hydroboration/hydrogenolysis steps in borane catalyzed hydrogenation of olefins. (b) Borane catalyzed hydrogenation of pyrene.

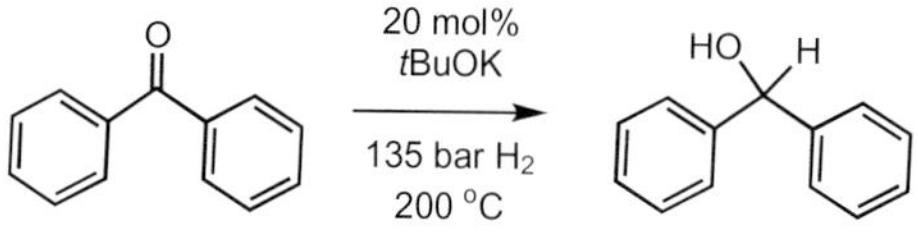

Scheme 11.2 Hydrogenation of benzophenone catalyzed by KO*t*-Bu.

More recent efforts to effect metal-free hydrogenations have fallen in the realm of organo-catalysts. For example, the hydrogenation of enones, imines, unsaturated α,β-aldehydes, and quinolines have been achieved without a metal, however, in these cases the stoichiometric source of H_2 was a Hantzsch ester (Scheme 11.3) [10–14]. Such a protocol can be applied to asymmetric reductions, as substituted enones can be reduced with reasonable enantiomer excesses between 73 and 91%. A variety of organocatalyst systems have been reviewed [15, 16].

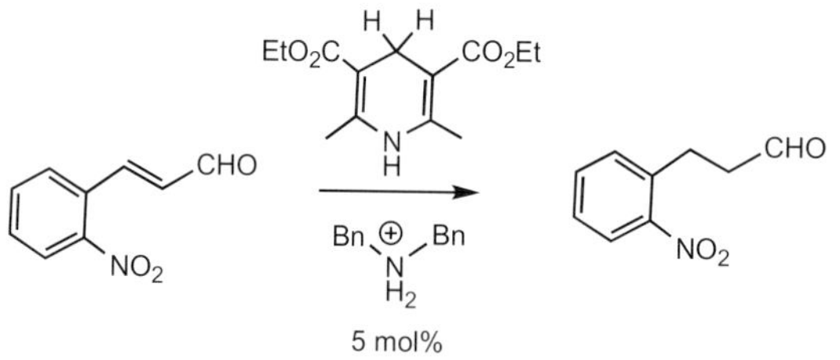

Scheme 11.3 Example of organocatalyst reduction of enone with the hantzsch ester.

The above systems notwithstanding, the notion of reducing organic substrates without the use of a metal remains a significant challenge. This difficulty is paralleled by the paucity of organic or main group systems that react easily and directly with H_2. Indeed, few such systems exist. Low-temperature matrix work has revealed main group elements will react with H_2 [17–20], and such intereactions have been supported by computational studies [21, 22]. Addition of H_2 to

Ge$_2$-alkyne analogs to give a mixture of Ge$_2$ and primary germane products (Scheme 11.4) has been described by Power and coworkers [23]. It is also noteworthy that a unique metal-free hydrogenase has been identified from *methanogenic archaea*. This enzyme catalyzes reactions with H$_2$ [24, 25], and it is suggested that a folate-like cofactor is important in the the reversible activation or liberation of H$_2$ [26, 27].

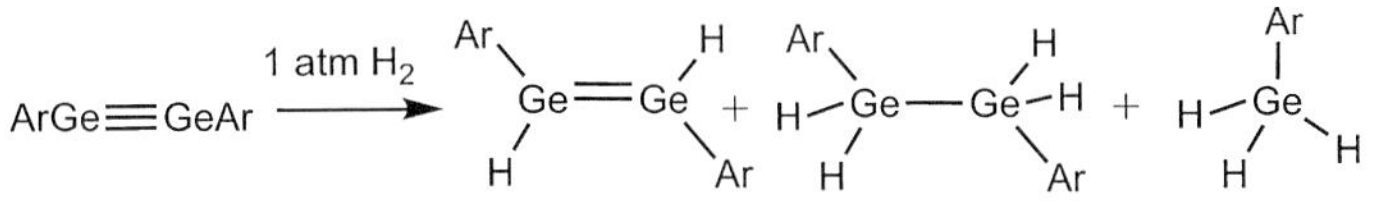

Scheme 11.4 Hydrogenation of digermyne with H$_2$.

A recent finding in our laboratories has revealed the first metal-free compound capable of reversible activation of H$_2$. We have investigated this system and shown subsequently that such materials are capable of metal-free hydrogenation catalysis under relatively mild conditions. These developments are the focus of this chapter.

11.1
Phosphine-Borane Activation of H$_2$

In studying the reactions of tertiary phosphines, R$_3$P (R = Cy, iPr) with the Lewis acid B(C$_6$F$_5$)$_3$, it was observed that sterically demanding phosphines that were too encumbered to form a P–B bond reacted at the *para*-carbon to provide the unique phosphonium borate salts of the formula [R$_3$P(C$_6$F$_4$)BF(C$_6$F$_5$)$_2$] (Scheme 11.5). In a similar fashion, bulky secondary phosphines gave analogous zwitterionic compounds of the form [R$_2$PH(C$_6$F$_4$)BF(C$_6$F$_5$)$_2$] (Scheme 11.5).

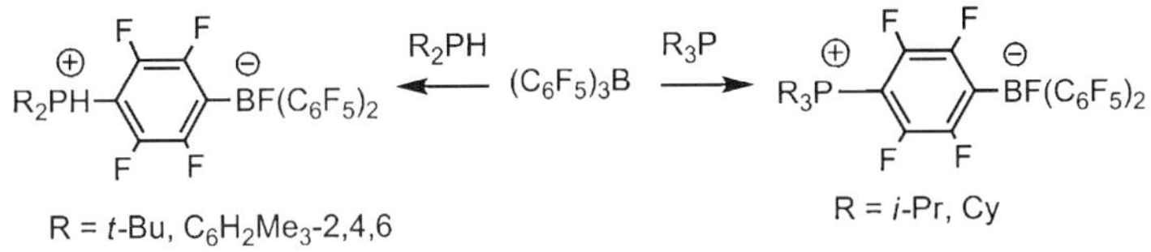

Scheme 11.5 Reactions of sterically demanding phosphines with B(C$_6$F$_5$)$_3$.

Subsequent reaction with Grignard reagents gave the neutral phosphino-boranes R$_2$P(C$_6$F$_4$)B(C$_6$F$_5$)$_2$ [28]. Alternatively, fluoride for hydride exchange at B was achieved by treatment with Me$_2$Si(H)Cl. This yielded the unusual salts [R'$_2$PH(C$_6$F$_4$)BH(C$_6$F$_5$)$_2$] which contain both protic and hydridic sites at the phosphonium and borate, respectively. Not surprisingly, thermolysis of one such latter species, specifically [(C$_6$H$_2$Me$_3$-2,4,6)$_2$PH(C$_6$F$_4$)BH(C$_6$F$_5$)$_2$] at 150 °C, resulted in the evolution of H$_2$ and the generation of (C$_6$H$_2$Me$_3$-2,4,6)$_2$P(C$_6$F$_4$)B(C$_6$F$_5$)$_2$ (Scheme 11.6) [29]. In this neutral phosphino-borane there is no interaction between the Lewis acidic B and the Lewis basic P center, rather this species remains monomeric in solution. Remarkably and unexpectedly, exposure of a solution of this red–orange phosphino-borane to 1 atm of H$_2$ at 25 °C resulted in the immediate and facile reformation of the zwitterionic and colorless salt [(C$_6$H$_2$Me$_3$-2,4,6)$_2$PH(C$_6$F$_4$)BH(C$_6$F$_5$)$_2$] (Scheme 11.6). This compound represents the first non-transition metal system that releases and takes up hydrogen. The significantly higher basicity of the P center in the analogous species tBu$_2$PH(C$_6$F$_4$)BH(C$_6$F$_5$)$_2$ prevents release of H$_2$ under similar thermolysis conditions.

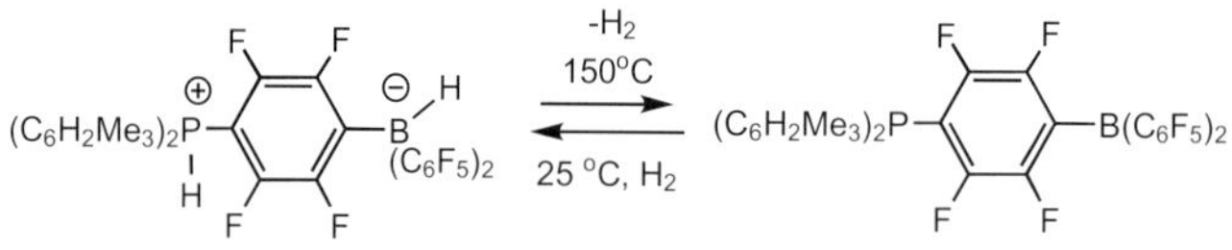

Scheme 11.6 Reversible metal-free activation of H$_2$.

11.2
"Frustrated Lewis Pairs"

The unique reactivity of the above system with H$_2$ appears to arise from the unquenched Lewis basicity and acidity of the respective donor P and the acceptor B centers. This inference prompted questions about the nature and reactivity of other phosphine-borane systems and, more broadly, of Lewis acid/base combinations. Is it necessary to have a link between the donor and acceptor sites? Could similar H$_2$ activation arise from combinations of donors and acceptors in which steric encumbrance frustrates Lewis acid–base adduct formation? If indeed such "frustrated Lewis pairs" could be uncovered, could one exploit them for the activation of small molecules and applications in catalysis?

In probing these questions, we began with a 1:1 combination of sterically demanding tertiary phosphines R$_3$P (R = t-Bu, C$_6$H$_2$Me$_3$) and the Lewis acid B(C$_6$F$_5$)$_3$. Such mixtures show no evidence of reaction, no adduct formation and no evidence of attack at the *para*-carbon of a fluorinated aryl ring. Nonetheless, addition of 1 atm of H$_2$ prompts immediate reaction to form the salt [R$_3$PH][HB(C$_6$F$_5$)$_3$] (Scheme 11.7a) [30]. This facile heterolytic activation of H$_2$ was also exhibited by the combination of tBu$_3$P and BPh$_3$, although the formation of [tBu$_3$PH][HBPh$_3$] proceeds much more slowly and the salt is obtained in much lower yield (33%). It is noteworthy that the combinations of (C$_6$H$_2$Me$_3$)$_3$P and BPh$_3$, (C$_6$F$_5$)$_3$P and B(C$_6$F$_5$)$_3$ or tBu$_3$P and B(C$_6$H$_2$Me$_3$)$_3$ appear to form frustrated Lewis pairs, in that no reaction is apparent from spectroscopic data. However, these combinations also show no reactivity with H$_2$ at 25 °C. Although the above salts

[R$_3$PH][HB(C$_6$F$_5$)$_3$] do not liberate H$_2$ upon heating to above 100 °C, adjustment of the Lewis acidity and basicity of the phosphine and borane does provide a system that reversibly activates H$_2$. Specifically, use of a borane that is slightly less Lewis acidic than B(C$_6$F$_5$)$_3$ and a slightly more protic phosphine as in B(p-C$_6$F$_4$H)$_3$ and (o-C$_6$H$_2$Me$_3$)$_3$P [31] affords the facile activation of H$_2$ to give [(o-C$_6$H$_4$Me)$_3$PH] [HB(p-C$_6$F$_4$H)$_3$], which under vacuum at 25 °C loses H$_2$, albeit slowly (85% after 9 days) (Scheme 11.7b). Alternatively, the same conversion proceeds at 80 °C after 12 h. Collectively these observations support the view that there is a threshold for the cumulative Lewis acidity and basicity required for these frustrated Lewis pairs (FLPs) to effect the activation and release of H$_2$.

Scheme 11.7 Activation of H$_2$ by phosphine and borane.

The mechanism of action of FLPs is the subject of considerable interest. Initial speculation noted the earlier computations for BH$_3$–H$_2$ adducts as well as the matrix isolation studies that demonstrated the interaction of phosphine with H$_2$. More recent DFT studies by Papaí and coworkers [32] infer a mechanism that proceeds via an "encounter complex" (Figure 11.1) in which the phosphine and borane approach but do not form a dative bond. Rather, the complex is stabilized by H·F interactions, yielding a "pocket" that is proposed to react with H$_2$. An analogous mechanism has been proposed for the C$_6$F$_4$-linked systems R$_2$P(C$_6$F$_4$)B(C$_6$F$_5$)$_2$ [33].

While the discovery of FLPs reveals a new strategy for reactivity, it is important to note that previous researchers had observed related systems where steric

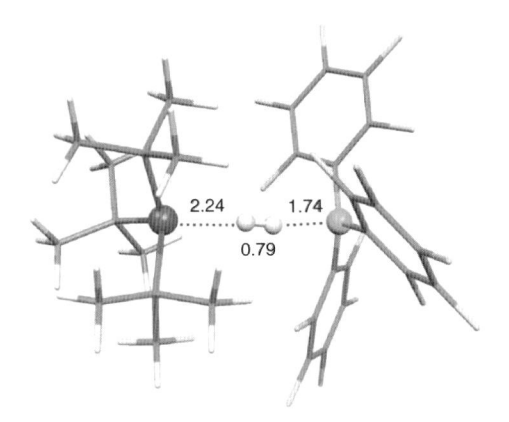

Figure 11.1 Calculated interaction of H$_2$ with phosphine and borane.

demands sculpt reactivity. For example, in 1942 H. C. Brown and coworkers [34] described the intervention of steric demands in the formation of simple donor–acceptor adducts of pyridines with BMe_3, specifically noting that lutidine failed to form an adduct. Some years later, in 1950, the research group of Wittig showed that the reaction of Ph_3CNa with $THF(BPh_3)$ [35] did not result in displacement of THF but rather THF ring-opening to give the anion $[Ph_3C(CH_2)_4OBPh_3]^-$ (Scheme 11.8). Similarly, in the 1960s, sterically encumbered amines were shown to react with trityl cation to result in proton abstraction to afford an iminium cation, rather than quaternization of the amine (Scheme 11.8) [36].

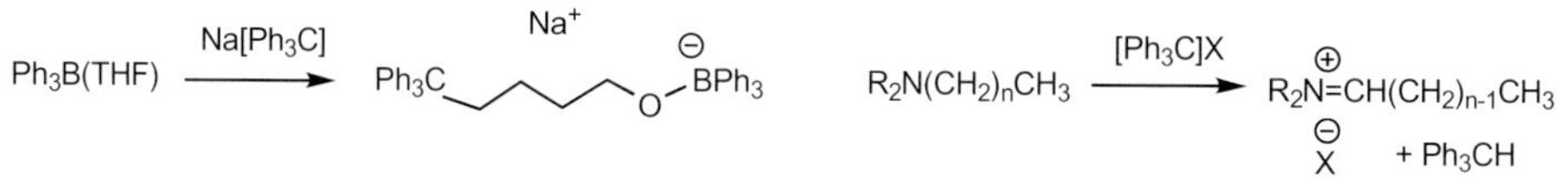

Scheme 11.8 Reactions of trityl anion and trityl cation.

In 1998, Erker and coworkers described the reaction of $B(C_6F_5)_3$ and a sterically encumbered ylide, Ph_3PCHPh. In this case, the simple ylide/borane adduct $(Ph_3PCHPh)B(C_6F_5)_3$ was shown to thermally rearrange to $(Ph_3PCHPh)(C_6F_4)BF(C_6F_5)_2$, inferring FLP-type reactivity (Scheme 11.9). In a similar fashion, we reported the reactions of sterically hindered phosphines with $[CPh_3][B(C_6F_5)_4]$ which proceed via para-attack to give species of the form $[(R_3PC_6H_5)CPh_2][B(C_6F_5)_4]$ (R = Cy, t-Bu) or $[(i-Pr_3PC_6H_4)Ph_2CH][B(C_6F_5)_4]$ (Scheme 11.9) [37]. Moreover, reactions of $(THF)B(C_6F_5)_3$ with bulky phosphines, proceed to effect nucleophilic ring opening of THF, yielding butoxy-tethered phosphonium-borates of the form $R'R_2P(C_4H_8O)B(C_6F_5)_3$ (Scheme 11.9) [38]. In all of these cases,

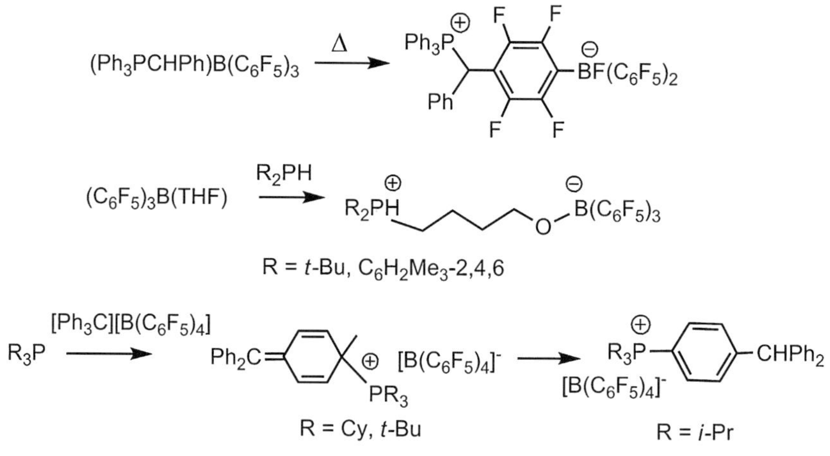

Scheme 11.9 FLP reactions of $B(C_6F_5)_3$ and trityl cation.

the presence of steric bulk alters conventional Lewis acid–base reactivity. A number of recent reports have explored the utility of FLPs in the activation of olefins [39], B–H [40] and N–H bonds. The role of carbenes as the base in FLP activation of H_2 and NH bonds has also been explored [41–43]. As these works are beyond the scope of the present chapter, they are not detailed herein.

11.3
Metal-Free Catalytic Hydrogenation

Conceptually, the heterolytic activation of H_2 is reminiscent of the Noyori systems [8, 9, 44], where heterolysis of H_2 affords a metal hydride and a protonated ligand. Extension of this analogy suggests that the ability to activate H_2 using a simple main group system opens the door to metal-free hydrogenation catalysis. Clearly what is required is a substrate that will receive a proton and hydride transferred from, for example, a phosphonium borate, yielding a phosphine and free borane, which would in turn further activate H_2 to regenerate the phosphonium borate. In early trials the stoichiometric reactions of the phosphonium borates (R_2PH)(C_6F_4)$BH(C_6F_5)_2$ (R = $C_6H_2Me_3$, tBu) with imines were probed. The result was the formation of the amine-adduct (R_2P)(C_6F_4)$B(C_6F_5)_2$($NHRCH_2R'$), inferring both proton and hydride transfer to the imine (Scheme 11.10).

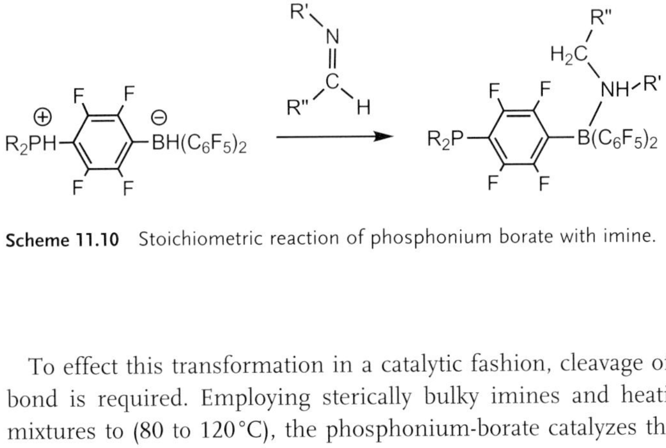

Scheme 11.10 Stoichiometric reaction of phosphonium borate with imine.

To effect this transformation in a catalytic fashion, cleavage of the B–N dative bond is required. Employing sterically bulky imines and heating the reaction mixtures to (80 to 120 °C), the phosphonium-borate catalyzes the hydrogenation of a series of imines to the corresponding amines (1–5 atm of H_2 and 5 mol% catalyst) [45]. For example, the imine $tBuN=C(H)Ph$ is reduced in 1 h at 1 atm H_2 and 80 °C (Table 11.1), while $PhSO_2N=C(H)Ph$ proceeds much more slowly (10.5–16 h) and requires higher H_2 pressure and higher temperature. These observations are attributed to the electron-rich and electron-deficient nature of the respective imines. Mechanistically, imine reduction proceeds via initial protonation of N to

Table 11.1 Catalytic hydrogenation of imines with phosphonium borate catalysts [45].

Substrate	Cat.	T (°C)	Time (h)	Yield (%)	Product
PhCH=NtBu	1[a]	80	1	79	PhCH$_2$NHtBu
PhCH=NtBu	2[a]	80	1	98	PhCH$_2$NHtBu
PhCH=NSO$_2$Ph	1	120	10.5	97	PhCH$_2$NHSO$_2$Ph
PhCH=NSO$_2$Ph	2	120	16	87	PhCH$_2$NHSO$_2$Ph
PhCH=NCHPh$_2$	1	140	1	88	PhPhCH$_2$NHCHPh$_2$
PhCH=NCH$_2$Ph	1	120	48	5[b]	PhCH$_2$NHCH$_2$Ph
PhCH=N(CH$_2$Ph)(B(C$_6$F$_5$)$_3$)	1	120	46	57	(PhCH$_2$NHCH$_2$Ph)(B(C$_6$F$_5$)$_3$)
MeCN(B(C$_6$F$_5$)$_3$)	1	120	24	75	MeCH$_2$NH$_2$(B(C$_6$F$_5$)$_3$)
PhCN(B(C$_6$F$_5$)$_3$)	1	120	24	84	PhCH$_2$NH$_2$(B(C$_6$F$_5$)$_3$)
(C$_6$F$_5$)$_3$BNC(CH$_2$)$_4$CNB(C$_6$F$_5$)$_3$	1[c]	120	48	99	(C$_6$F$_5$)$_3$BH$_2$N(CH$_2$)$_6$NH$_2$B(C$_6$F$_5$)$_3$
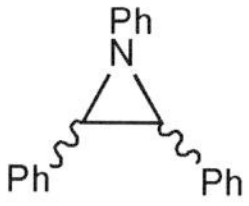 Ph-N(triangle with Ph, Ph)	1[c]	120	1.5	98	PhCH$_2$C(H)PhN(H)Ph

1 = (C$_6$H$_2$Me$_3$)$_2$PH(C$_6$F$_4$)BH(C$_6$F$_5$)$_2$, 2 = tBu$_2$PH(C$_6$F$_4$)BH(C$_6$F$_5$)$_2$. Standard Conditions: 5 mol% catalyst, 4 mL toluene, ca. 5 atm H$_2$.
a) 1 atm. H$_2$.
b) Determined by ^{1}H NMR spectroscopy.
c) 10 mol% catalyst.

afford the activated iminium cation, which is then attacked by the borohydride moiety to give the amine (Scheme 11.11). Coordination of the amine to the borane occurs, but at the temperature of reduction, thermally encouraged dissociation of amine allows the free phosphino-borane to again activate H$_2$ and thus re-enter the catalytic cycle. Initial protonation was unambiguously confirmed by the observation that treatment of (Cy$_3$P)(C$_6$F$_4$)BH(C$_6$F$_5$)$_2$ with the imine tBuN=CPh(H) resulted in no reaction.

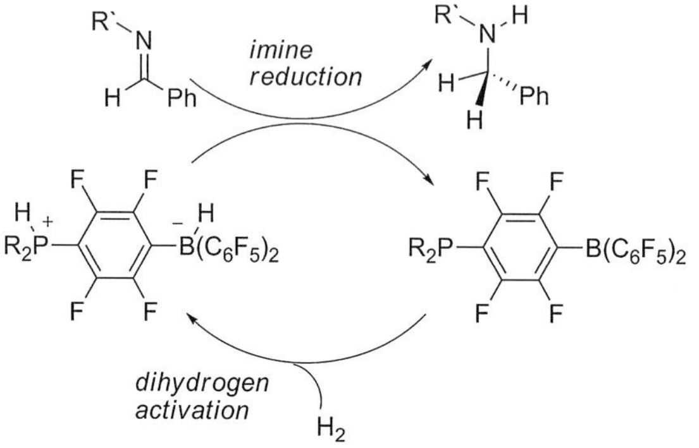

Scheme 11.11 Catalytic hydrogenation of imine using phosphonium borate catalyst.

The impact of steric efforts on this reduction process is evident from the data. Sterically less encumbered imines are reduced more slowly, which in the case of the imine $PhCH=NCH_2Ph$ leads to it being reduced only stoichiometrically [45]. A strategy to overcome this steric requirement is to employ a bulky protecting group. Thus, reduction of the adduct $(PhCH=NCH_2Ph)(B(C_6F_5)_3)$ proceeds catalytically. Similarly, this strategy was used to reduce the benzylnitrile- $B(C_6F_5)_3$ and the adiponitrile-bis-$B(C_6F_5)_3$ adducts to the corresponding primary amine-borane species (Table 11.1) [45]. On the other hand, this metal-free reduction catalysis was used effectively to hydrogenate the cis-1,2,3-triphenylaziridine, (PhCHCHPhNPh) to the corresponding ring-opened amine [45].

Attempts to reduce benzaldehyde gave only the stoichiometric product $R_2PH(C_6F_4)B(C_6F_5)_2OCH_2Ph$ (R = 2,4,6-$Me_3C_6H_2$, tBu) (Scheme 11.12), presumably a result of the strength of the B–O bond that is formed. In developing new FLP systems, the Erker group prepared $(C_6H_2Me_3)_2P(C_2H_4)B(C_6F_5)_2$ via hydroboration of the corresponding vinylphosphine [46]. Activation of H_2 gave $(2,4,6-Me_3C_6H_2)_2PH(C_2H_4)BH(C_6F_5)_2$, which reacted with benzaldehyde to give $(2,4,6-Me_3C_6H_2)_2PH(C_2H_4)B(C_6F_5)_2(OCH_2Ph)$. Thus, it appears that the oxophilicity of B is the limit to the potential of these systems for hydrogenation of carbonyl groups.

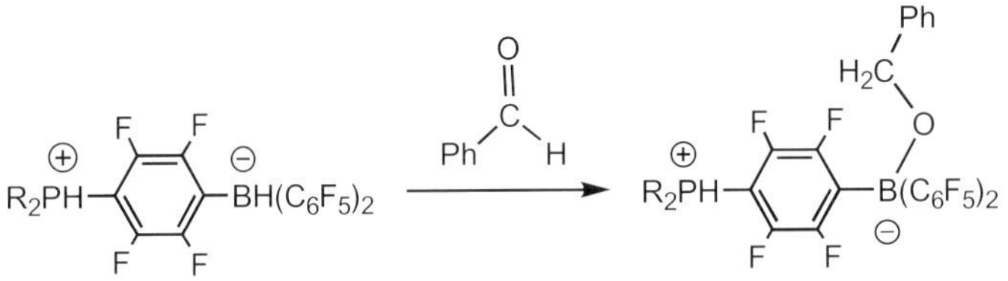

Scheme 11.12 Reaction of phosphonium borate with benzaldehyde.

Recognizing that FLP activation of H_2 requires a bulky base and acid combination, we probed the question of whether a bulky imine could act both as the base component of an FLP and as the substrate for catalytic reduction. Indeed, a number of aldimines and ketimines could be hydrogenated with a catalytic amount of $B(C_6F_5)_3$ under H_2 in a facile manner (Table 11.2, Scheme 11.13) [47]. As with the phosphino-borane catalyst, cis-triphenylaziridine is also reduced under these conditions to N-(1,2-diphenylethyl)aniline (Table 11.2) [47].

It is also noteworthy that addition of extra equivalents of imine substrate prompts ongoing reduction, demonstrating the "living" nature of the catalysis. In addition, it was shown that the resulting amine in combination with $B(C_6F_5)_3$ acts to split H_2, as the resting catalyst is $[R'RNH_2][HB(C_6F_5)_3]$ [47]. A paper published subsequently described the heterolytic activation of H_2 by sterically encumbered amines and $B(C_6F_5)_3$ [48]. In the case of the extremely bulky $(C_6H_3(i-Pr)_2)N=CMe(t-Bu)$ with $B(C_6F_5)_3$ under H_2, the imine is not reduced. Instead, the reaction gave only the ion pair $[(C_6H_3(i-Pr)_2)N(H)=CMe(t-Bu)][HB(C_6F_5)_3]$, consistent with the mecha-

Table 11.2 Catalytic hydrogenation of imines B(C$_6$F$_5$)$_3$/H$_2$ [47].

Substrate	Time (h)	Yield (%)	Product
PhCH=NtBu	2$^{a)}$	89	PhCH$_2$NHtBu
PhCH=NCH$_2$Ph	1	99	PhCH$_2$NHCH$_2$Ph
PhCH=NSO$_2$Ph	41	94	PhCH$_2$NHSO$_2$Ph
Ph$_2$C=NtBu	1	98	Ph$_2$CHNHtBu
Ph(Me)C=N(C$_6$H$_2$(iPr)$_2$)	8	94	Ph(Me)CHNH(C$_6$H$_2$(iPr)$_2$)
Ph(tBu)C=N(C$_6$H$_2$(iPr)$_2$)	48	0	Ph(tBu)CHNH(C$_6$H$_2$(iPr)$_2$)
PhCH=NCH$_2$Ph(B(C$_6$F$_5$)$_3$)	46	57	PhCH$_2$NHCH$_2$Ph(B(C$_6$F$_5$)$_3$)
PhCHCHPhNPh	2	95	PhCH$_2$C(H)PhN(H)Ph

Standard Conditions: 5 mol% B(C$_6$F$_5$)$_3$, 120 °C, 1 atm H$_2$.
a) 80 °C.

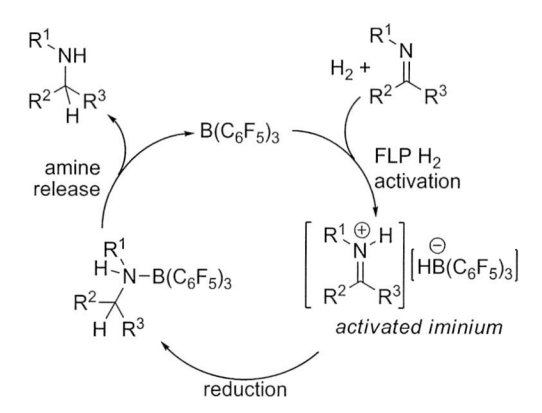

Scheme 11.13 Catalytic reduction of imines by B(C$_6$F$_5$)$_3$/H$_2$.

nism proposed for the phosphonium-borate catalyst involving protonation of the imine. Obviously in this case hydride attack of the iminium cation is sterically precluded.

The proposed mechanism is further supported by the observation that the reduction of PhSO$_2$N=CPh(H) by B(C$_6$F$_5$)$_3$ and H$_2$ is slow, but addition of 5 mol% P(C$_6$H$_2$Me$_3$)$_3$ accelerates the reduction [47]. This rate acceleration is attributed to the ability of P(C$_6$H$_2$Me$_3$)$_3$/B(C$_6$F$_5$)$_3$ to rapidly activate H$_2$, giving the phosphonium salt [(C$_6$H$_2$Me$_3$)$_3$PH][HB(C$_6$F$_5$)$_3$], which then effects imine reduction. Similarly, (RCN)B(C$_6$F$_5$)$_3$ (R = Me, Ph) with 5 mol% B(C$_6$F$_5$)$_3$ alone and H$_2$ was not reduced, but addition of 5 mol% P(C$_6$H$_2$Me$_3$)$_3$ prompted clean hydrogenation to the amine-borane adducts (Scheme 11.14) [47]. This mechanism parallels that proposed by Piers and coworkers for the B(C$_6$F$_5$)$_3$ catalyzed hydrosilylation of imines, ketones, enones and silyl enol ethers [49–53].

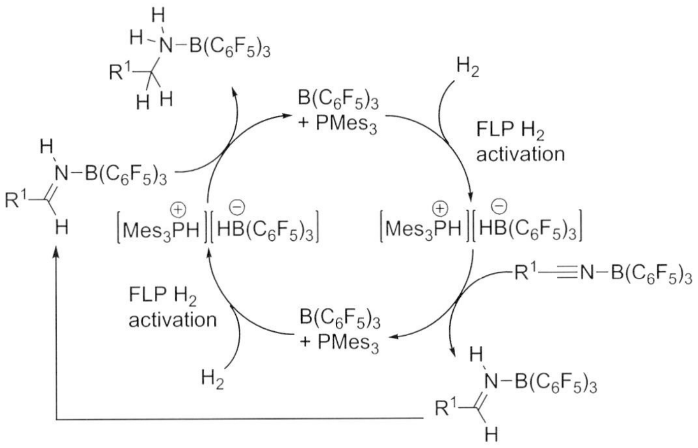

Scheme 11.14 Catalytic reduction of nitrile-borane adducts by $R_3P/B(C_6F_5)_3/H_2$. (Mes = $C_6H_2Me_3$).

In a related and subsequent report by Chen and Klankermayer [54], $B(C_6F_5)_3$ and H_2 were employed to catalytically reduce acyclic and cyclic imines in a fashion similar to that described above. In addition these authors also reported the use of optically impure (α-pinanyl)$B(C_6F_5)_2$ [55] under a H_2 atmosphere (Scheme 11.15) to effect the reduction of PhN=CPh(Me) with some asymmetric bias, obtaining the corresponding amine with a 13% enantiomeric excess [54].

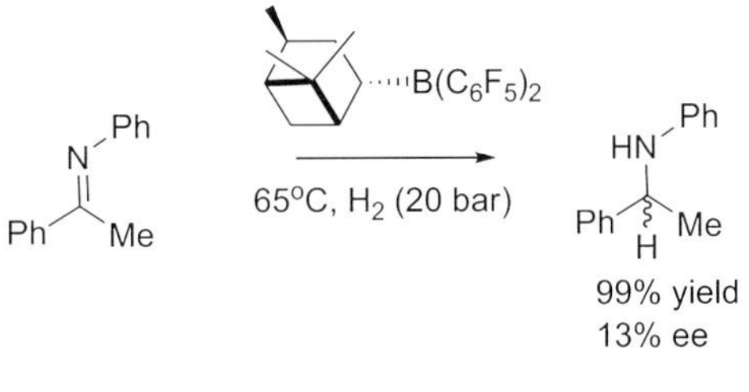

Scheme 11.15 Assymetric reduction of imine using a chiral borane.

In recent efforts, Erker and coworkers have prepared the olefin-linked phosphine-borane $(C_6H_2Me_3)_2P(CH=CR)B(C_6F_5)_2$ [56], which also activates H_2. This species also effects the catalytic reduction of the imine, tBuN=CH(Ph) and enamines at room temperature (Scheme 11.16).

Using a similar strategy, the research groups of Repo and Rieger reported the catalytic hydrogenation of imines and enamines using a linked amine-borane species derived from 2,2,6,6-tetramethylpiperidine [57]. In this case, the link

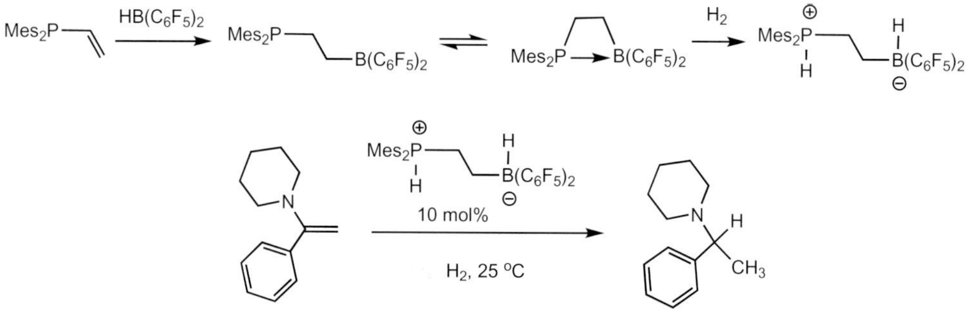

Scheme 11.16 Synthesis and reduction of enamine by ethylene-linked phosphonium borate.

between the B and piperidine units is thought to provide a more crowded environment about B, thus permitting the reduction of sterically less encumbered imines (Scheme 11.17).

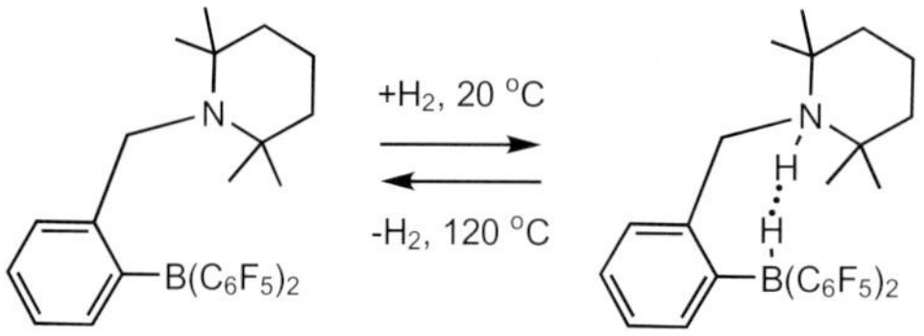

Scheme 11.17 Amino-borane activation of H_2: "molecular tweezers".

In most recent efforts, Erker and colleagues have developed an interesting new FLP based on the 1,8-diphosphino-napthalene 1,8-$(Ph_2P)_2C_{10}H_6$ [58]. This diphosphine with $B(C_6F_5)_3$ activates H_2. Perhaps, most interestingly however, this FLP catalytically reduces a series of silylenol ethers at 25 °C under H_2 pressure of 2 bar (Scheme 11.18).

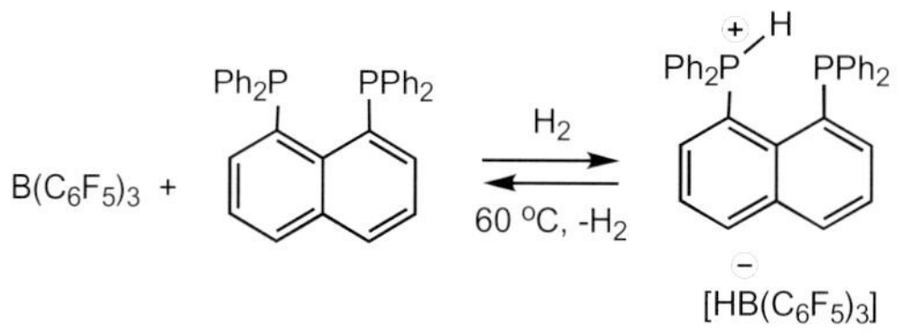

Scheme 11.18 Activation of H_2 by 1,8-diphosphino-napthalene and $B(C_6F_5)_3$.

11.4
Future Considerations

The discovery of the ability of FLPs to activate H_2 is only a few years old. Nonetheless, it brings a new perspective to the realm of catalysis, offering the potential for metal-free hydrogenation. While the breadth of substrates is limited at this point, the promise of cheap catalysts in comparison to those derived from precious metals, as well as metal-free products, has stimulated interest from commercial enterprises. Aside from a broadened range of substrates and developing strategies to generate functional group selectivity, adaptation of FLPs for enantioselective synthesis is an extension that would generate interest from both the organic community and industrial researchers. It is clear that this finding creates a new opportunity for new forms of catalysis. To what extent this approach will impact on commonly used methodologies remains to be seen.

Acknowledgements

The contributions and helpful discussions with a number of very talented students and postdoctoral fellows in my group is gratefully acknowledged In particular, the contributions and discussions of Dr. Preston Chase, Dr. Greg Welch, Dr. Jenny McCahill, Lourisa Cabrera, Dr. Emily Hollink, Dr. Matthias Ullrich, Meghan Dureen, Rebecca Neu, Xiaoxi Zhao, Dr. Alberto Ramos, Kelvin Seto, Dr. Zach Heiden, Dr. Consuelo Herrera and Dr. Edwin Otten were invaluable. Financial support from NSERC of Canada is gratefully acknowledged.

References

1 DeWitt, E.J., Ramp, F.L., and Trapasso, L.E. (1961) *J. Am. Chem. Soc.*, **83**, 4672.
2 Haenel, M.W., Narangerel, J., Richter, U.B., and Rufińska, A. (2006) *Angew. Chem. Int. Ed.*, **45**, 1061.
3 Koester, R., Bruno, G., and Binger, P. (1961) *Liebigs Ann. Chem.*, **644**, 1.
4 Ramp, F.L., DeWitt, E.J., and Trapasso, L.E. (1962) *J. Org. Chem.*, **27**, 4368.
5 Walling, C., and Bollyky, L. (1964) *J. Am. Chem. Soc.*, **86**, 3750–3752.
6 Walling, C., and Bollyky, L. (1961) *J. Am. Chem. Soc.*, **83**, 2968–2969.
7 Berkessel, A., Schubert, T.J.S., and Mueller, T.N. (2002) *J. Am. Chem. Soc.*, **124**, 8693.
8 Noyori, R., and Hashiguchi, S. (1997) *Acc. Chem. Res.*, **30**, 97.
9 Noyori, R., Kitamura, M., and Ohkuma, T. (2004) *Proc. Natl. Acad. Sci. USA*, **101**, 5356.
10 Adolfsson, H. (2005) *Angew. Chem. Int. Ed.*, **44**, 3340.
11 Dalko, P.I., and Moisan, L. (2004) *Angew. Chem. Int. Ed.*, **43**, 5138.
12 Rueping, M., Antonchick, A.P., and Theissmann, T. (2006) *Angew. Chem. Int. Ed.*, **45**, 3683.
13 Tuttle, J.B., Ouellet, S.G., and MacMillan, D.W.C. (2006) *J. Am. Chem. Soc.*, **128**, 12662.
14 Yang, J.W., Hechavarria Fonseca, M.T., and List, B. (2004) *Angew. Chem. Int. Ed.*, **43**, 6660.
15 (a) Lelais, G., and MacMillan, D.W.C. (2006) *Aldrichchim. Acta*, **39**, 79–87; (b) Ouellet, S.G., Walji, A.M., and

Macmillan, D.W.C. (2007) *Acc. Chem. Res.*, **40**, 1327–1339.

16 MacMillan, D.W.C. (2008) *Nature*, **455**, 304–308.

17 Aldridge, S., and Downs, A.J. (2001) *Chem. Rev.*, **101**, 3305.

18 Himmel, H.J. (2003) *Dalton Trans.*, 3639.

19 Himmel, H.J., and Vollet, J. (2002) *Organometallics*, **21**, 11626.

20 Xiao, Z.L., Hauge, R.H., and Margrave, J.L. (1993) *Inorg. Chem.*, **32**, 642.

21 Kulkarni, S.A. (1998) *J. Phys. Chem. A*, **102**, 7704.

22 Kulkarni, S.A., and Srivastava, A.K. (1999) *J. Phys. Chem. A*, **103**, 2836.

23 Spikes, G.H., Fettinger, J.C., and Power, P.P. (2005) *J. Am. Chem. Soc.*, **127**, 12232.

24 Pilak, O., Mamat, B., Vogt, S., Hagemeier, C.H., Thauer, R.K., Shima, S., Vonrhein, C., Warkentin, E., and Ermler, U. (2006) *J. Mol. Biol.*, **358**, 798.

25 Shima, S., Lyon, E.J., Thauer, R.K., Meinert, B., and Bill, E. (2005) *J. Am. Chem. Soc.*, **127**, 10430.

26 Scott, A.P., Golding, B.T., and Radom, L. (1998) *New J. Chem.*, **22**, 1171.

27 Teles, J.H., Brode, S., and Berkessel, A. (1998) *J. Am. Chem. Soc.*, **120**, 1345.

28 Welch, G.C., Cabrera, L., Chase, P.A., Hollink, E., Masuda, J.D., Wei, P., and Stephan, D.W. (2007) *Dalton Trans.*, 3407.

29 Welch, G.C., SanJuan, R.R., Masuda, J.D., and Stephan, D.W. (2006) *Science*, **314**, 1124.

30 Welch, G.C., and Stephan, D.W. (2007) *J. Am. Chem. Soc.*, **129**, 1880–1881.

31 Ullrich, M., Lough, A.J., and Stephan, D.W. (2009) *J. Am. Chem. Soc.*, **131**, 52–53.

32 Rokob, T.A., Hamza, A., Stirling, A., Soós, T., and Pápai, I. (2008) *Angew. Chem. Int. Ed.*, **47**, 2435.

33 Guo, Y., and Li, S. (2008) *Inorg. Chem.*, **47**, 6212.

34 Brown, H.C., Schlesinger, H.I., and Cardon, S.Z. (1942) *J. Am. Chem. Soc.*, **64**, 325.

35 Wittig, G., and Ruckert, A. (1950) *Liebigs Ann. Chem.*, **566**, 101.

36 Damico, R., and Broadus, C.D. (1966) *J. Org. Chem.*, **31**, 1607.

37 Cabrera, L., Welch, G.C., Masuda, J.D., Wei, P., and Stephan, D.W. (2006) *Inorg. Chim. Acta*, **359**, 3066.

38 Welch, G.C., Masuda, J.D., and Stephan, D.W. (2006) *Inorg. Chem.*, **45**, 478–480.

39 McCahill, J.S.J., Welch, G.C., and Stephan, D.W. (2007) *Angew. Chem. Int. Ed.*, **46**, 4968.

40 Dureen, M.A., and Stephan, D.W. (2008) *Chem. Commun.*, 4303.

41 Frey, G.D., Lavallo, V., Donnadieu, B., Schoeller, W.W., and Bertrand, G. (2007) *Science*, **316**, 439.

42 Chase, P.A., and Stephan, D.W. (2008) *Angew. Chem. Int. Ed.*, **47**, 7433.

43 Holschumacher, D., Bannenberg, T., Hrib, C.G., Jones, P.G., and Tamm, M. (2008) *Angew. Chem. Int. Ed.*, **47**, 7428.

44 Casey, C.P., Bikzhanova, G.A., and Guzei, I.A. (2006) *J. Am. Chem. Soc.*, **128**, 2286.

45 Chase, P.A., Welch, G.C., Jurca, T., and Stephan, D.W. (2007) *Angew. Chem. Int. Ed.*, **49**, 8050.

46 Spies, P., Erker, G., Kehr, G., Bergander, K., Fröhlich, R., Grimme, S., and Stephan, D.W. (2007) *Chem. Commun.*, 5072–5074.

47 Chase, P.A., Jurca, T., and Stephan, D.W. (2008) *Chem. Commun.*, 1701.

48 Sumerin, V., Schulz, F., Nieger, M., Leskela, M.T., Rieger, R., and Blackwell (2008) *Angew. Chem. Int. Ed.*, **47**, 6001.

49 B., J.M., Foster, K.L., Beck, V.H., Piers, and W.E. (1999) *J. Org. Chem.*, **64**, 4887–4892.

50 Blackwell, J.M., Morrison, D.J., and Piers, W.E. (2002) *Tetrahedron*, **58**, 8247–8254.

51 Blackwell, J.M., Sonmor, E.R., Scoccitti, T., and Piers, W.E. (2000) *Org. Lett.*, **2**, 3921.

52 Parks, D.J., Blackwell, J.M., and Piers, W.E. (2000) *J. Org. Chem.*, **65**, 3090.

53 Parks, D.J., and Piers, W.E. (1996) *J. Am. Chem. Soc.*, **118**, 9440–9441.

54 Chen, D., and Klankermayer, J. (2008) *Chem. Commun.*, 2130.

55 Parks, D.J., Piers, W.E., and Yap, G.P.A. (1998) *Organometallics*, **17**, 5492.

56 Spies, P., Schewendemann, S., Lange, S., Kehr, G., Fröhlich, R., and Erker, G. (2008) *Angew. Chem. Int. Ed.*, **47**, 7543.

57 Sumerin, V., Schulz, F., Atsumi, M., Wang, C., Nieger, M., Leskela, M., Repo, T., Pyykkö, P., and Rieger, B. (2008) *J. Am. Chem. Soc.*, **130**, 14117–14119.

58 Wang, H., Fröhlich, R., Kehr, G., and Erker, G. (2008) *Chem. Commun.*, 5966–5968.

Index

a
abiological catalysts 25
acceptors, hydride 171
acetate, ethyl 78
acetophenone 63
– hydrosilylation 72
acetylene
– complexes 100
– terminal 224–225
acids
– acid-catalyzed condensation 112
– (+)-α-allokainic 190
– ionic hydrogenation 59–64
– kainoid amino 190
– Lewis acid–base reactivity 267
– sulfinic 229
acryl amide 77
acrylonitrile 77
activated methylene compounds 225–227
activation
– amino-borane 272
– C–H/C–N 143–164
– dihydrogen 268
– phosphine-borane 262–263
– precatalysts 115–125
active catalysts 125
activity, structure–activity profiling 252
acyclic 1,3-dienes 193
acyl chlorides 57
acyl complexes 73
addition
– cyclo-, see cycloaddition
– Prins 181
adducts
– bis-phosphine 203
– nitrile-borane 271
adhesives, copper-binding 253–255
agents, reducing 181, 192
agricultural chemicals 222

alchemy, modern 83–110
alcohols
– aliphatic 223–224
– complexes 67
– isopropyl 75
– silylated allylic 204
aldehydes 251
– aliphatic/aromatic 199
– α-branched 199
– coupling 191
– hydrosilylation 102
– pre-equilibrium protonation 62
– simple couplings 194–196
aliphatic alcohols 223–224
aliphatic aldehydes 199
aliphatic amines 213–215
alkaloids, allopumiliotoxin 205
alkenes 96
– carbonyl-catalyzed isomerization 89
– carboxylated 97
– electron-deficient 183
– hydrogenation 86
– tri-/tetra-substituted 182
alkenyl sulfides 229
alkoxide, chromium 59
alkoxysilanes 71–72
alkyl complexes 91
alkyl groups, transfer 182–184, 192–193
alkyl halides 57
alkyl model system 129
alkyl radicals, secondary 5
alkylation 129
alkylative couplings 205
alkyls, cobalt/iron 129–132
alkynes 96
– coupling 182
– cycloaddition 235–260
– disubstituted 157
– hydrogenation 91–92

– immobilized   240
– insertion   186
– mono-aryl internal   197
– polyvalent   255
– simple couplings   194–196
– terminal   251
alkynyl halides   220–221
(+)-α-allokainic acid   190
allopumiliotoxin alkaloids   205
allyl benzene, hydrogenation   61
allylcobalttricarbonyl   151
allylic alcohol   204
allyls   152
alternative nitrogenases   25
amidation   218–221
amides, cross-coupling   219–221
amido nitrogen
– protonation   34
– silylated   28
amido p-orbitals, in-plane   26
amines   97, 157
– arylation   213–216
– sterically encumbered   266
– tris(1,2,3-triazolyl)methyl   242–243
amino acids, kainoid   190
amino-borane activation   272
ammonia   25–50
– arylation   215–216
– cationic complexes   37
– displacement   47
– ^{15}N-labeled   40
– substoichiometric formation   39
amphidinolide   196, 207
ancillary ligands   247
anionic chromium alkoxide   59
anionic metal hydrides
– carbonyl   56
– catalytic hydrogenation   58–59
– complexes   56–58
anions, cation-anion pairs   131
annulation, hetero-   143
anthracene, reduction   7
antitumor activity   73
applications
– CuAAC reaction   252
– iron/cobalt complexes   133
– natural product synthesis   189–191, 205–209
"aqueous ascorbate" method   255
aqueous buffer   255–256
arenes   99
argon   45
"arm-off" species   48
aromatic aldehydes   199

aromatic amidation   218–221
aromatic thiols   154
aryl amines   215–216
aryl azides   244
N-aryl groups   124
aryl halides
– coupling   217–218
– cross-coupling   219–220, 223–224
aryl imines   148
aryl nitriles   228
aryl ring, fluorinated   264
aryl sulfones   229
arylalkyl-thioethers   228
arylation
– activated methylene compounds   225–227
– amines   213–216
– ammonia   215–216
– N-heterocycles   217
aryloxylation   223
arylphosphinates   230
ascorbate
– "aqueous ascorbate" method   255
– sodium   239–240
asymmetric reduction   271
asymmetric ketone hydrogenation   261
asymmetric variants   199–200
atom transfer, hydrogen   184–186, 193–200
azadithiolate   167
azidation   221
azide-alkyne cycloaddition   235–260
azides
– aryl   244
– benzyl   248
– *in situ* generated   244
– immobilized   240
– organic   235
– polyvalent   255
– sulfonyl   250–251
– vinyl   244
azomethine imines   252

b

backbonding   32, 46
bases
– Lewis acid–base reactivity   267
– pendant nitrogen   166
benzaldehyde   269
benzene   39
benzimidazoles   217–218
benzyl azide   248
N-benzylmethylamine   155
Bergman cycloaromatization   4
Bianchini's catalyst precursors   90

bidentate compounds 213
bifunctional catalysts
– iron 103
– Noyori's 84
bifunctional complexes 103–105
bimetallic chemistry 29
bimetallic complexes 59
binuclear complexes 249
biocatalytic reactions 85
bioconjugation 242, 250
– aqueous buffer 255–256
biological screening 252–253
biomolecules 255–256
biphenylene 157–158
bipyramidal structure 115
bis-phosphine adduct 203
bisaryl-thioethers 228
2,6-bis(arylimino)pyridines 112
bis(diphosphine) metal hydrides 53
bis(hydrazone)pyridine iron complexes 119
κ^2-bis(imino)pyridine chelates 98
bis(imino)pyridine iron complexes 93–99
bis(imino)pyridine ligands 111–141
blending, reactor 134
bonds
– C–C 7–11, 224–228
– C–H 143–164
– C–N 143–164, 213–221, 249
– C–O 222–224
– C=O 62–63
– C–P 230
– C–S 228–229
– dissociation energy 1–3, 10
– M–H, see M–H bonds
– N–H 104
– P–S/P–O 74–75
borane
– activation 262–263
– amino- 272
– chiral 271
– nitrile-borane adducts 271
– phosphino- 264
borate, phosphonium 267–269, 272
bound ketones 51
α-branched aldehydes 199
bridge, dithiolate 173
bridging hydride 59
bromides
– cetyltrimethylammonium 153
– vinyl 218
buffer, aqueous 255–256
Burk's DuPhos Rh catalyst 84
2-butanone 75

c

C_2-bridged catalysts 69–70
C–C bonds
– formation 224–228
– H· transfer reactions 7–11
C–H bonds
– activation reactions 143–164
– β-elimination 148
C–N bonds
– activation reactions 143–164
– formation 213–221, 249
C–O bonds, formation 222–224
C–O cleavage, competing 97
C=O bonds 62–63
C–P bonds, formation 230
C–S bonds, formation 228–229
carbamates 219–221
carbenes
– ligands 70–71
– Ni-carbene intermediates 161
carbonyl acyl complexes 73
carbonyl catalysts, alkene isomerization 89
carbonyl complexes 86–89
carbonyl hydrides 53, 56
carbonyl hydrogenation 101
carbonyls, α,β-unsaturated 182–191
carboxylated alkenes 97
cascade cyclization
– 6-endo/6-exo 18
– reductive 5
catalysis
– homogeneous 83
– metal-free strategy 267–272
– olefin oligomerization 122
– olefin polymerization 115–121
– radical cyclization 15–19
– tandem 134
catalysts
– abiological 25
– active 125
– Bianchini's precursors 90
– bifunctional 103
– Burk's DuPhos Rh 84
– C_2-bridged 69–70
– carbonyl 89
– co- 117
– cobalt 143–153
– copper 213–233
– Cp_2Mo 73–78
– Crabtree's 84
– cycloaddition 238–243
– design 166–169
– electro- 165–180

– inhibition   177
– (^{iPr}PDI)Fe(N$_2$)$_2$ 94–95
– iron   83–110
– ketone hydrosilylation   71–73
– Liebeskind   220
– lifetime   69–70
– MAO, see MAO
– molecular   173–174
– molybdenum   25–81
– molybdenum phosphine   69–70
– N-heterocyclic carbene ligands   70–71
– nickel, see nickel catalysts
– Noyori's   84
– organo-   262
– performance   111
– phosphonium borate   267–269
– pre-, see precatalysts
– precious metal   84
– ruthenium   237, 261
– Schrock-Osborn   84
– SHOP   122
– Shvo-type iron   103
– Shvo's   84
– synthetic   171
– tethering   133
– tungsten   51–81
– Wilkinson's   84
catalytic chain transfer (CCT)   11–15
catalytic core   170
catalytic cycle   176
– CuAAC reaction   247
catalytic epoxide opening   4
catalytic evaluation   116, 122
catalytic hydroformylation   52
catalytic hydrogenation   58–59, 65–69
– alkenes and carbonyls   83–110
– unactivated olefins   95
catalytic intermediates, energy   171–173
catalytic ionic hydrogenation   65–69
catalytic reduction   25–50, 270
catalytic resting state   89
catalytically competent intermediates   96
cations
– cation-anion pairs   131
– cationic alkyl model system   129
– cationic ammonia complexes   37
– cationic monochloride complex   127
– iminium   270
– phosphonium   69
– trityl   266
CCT (catalytic chain transfer)   11–15
cetyltrimethylammonium bromide   153
chain
– growth on zinc   125

– propagation   129
– proton transport   168
– ring-chain isomerization   251
chain-carrying radical   12
chain transfer
– catalytic   11–15
– pathways   127
– rate   124
channels   167
"Cheap Metals for Noble Tasks"   51
chelates   98, 243
chelating diphosphine ligands   175
chemical reactions, see reactions
chemical warfare   74
chiral borane   271
chirality, transfer of   197–198
chlorides, acyl   57
4-chlorobenzenethiol   154–156
chromium alkoxide   59
chromium hydrides   2–3
cleavage
– competing C–O   97
– modes   52
"clicking"   236
co-catalysts   117
co-crystallization   33
coal liquefaction   261
cobaloxime   14
cobalt
– alkyls   129–132
– complexes   111–141, 166
– dialkyl species   127
– dicobalt complexes   9
– halides   113
cobalt catalysts
– active   127
– C–H activation reactions   143–153
– tethering   133
COD (1,5-cyclooctadiene)   182, 194–196
collidinium   40–41
competing C–O cleavage   97
competing coordination and deactivation   99
competing methods   19–20
complexation with iron/cobalt   113–115
complexes
– acetylene   100
– alcohol   67
– anionic metal hydrides   56–58
– bifunctional   103–105
– bimetallic   59
– binuclear   249
– bis(hydrazone)pyridine iron   119

– bis(imino)pyridine iron   93–99
– cationic ammonia   37
– cationic monochloride   127
– cobalt   111–141, 166
– dicobalt   9
– dihydrogen   60
– α-diimine iron   99–101
– "encounter"   265
– iron   111–141, 174
– iron(II) alkyl   91
– iron carbonyl   86–89
– kinetically stabilized   62
– manganese carbonyl acyl   73
– metallocene   73
– nickel   166
– olefin   100
– phosphine   66
– pseudo-tetrahedral   91
– transition-metal hydride   1
– triamidoamine   26–30
compound libraries   252–253
condensation, acid-catalyzed   112
Conrad–Limpach reaction   143
coordinated nitrile   78
coordination, arenes   99
coordination sphere   171–178
copper, nanoparticles   241
copper-binding adhesives   253–255
copper catalysts   213–229
– azide-alkyne cycloaddition   235–260
– couplings   213–233
copper salts   238
core, catalytic   170
coupling
– aldehydes   191
– alkylative   205
– alkynes   182
– aryl halides   217–218
– byproducts   240
– copper-catalyzed   213–233
– cross-, *see* cross-coupling
– Glaser   239
– ligand promoted   213–233
– multi-component   210
– olefin-directed   196
– oxidative   240
– reductive   181–212
– Sonogashira   225
– spin–orbit   94
– three-component   182–184, 192–193
– Ullmann-type   213–233
– vinyl bromides   218
Cp$_2$Mo catalysts   73–78
Crabtree's catalyst   84

cross-coupling
– aliphatic alcohols   223–224
– aryl halides   219–220, 223–224
– selective   148
– with terminal acetylene   224–225
crossover experiments   201
CuAAC (copper-catalyzed azide-alkyne cycloaddition)   235–260
Cu . . . *see also* copper
cyanation   227–228
cycle, catalytic   176, 247
cyclic 1,3-dienes   193
cyclic voltammetry   35
cyclization
– dienes   19–20
– 6-endo/6-exo cascade   18
– macro-   194
– oxocarbenium ion/vinyl silane   207
– radicals   15–19
– reductive   5, 181–212
cycloaddition
– azide-alkyne   235–260
– Diels–Alder   181
– 1,3-dipolar   235
– thermal   237
cycloalkyl substituents   118
cycloaromatization   204–205
– Bergman   4
cyclohexadiene   4
cyclohexanone   58
1,5-cyclooctadiene (COD)   182, 194–196
cyclopentanol framework   206

d

deactivation, arenes   99
decomposition pathway   132
derivatives
– quinoline   146–147
– TREN   28–29
design, catalysts   166–169
deuteration   57
– norbornene   92
deuterium, H/D exchange   43
DFT calculations, ketone hydrogenation   104–105
dialkyl species   127
diallylanilines   144
diamagnetic intermediates   36
diamines, vicinal   219
diaryl ethers   222–223
diastereoselective variants   197–198
diazenido proton resonance   33
1,3-diazide   246
dibutyl phosphite   230

dicarboxylate esters   237
dicobalt complexes   9
Diels–Alder cycloadditions   181
dienes
– COD   182, 194–196
– cyclization   19–20
– cyclohexa-   3
– heterocycle-substituted   19
1,3-dienes, (a)cyclic   193
diethyl ether   36
diethyl malonate   226
differential pulse voltammetry   38
differentiated π-systems   210
digermyne   263
dihalides, iron   117
dihydrides
– intermediate   177
– ionic hydrogenations   64–65
– regeneration   67
dihydrogen   96
– activation   268
– complexes   60
– see also hydrogen
α-diimine iron complexes   99–101
dimerization, head-to-head   125
dimers
– molybdenum sulfur   166
– oxygen-bound   202
dinitrogen   25–50
diphenylacetylene   158
9,10-diphenylphenanthrene   158–160
diphosphine
– bis(diphosphine) metal hydrides   53
– ligands   166, 175
1,8-diphosphino-napthalene   272
1,3-dipolar cycloaddition   235
dipolar species   250–252
dippe   157, 160–161
directed processes   196–197
displacement, ammonia   47
dissociated dinitrogen   45
dissociation energy, bond   1–3, 10
disubstituted alkynes   157
3,5-disubstituted isoxazoles   251
1,4-disubstituted-1,2,3-triazoles   237
dithiolate bridge   173
ditriazole   246
Doebner–Von Miller reaction   143
donor ligands
– "hard"   241
– weak-/strong-donor conditions   249
donors, hydride   54
driving forces, thermodynamic   175
DuPhos Rh catalyst, Burk's   84

e
effects
– electronic   116–119
– inverse isotope   8
– solvent   145
– steric   9, 116–119
– substituent   122–124
– Thorpe–Ingold   17
electrocatalysts, molecular   165–180
electrochemical reduction   37
electron-deficient alkenes   183
π electron system in Mo(N$_2$)$^-$ 31
electron transfer reactions   2
electronic effects   116–119
electronic energy barriers   174
electrophiles   185
elimination, reductive   160, 184
β-elimination, C–H bonds   148
ellipsoids, thermal   47
enamine   272
enantioselectivity   227
"encounter complex"   265
6-endo/6-exo cascade cyclization   18
ene reactions   181
energy
– bond dissociation   1–3, 10
– catalytic intermediates   171–173
– free energy landscape   172–173
– matching strategies   172
– steric/electronic barriers   174
enoates   185
Z-enol silane   189
enolates   187
enones   262
– α,β-unsaturated   183
enyne   202
1,6-enyne   198
enzymes
– HIV protease   254
– hydrogenase   166–169
– transferase   253
epimeric natural products   190
epoxide opening   4
esters
– β-keto   223
– D-serine methyl   189–191
– dicarboxylate   237
– ethyl   16
– Hantzsch   262
– trifluoroacetate   61
ethers   97
– diaryl   222–223
– diethyl   36
– thio-   228

ethyl acetate, hydrolysis 78
ethyl benzene 63
ethyl ester 16
2-ethyl-3-methylquinoline 148
ethyl 2-oxocyclohexanecarboxylate 218
ethylene
– gas-phase hydrogenation 88
– HDPE 111, 116
– -linked phosphonium borate 272
– oligomerization 112
– polymerization 112, 116
evaluation, catalytic 116, 122
exchange
– H/D 43
– intramolecular 174
extrusion, hydrogen gas 204–205

f

favored ligand 249
Fe, see iron
ferrous oxidation state 130
first coordination sphere 171–173
first-row transition metals 1–24
Flory distribution, Schulz– 124
fluorinated aryl ring 264
fluorocarbons 97
fouling, reactors 133
framework, cyclopentanol 206
free energy landscape 172–173
Friedlaender reaction 143
"frustrated Lewis pairs" 264–267
fucosyl transferase inhibitors 253
fuel cells 165
functional group tolerance 97
functionalized 1,3-dienes 193
functionalized hydroxylamines 219

g

gas extrusion, hydrogen 204–205
gas-phase hydrogenation 88
generated azides, in situ 244
"generic" model system, cationic alkyl 129
Glaser coupling 239
glyoxime ligands 166
Goldberg reaction 218
"green" media 162
Grignard reagents 263
group 10 metals, zero-valent 161

h

H2, see hydrogen
H· transfer reactions 1–24
– C–C bonds 7–11

– ligands 4–7
– M–H bonds 2–4, 7–11
– olefins 11
– organic radicals 2–7
– see also proton transfer reactions
halide-containing precatalysts 129
halides
– alkyl 57
– alkynyl 220–221
– aryl 217–220
– iron/cobalt 113
– vinyl 220–223
– see also bromides, chlorides
halogen-exchanging process 228
Hantzsch ester 262
"hard" donor ligands 241
HAT (hydrogen atom transfer), see H· transfer reactions
head-to-head dimerization 125
heptane 39–40
hetero-annulation 143
heterocycle-substituted dienes 19
N-heterocycles 143
– arylation and vinylation 217
– couplings 217–218
heterocyclic chelates 243
N-heterocyclic carbene ligands 70–71
N-heterocyclic carbenes 161
hexaphenylbenzene 160
1-hexene 92
high density polyethylene (HDPE) 111, 116
[HIPTN$_3$N]Mo intermediates 30–38
HIV protease inhibitors 254
homogeneous catalysis 83
Hurtly reaction 225
"hybrid" heterocyclic chelates 243
hydration 73–78
hydrazine 25
hydrazone 119
"hydricities" 1
hydride acceptors 171
hydride complexes, transition-metal 1
hydride donors 54
hydride transfer reactions 54–56
– kinetics 55
hydride transfer reactivity 56–58
hydrides
– bond dissociation energy 2–3, 10
– bridging 59
– di-, see dihydrides
– metal, see metal hydrides
– metal carbonyl 53
hydroboration 262

hydroformylation 52
hydrogen
– amino-borane activation 272
– atom transfer 184–186, 193–200
– gas extrusion 204–205
– H/D exchange 43
– oxidation and production 165–180
– phosphine-borane activation 262–263
– see also dihydrogen
β-hydrogen-bearing organozincs 200
hydrogenase 166–169
hydrogenation 73–78
– alkenes 86
– alkynes 91–92
– allyl benzene 61
– asymmetric ketone 261
– carbonyl 101
– catalysis 267–272
– catalytic, see catalytic hydrogenation
– catalytically competent intermediates 96
– DFT calculations 104–105
– gas-phase 88
– ionic 1, 59–69
– molybdenum and tungsten catalysts 51–81
– N-heterocyclic carbene ligands 70–71
– olefins 86
– photocatalytic 87
hydrogenolysis 262
hydrolysis 73–78
– ethyl acetate 78
hydrophobic tunnels 168
hydrosilylation
– acetophenone 72
– aldehydes 102
– catalytically competent intermediates 96
– iron catalysts 101–103
– ketones 71–73, 102
– photocatalytic 87
hydroxylamines 251
– functionalized 219

i

imidazoles 217–218
imidazolium salts 162
imines 149–151, 267–268
– asymmetric reduction 271
– azomethine 252
iminium cation 270
imino-carbon substituents 115
imino-nitrogen position 113
iminoaryl units 117, 123
immobilization 133–134

immobilized alkynes/azides 240
in-plane amido p-orbitals 26
in situ generated azides 244
indolidizine skeleton 201
inhibition, catalysts 177
inhibitors
– fucosyl transferase 253
– HIV protease 254
insertion
– alkynes 186
– bound ketones 51
interconversion, $Mo(NH_3)$ and $Mo(N_2)$ 38–39
intermediates
– catalytic 96, 171–173
– diamagnetic 36
– dihydrides 177
– $[HIPTN_3N]Mo$ 30–38
– ketenimine 250
– Ni-carbene 161
– six-coordinate 48
intermolecular allyl transfer 152
intermolecular reductive couplings 185
internal alkynes 197
intramolecular exchange 174
inverse isotope effect 8
ionic hydrogenation 1, 59–64
– catalytic 65–69
– dihydrides 64–65
ionic liquids 162
iron alkyls 129–132
iron catalysts 83–110
– active 126–127
– bifunctional 103
– olefin hydrogenation 86
– Shvo-type 103
– tethering 133
iron complexes
– acetylene 100
– alkyl 91
– bis(hydrazone)pyridine 119
– bis(imino)pyridine 93–99
– carbonyl 86–89
– α-diimine 99–101
– olefin 100, 111–141
– pendant nitrogen bases 174
iron dihalides 117
iron(II) halides 113
iron phosphine compounds 89–92
isobutyraldehyde 62
isolated yields 144, 149–151
isolobal relationship 93
Z-isomer stereoselectivity 188

isomerization   16–17
– alkenes   89
– photocatalytic   87
– ring-chain   251
isopropyl alcohol   75
isotope effect, inverse   8
isotopic labeling experiments   95
isoxazoles   256
– 3,5-disubstituted   251

k

kainoid amino acid   190
ketenimine intermediates   250
β-keto ester   223
ketones   97
– asymmetric hydrogenation   261
– catalytic hydrogenation   58–59
– DFT calculations   104–105
– hydrosilylation   71–73, 102
– insertion   51
– ionic hydrogenation   59–64
kinetically stabilized complex   62
kinetics, hydride transfer   55

l

labilization, ligands   44
layered approach to catalyst design   170–171
Lewis acid–base reactivity   267
Lewis pairs, "frustrated"   264–267
libraries, compound   252–253
Liebeskind catalyst   220
lifetime, catalysts   69–70
ligands
– ancillary   247
– azadithiolate   167
– bidentate   213
– bis(imino)pyridine   111–141
– chelating diphosphine   175
– cycloaddition   238–243
– favored   249
– H· transfer reactions   4–7
– "hard" donor   241
– labilization   44
– ligand promoted couplings   213–233
– monodentate phosphine   182, 194
– N-heterocyclic carbene   70–71
– pentabenzylcyclopentadienyl   68
– preparation   112–113
– "soft"   241
– tetradentate   104
– [TMSN$_3$N]$^{3-}$ 27
– trianionic   46
– variations   44–47

linear low density polyethylene (LLDPE)   122, 134–135
liquefaction of coal   261
liquids, ionic   162
LLDPE (linear low density polyethylene)   122, 134–135
low density polyethylene, linear   122, 134–135
lutidinium   33–34

m

M–H bonds
– H· transfer reactions   2–4, 7–11
– ketone insertion   51
M–H cleavage modes   52
macrocyclization   194
macrolides   208
malonate   226
manganese carbonyl acyl complexes   73
MAO (methylaluminoxane)   93, 111, 119–120
matching strategies, energy   172
mechanistic pathway, metallacycle-based   186–189
mesityl ring   71
metal hydrides
– anionic   56–59
– bis(diphosphine)   53
– hydride transfer reactions   54–56
– ionic hydrogenation   59–64
– proton transfer reactions   52–54
metal-free hydrogenation strategy   267–272
metallacycle-based mechanistic pathway   186–189
metallacycles, oxa-   200
metallocenes   35
– complexes   73
metals
– anionic carbonyl hydrides   56
– carbonyl hydrides   53
– "Cheap Metals for Noble Tasks"   51
– transition, see transition metals
– variations   44–47
– zero-valent   161
methacrylate, methyl   14
methods
– "aqueous ascorbate"   255
– competing   19–20
– development   182–186, 192–194
– stopped-flow   55
methyl ester   189–191
methyl methacrylate   14
methyl sorbate   86

methylaluminoxane (MAO) 93, 111, 119–120
methylene
– activated compounds 225–227
– group 46
migration reactions 27
modality, polymers 120
model, space filling 169
model system, cationic alkyl 129
modern alchemy 83–110
modified methylalumoxane (MMAO) 112, 116, 121
modular approach to catalyst design 170–171
molecular catalysts 173–174
molecular electrocatalysts 165–180
molecular nitrogen 30–38
"molecular tweezers" 272
molybdenum catalysts 25–50
– catalytic ionic hydrogenation 65–69
– hydrogenation 51–81
– phosphine 69–70
molybdenum sulfur dimers 166
$Mo(NH_3)/Mo(N_2)$, interconversion 38–39
mono-aryl internal alkynes 197
monochloride complex 127
monodentate phosphine ligands 182, 194
monomers, α-olefin 121, 125
Morita–Baylis–Hillman reaction 18
movement of protons 173–174
multi-component coupling 210
multi-electron/multiproton processes 169

n

N–H bonds, tetradentate ligands 104
^{15}N-labeled ammonia 40
Na, see sodium
nanoparticles, copper 241
naphthalene, 1,8-diphosphino- 272
natural products
– epimeric 190
– synthesis 189–191, 205–209
Ni-carbene intermediates 161
nickel
– complexes 166
– enolate 187
– vinyl nickel species 204
nickel catalysts 153–163
– reductive couplings and cyclizations 181–212
– zero-valent 182
nitrile, coordinated 78

nitrile-borane adducts 271
nitrogen
– imino-nitrogen position 113
– molecular 30–38
– pendant bases 166
– silylated amido 28
nitrogenases, alternative 25
non-bonding orbital 26
norbornene 92
norcarane 6
Noyori's bifunctional catalyst 84
nucleophiles, ynoates 185

o

olefin complexes 100
olefin-directed couplings 196
olefinic units 261
olefins
– H· transfer reactions 11
– hydrogenation 86
– imidazolium salts 162
– α-olefin monomers 121, 125
– oligomerization 111–141
– polymerization 111–141
– unactivated 95
oligomerization
– ethylene 112
– olefins 111–141
– precatalysts 124
one-pot synthesis 106
– MoH 42
– triazoles 245
opening, epoxide 4–5
orbit, spin–orbit coupling 94
orbitals, non-bonding 26
organic azides 235
organic radicals, H· transfer reactions 2–7
organocatalyst reduction 262
organometallic chemistry 83
organophosphate toxins 74
organozinc 192
– β-hydrogen-bearing 200
ortho-substituents 115
overpressure 41
oxazolidinones 221
oxidation, hydrogen 165–180
oxidation state, ferrous 130
oxidative coupling byproducts 240
oxocarbenium ion/vinyl silane cyclization 207
2-oxocyclohexanecarboxylate 218
oxygen, rebound mechanism 6
oxygen-bound dimer 202

p

P–S/P–O bonds   74–75
π electron system in Mo(N$_2$)$^-$ 31
π-systems, differentiated   210
pairs
– cation-anion   131
– Lewis   264–267
pathways
– chain transfer   127
– decomposition   132
– metallacycle-based mechanistic   186–189
PCET (proton-coupled electron transfer reactions)   2
(^{iPr}PDI)Fe(N$_2$)$_2$ 94–95
pendant nitrogen bases   166
pentabenzylcyclopentadienyl ligand   68
performance, catalysts   111
Pfitzinger reaction   143
pharmaceuticals   222
phenanthrenes   158
1,10-phenanthroline   217
phenols   223
phenyl acetylene   248
phosphate esters   74
phosphines
– bis-phosphine adduct   203
– catalysts   69–70
– complexes   66
– iron compounds   89–92
– monodentate ligands   182, 194
– phosphine-borane activation   262–263
– sterically demanding   263
– tertiary   263
– tricyclopentyl-   198
phosphite, dibutyl   230
phosphonium borate   267–269
– ethylene-linked   272
phosphonium cations   69
photocatalytic reactions   87–88
Poisson distribution   124
polyacrylamide   77
polyethylene
– HDPE   111, 116
– LLDPE   122, 134–135
– ULDPE   134
polymerization
– CCT   12
– ethylene   112
– methyl methacrylate   14
– olefins   111–141
– propylene   121
– radicals   11

polymers
– modality   121
– yield   121
polyvalent azides/alkynes   255
POV-Ray rendering   47
pre-equilibrium protonation   62
precatalysts   41
– activation   115–125
– halide-containing   129
– oligomerization   124
– synthesis   112–115
precious metal catalysts   84
precoordination   202
precursors, catalyst   90
pressure
– effect on olefin polymerization   120–121
– over-   41
primary amines   213–215
Prins addition reactions   181
processes
– directed   196–197
– halogen-exchanging   228
production of hydrogen   165–180
profiling, structure–activity   252
L-proline   214, 229
propagation   127
– chain   129
propionaldehyde   64
propylene, polymerization   121
protease, HIV   254
proteolysis   248
"proton-catalyzed" reduction   33
proton-coupled electron transfer (PCET) reactions   2
proton relays   175
proton transfer reactions   52–54
– see also H• transfer reactions
proton transport chain   168
protonation
– amido nitrogen   34
– pre-equilibrium   62
protons, movement   173–174
pseudo-tetrahedral iron(II) alkyl complexes   91
pulse voltammetry, differential   38
pyramidal structure   115
pyrazoles   217
pyrene   262
pyridine
– 2,6-bis(arylimino)pyridines   112
– bis(imino)pyridine iron complexes   93–99
– bis(imino)pyridine ligands   111–141
4-pyridyl position   134

q

quinolines 143–152
– derivatives 146–147

r

radicals
– chain-carrying 12
– cyclization 15–19
– organic 2–7
– polymerization 11
– secondary alkyl 5
– tris(*p-tert*-butylphenyl)methyl 3
rate constants, hydride transfer 56
rate-determining step 177
reactions
– aromatic amidation 218–221
– arylation 213–216, 225–227
– aryloxylation 223
– azidation 221
– β-elimination 148
– β scission 15
– Bergman cycloaromatization 4
– biocatalytic 85
– C–H/C–N activation 143–164
– condensation 112
– Conrad–Limpach 143
– cross-coupling, *see* cross-coupling
– cyanation 227–228
– cyclization, *see* cyclization
– cycloaddition 181, 235–260
– cyclocondensation 204–205
– deuteration 92
– dimerization 125
– 1,3-dipolar cycloaddition 235
– Doebner–Von Miller 143
– ene 181
– Friedlaender 143
– Goldberg 218
– H• transfer, *see* H• transfer reactions
– H/D exchange 43
– hetero-annulation 143
– Hurtly 225
– hydroformylation 52
– hydrogenation, *see* hydrogenation
– hydrolysis 73–78
– hydrosilylation 71–73
– ionic hydrogenation 1
– isomerization 16–17, 251
– migration 27
– Morita–Baylis–Hillman 18
– oxidation 165–180
– Pfitzinger 143
– photocatalytic 87–88
– polymerization 11–13

– Prins addition 181
– proteolysis 248
– protonation 34, 62
– reduction 7
– reductive coupling, *see* reductive coupling
– reductive elimination 160, 184
– Skraup 143
– sulfonyl azides 250–251
– transfer, *see* transfer reactions
reactivity
– Lewis acid–base 267
– transfer 56–58
reactor blending 134
reactor fouling 133
rearrangement, norcarane 6
rebound mechanism, oxygen 6
reducing agents 181
– organozinc 192
reduction
– anthracene 7
– asymmetric 271
– benzaldehyde 269
– catalytic 25–50, 270
– electrochemical 37
– organocatalyst 262
– "proton-catalyzed" 33
reductive couplings 181–212
– hydrogen atom transfer 184–186, 193–200
– intermolecular 185
reductive cyclization 181–212
reductive cyclization cascade 5
reductive elimination 160, 184
regeneration, dihydrides 67
regioselectivity 145, 186–187
resin-immobilized TBTA 242
resorcinylic macrolides 208
resting state, catalytic 89
rhodium, vinyl rhodium species 20
rings
– fluorinated aryl 264
– mesityl 71
– ring-chain isomerization 251
ruthenium catalysts 4, 261
ruthenium catalyzed azide-alkyne cycloaddition (RuAAC) 237

s

salts
– copper 238
– imidazolium 162
– sulfinic acid 229
– zwitterionic 264
saturated N-heterocyclic carbenes 161

Schrock-Osborn catalyst 84
Schulz–Flory distribution 124
β scission 15
screening, biological 252–253
second coordination sphere 173–174
secondary alkyl radical 5
secondary amines 213–215
selective cross-coupling 148
D-serine methyl ester 189–191
sesterterpene 209
SHOP catalyst 122
Shvo-type iron catalysts 103
Shvo's catalyst 84
silacycles 204
silanes 96, 194
– Z-enol 189
– vinyl 207
silica 134
silylated allylic alcohol 204
silylated amido nitrogen 28
six-coordinate intermediates 48
Skraup reaction 143
sodium ascorbate 239–240
"soft" ligands 241
solvent effects 145
Sonogashira coupling 225
sorbate, methyl 86
space filling model 169
sphere, coordination 171–178
spin doublet 32
spin–orbit coupling 94
square pyramidal structure 115
stabilized complex, kinetically 62
stereodivergent approach 190
stereoselectivity 186–187
– Z-isomer 188
steric effects 9, 116–119
steric energy barriers 174
steric variations 123
sterically demanding phosphines 263
sterically encumbered amines 266
stoichiometric hydride transfer reactivity 56–58
stopped-flow methods 55
strong-donor conditions 249
structure–activity profiling 252
styrene 10
substituents
– cobaloxime 14
– cycloalkyl 118
– effects 122–124
– imino-carbon 115
– ortho- 115
substoichiometric formation, ammonia 39

sulfinic acid salts 229
sulfones, aryl 229
sulfonyl azides 250–251
sulfur, molybdenum sulfur dimers 166
sustainability 85
symmetrical 2,6-bis(arylimino)pyridines 112
σ symmetry 31
synthesis
– alkenyl sulfides 229
– diaryl ethers 222–223
– natural products 189–191, 205–209
– one-pot 42, 106, 245
– precatalysts 112–115
synthetic catalysts 171

t
tandem catalysis 134
tandem catalytic epoxide opening 4
TBTA (tris(1,2,3-triazolyl)methyl amine) 242–243
temperature, effect on olefin polymerization 120–121
terminal acetylene 224–225
terminal alkynes 251
(−)-terpestacin 209
tertiary phosphines 263
tethering, iron/cobalt catalysts 133
tetradentate ligands, N–H bonds 104
therapeutic agents 73
thermal cycloaddition 237
thermal ellipsoids 47
thermodynamic driving forces 175
THF 30–31
thietan-2-one 153
thiirane 153
thiocarbamate 155–156
thioethers 228
thiols 157, 229
– aromatic 154
thiophosphinate 75
Thorpe–Ingold effect 17
three-component couplings, alkyl group transfer 182–184, 192–193
[TMSN$_3$N]$^{3-}$ ligand 27
toxins, organophosphate 74
trans-cinnamaldehyde 256
transfer of chirality 197–198
transfer reactions
– alkyl group 182–184, 192–193
– CCT 11–15
– H•, see H• transfer reactions
– hydride 54–56
– hydrogen atom 184–186, 193–200

– intermolecular allyl   152
– PECT   2
– proton   52–54
transfer reactivity, stoichiometric hydride   56–58
transferase, fucosyl   253
transition metals
– first-row   1–24
– hydride complexes   1
transport chain, proton   168
trapping   45
TREN derivatives   28–29
tri-/tetra-substituted alkenes   182
triamidoamine complexes   26–30
trianionic ligand   46
triazoles, one-pot synthesis   245
1,2,3-triazoles   237, 255
tributylphosphine   184
tricyclopentylphosphine   198
triethylenetetramine   27
trifluoroacetate esters   61
trigonal bipyramidal structure   115
tris(1,2,3-triazolyl)methyl amine (TBTA)   242–243
tris(p-tert-butylphenyl)methyl radical   3
trityl cation   266
tungsten catalysts   51–81
– catalytic ionic hydrogenation   65–69
– N-heterocyclic carbene ligands   70–71
tungsten trihydride species   42
tunnels, hydrophobic   168
tweezers, molecular   272

u
Ullmann-type couplings   213–233
ultra low density polyethylene (ULDPE)   134
unactivated olefins   95
units
– iminoaryl   117, 123
– olefinic   261
α,β-unsaturated carbonyls   182–191
unsaturated N-heterocyclic carbenes   161

v
vanadium hydrides   2–3
variants
– asymmetric   199–200
– diastereoselective   197–198
– steric   123
vicinal diamines   219
vinyl azides   244
vinyl bromides   218
vinyl halides   220–223
– aryloxylation   223
vinyl nickel species   204
vinyl rhodium species   20
vinyl silane cyclization, oxocarbenium   207
vinylation, N-heterocycles   217
voltammetry
– cyclic   35
– differential pulse   38

w
W (tungsten), see tungsten
warfare, chemical   74
weak-donor conditions   249
well-defined, alkyls   129–132
Wilkinson's catalyst   84

y
yield
– isolated   144, 149–151
– polymers   121
ynoates   185

z
Z-enol silane   189
Z-isomer stereoselectivity   188
zero-valent nickel, 1,5-cyclooctadiene (COD)-stabilized   182
zero-valent group 10 metals   161
zinc
– chain growth   125
– organo-   192, 200
zwitterionic cobalt dialkyl species   127
zwitterionic salts   264

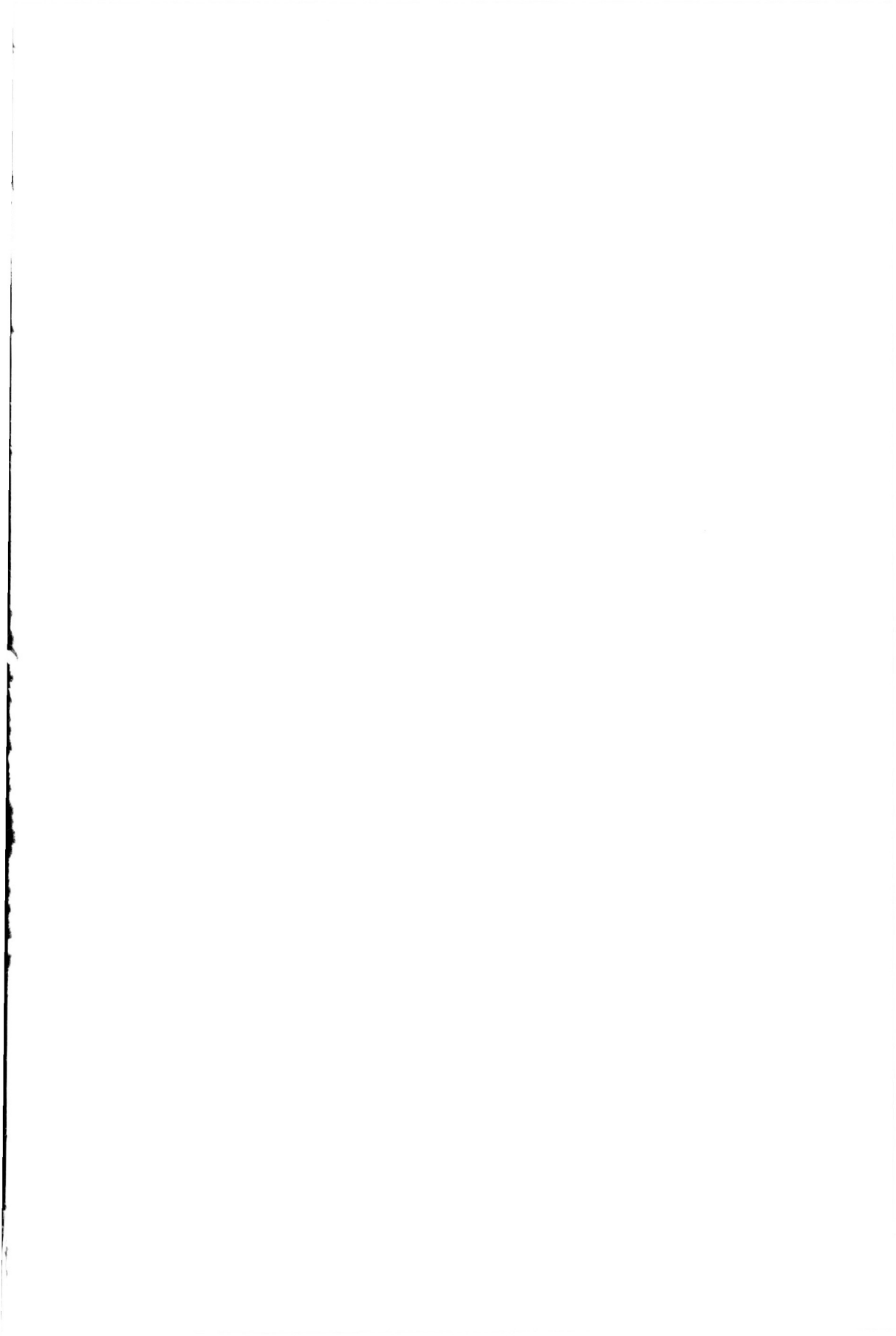